# Photocured Materials

# RSC Smart Materials

*Series Editors:*
Hans-Jörg Schneider, Saarland University, Germany
Mohsen Shahinpoor, University of Maine, USA

*Titles in this Series:*
1: Janus Particle Synthesis, Self-Assembly and Applications
2: Smart Materials for Drug Delivery: Volume 1
3: Smart Materials for Drug Delivery: Volume 2
4: Materials Design Inspired by Nature
5: Responsive Photonic Nanostructures: Smart Nanoscale Optical Materials
6: Magnetorheology: Advances and Applications
7: Functional Nanometer-Sized Clusters of Transition Metals: Synthesis, Properties and Applications
8: Mechanochromic Fluorescent Materials: Phenomena, Materials and Applications
9: Cell Surface Engineering: Fabrication of Functional Nanoshells
10: Biointerfaces: Where Material Meets Biology
11: Semiconductor Nanowires: From Next-Generation Electronics to Sustainable Energy
12: Supramolecular Materials for Opto-Electronics
13: Photocured Materials

*How to obtain future titles on publication:*
A standing order plan is available for this series. A standing order will bring delivery of each new volume immediately on publication.

*For further information please contact:*
Book Sales Department, Royal Society of Chemistry, Thomas Graham House, Science Park, Milton Road, Cambridge, CB4 0WF, UK
Telephone: +44 (0)1223 420066, Fax: +44 (0)1223 420247
Email: booksales@rsc.org
Visit our website at www.rsc.org/books

# *Photocured Materials*

Edited by

**Atul Tiwari**
*University of Hawaii at Manoa, Honolulu, Hawaii, USA*
*Email: tiwari@hawaii.edu*

**Alexander Polykarpov**
*AkzoNobel Inc., Strongsville, Ohio, USA*
*Email: alexander.polykarpov@akzonobel.com*

RSC Smart Materials No. 13

Print ISBN: 978-1-78262-001-3
PDF eISBN: 978-1-78262-007-5
ISSN: 2046-0066

A catalogue record for this book is available from the British Library

© The Royal Society of Chemistry 2015

*All rights reserved*

*Apart from fair dealing for the purposes of research for non-commercial purposes or for private study, criticism or review, as permitted under the Copyright, Designs and Patents Act 1988 and the Copyright and Related Rights Regulations 2003, this publication may not be reproduced, stored or transmitted, in any form or by any means, without the prior permission in writing of The Royal Society of Chemistry or the copyright owner, or in the case of reproduction in accordance with the terms of licences issued by the Copyright Licensing Agency in the UK, or in accordance with the terms of the licences issued by the appropriate Reproduction Rights Organization outside the UK. Enquiries concerning reproduction outside the terms stated here should be sent to The Royal Society of Chemistry at the address printed on this page.*

The RSC is not responsible for individual opinions expressed in this work.

The authors have sought to locate owners of all reproduced material not in their own possession and trust that no copyrights have been inadvertently infringed.

Published by The Royal Society of Chemistry,
Thomas Graham House, Science Park, Milton Road,
Cambridge CB4 0WF, UK

Registered Charity Number 207890

For further information see our web site at www.rsc.org

Printed and bound by CPI Group (UK) Ltd, Croydon, CR0 4YY

# Preface

Synthetic polymers have been conventionally hardened by heating them at elevated temperatures. The properties and service life span of such synthetic materials largely depends on the degree of cure and crosslinking. In some cases, hardening of polymeric materials can be achieved in ambient conditions with the aid of catalysts. However, postcuring of such materials at higher temperature imparts better strength and resistance to harsh environmental conditions. The heating or hardening of such materials at higher temperature demands a high level of safety, excessive energy, and often a large working space.

The invention of photocuring technology is deemed to revolutionize the materials industry. This technique is considered as one of the most effective ways to rapidly transform a liquid polymeric resin into a dense crosslinked stable product without the release of harmful volatile organic components. Curing and crosslinking reactions triggered by light occur extremely rapidly that saves significant time required in conventional hardening of the materials. The ability to rapidly cure at low temperature, high spatial resolution, and use of stable single-component systems without the need for solvents continue to fuel industrial and academic interest in photocured materials. The past decade has noticed significant research and development activities in the curing technology as well as chemical ingredients. A wide array of photoinitiators, polymer precursors, fillers, and light devices has been discovered for the newer developments. Market survey studies have suggested that interest in photocured materials is gaining wide acceptance in academia as well as industry. It is thought that the market of photocured materials such as acrylic resins would grow at a rapid pace and reached approximately US$ 4.94 billion in year 2012. It clearly demonstrates the applicability and usefulness of the materials and technology. This staunch effort could help educating young scholars about the unforeseen applications that may originate from the interdisciplinary approach.

This book is a collection of excellent chapters written by the experts utilizing the technology in various innovative areas of materials science and engineering. Chapters dedicated to the synthesis of new monomers and resulting polymeric precursors are included to lay the foundation for novice. The utility of photocuring in coatings is the fastest growing area and has been extensively highlighted in the book. Besides traditional photocuring using UV lamps, LED and lasers a two-photon curing technique will impart new research directions for innovative developments. Chapters on the utility of photocuring technology in obtaining complex 3D structures, composite curing, and functional polymers will be of enormous interest to the readers of many research areas. The utilization of commercially successful polymers in microscale structures fabrication is well demonstrated in three inter-related chapters. The chapter on migration from cured coatings in food-contact applications reminds the researchers to consider the levels of residual photoinitiators and other low molecular weight species while designing their materials for faster curing and better properties.

We are confident that this book will be of interest to readers from diverse backgrounds in chemistry, physics, biology, materials science and engineering, and chemical engineering. It can serve as a reference book for students and research scholars and as a unique guide for the industrial technologists.

Atul Tiwari<br>
Alexander Polykarpov

# Contents

**Chapter 1  Photocured Materials: A General Perspective**  1
*Alexander Polykarpov and Atul Tiwari*

    1.1  Current Trends and Future Avenues  1
        1.1.1  Photocured Applications  1
        1.1.2  Graphic Arts  2
        1.1.3  Adhesives and Sealants  2
        1.1.4  Barrier Coatings  3
        1.1.5  Release Coatings  3
        1.1.6  UV Powder Coatings  3
        1.1.7  Corrosion-Protection Coatings  4
        1.1.8  Automotive Refinish  4
        1.1.9  Aircraft Coatings  4
        1.1.10  Coatings for Plastics  5
        1.1.11  Can Coatings  5
        1.1.12  Wood Coatings  6
        1.1.13  Concrete Coatings  6
        1.1.14  Alkyd Paints  6
        1.1.15  Photochromic and Imaging Materials  6
        1.1.16  Photoresists  7
        1.1.17  Stereolithography  7
        1.1.18  Optics and Electronics  7
        1.1.19  Photovoltaics  8
        1.1.20  EB Crosslinked Polymers and Composites  8
        1.1.21  Nail Polish  8
        1.1.22  Dental Work  8
        1.1.23  Biomaterials  8
    1.2  An Overview of Chapter Contributions  9
    References  10

---

RSC Smart Materials No. 13
Photocured Materials
Edited by Atul Tiwari and Alexander Polykarpov
© The Royal Society of Chemistry 2015
Published by the Royal Society of Chemistry, www.rsc.org

| Chapter 2 | UV-Curable Coating Technologies | 15 |
|---|---|---|
| | Mark D. Soucek and Xiaofeng Ren | |

| | 2.1 | Introduction to UV Curing | 15 |
|---|---|---|---|
| | | 2.1.1 Advantages | 16 |
| | | 2.1.2 Disadvantages | 17 |
| | 2.2 | UV Energy and its Properties | 17 |
| | 2.3 | Equipment | 18 |
| | | 2.3.1 Light Sources | 19 |
| | | 2.3.2 Coating Methods | 21 |
| | 2.4 | Components | 21 |
| | | 2.4.1 Photoinitiator | 21 |
| | | 2.4.2 Oligomers | 23 |
| | | 2.4.3 Reactive Diluents | 27 |
| | | 2.4.4 Additives | 31 |
| | 2.5 | Free-Radical Photopolymerization | 31 |
| | | 2.5.1 Free-Radical Photoinitiator | 31 |
| | 2.6 | Cationic Photopolymerization | 34 |
| | | 2.6.1 Cationic Photoinitiators | 34 |
| | | 2.6.2 Cationic Oligomers | 36 |
| | | 2.6.3 Comparison of Cationic with Free Radicals | 39 |
| | 2.7 | Evaluation of Cure Extent | 40 |
| | | 2.7.1 Real-Time Infrared (RT-IR) Spectroscopy | 40 |
| | | 2.7.2 Confocal Raman Microscopy | 41 |
| | | 2.7.3 Photo-DSC | 41 |
| | 2.8 | Development and Prospect | 42 |
| | | 2.8.1 Oil-Based Coatings | 42 |
| | | 2.8.2 Water-Reducible UV Coatings | 42 |
| | | 2.8.3 UV Powder Coatings | 44 |
| | | 2.8.4 UV-Moisture Dual-Curable Coatings | 45 |
| | | 2.8.5 UV-Thermal Dual-Cure Coatings | 45 |
| | 2.9 | Conclusion | 45 |
| | References | | 46 |

| Chapter 3 | Newly Synthesized Photocrosslinkable Liquid-Crystalline Polymers and their Properties | 49 |
|---|---|---|
| | G. Kumar and K. Subramanian | |

| | 3.1 | Introduction | 49 |
|---|---|---|---|
| | 3.2 | Synthesis of Monomers and Polymers | 50 |
| | 3.3 | Structural Conformations by Proton and $H^1$-NMR Spectroscopy | 50 |
| | 3.4 | Structural Conformations by Proton and $^{13}C$-NMR Spectroscopy | 52 |
| | 3.5 | Molecular Weight of Polymers | 52 |

|  |  |  |
|---|---|---|
| 3.6 | Thermal Stability | 53 |
| 3.7 | Photocrosslinking Properties | 55 |
| | 3.7.1 UV Spectral Studies | 55 |
| | 3.7.2 Fluorescence Studies | 61 |
| 3.8 | Scanning Electron Microscopy (SEM) | 66 |
| 3.9 | Liquid-Crystalline Properties of Polymers | 67 |
| 3.10 | Conclusion | 72 |
| | References | 73 |

**Chapter 4 Efficient Photoinitiators for Two-Photon Polymerization** — 75
*Jan Torgersen*

|  |  |  |
|---|---|---|
| 4.1 | Introduction | 75 |
| 4.2 | Photopolymerizable Formulation | 76 |
| 4.3 | Photoinitiators and Two-Photon Efficiency | 78 |
| 4.4 | Initiator Design | 79 |
| 4.5 | Structuring Results | 80 |
| 4.6 | Discussion | 81 |
| 4.7 | Conclusion | 84 |
| | Acknowledgements | 84 |
| | References | 85 |

**Chapter 5 Inhomogeneous Photopolymerization in Multicomponent Media** — 87
*A. Veltri, A. V. Sukhov, R. Caputo, L. De Sio, M. Infusino and C. P. Umeton*

|  |  |  |
|---|---|---|
| 5.1 | Introduction | 87 |
| 5.2 | Photochemistry | 88 |
| 5.3 | Mass Transport | 92 |
| 5.4 | The System of Equations | 94 |
| 5.5 | Analytical Solutions | 94 |
| 5.6 | Numerical Solutions | 97 |
| 5.7 | Conclusions | 100 |
| | References | 101 |

**Chapter 6 Microfabrication Processes and Applications of Liquid Photosensitive Materials** — 103
*Chin-Tai Chen*

|  |  |  |
|---|---|---|
| 6.1 | Introduction | 103 |
| | 6.1.1 Fusion of Liquid and Light | 104 |
| | 6.1.2 Demand and Development of Technologies | 105 |

|  |  |  |  |
|---|---|---|---|
| | 6.2 | Fundamental Principles | 105 |
| | | 6.2.1 Characterization of Materials | 106 |
| | | 6.2.2 Photoinduced Polymerization | 106 |
| | | 6.2.3 Miniaturization of Objects | 107 |
| | 6.3 | Microfabrication Schemes | 108 |
| | | 6.3.1 Photolithography (PL) | 108 |
| | | 6.3.2 Soft Lithography (SL) | 109 |
| | | 6.3.3 Light Stereolithography (LS) | 110 |
| | | 6.3.4 Inkjet Print (IJP) | 111 |
| | 6.4 | Applications in Microdevices | 112 |
| | | 6.4.1 Microactuators | 112 |
| | | 6.4.2 Microsensors | 113 |
| | | 6.4.3 Microfluidic Components | 114 |
| | | 6.4.4 Optical Components | 114 |
| | | 6.4.5 Medical Devices | 116 |
| | | 6.4.6 Miscellaneous | 116 |
| | 6.5 | Conclusion and Outlook | 117 |
| | References | | 118 |

**Chapter 7   UV-Cured Functional Coatings** — 121
*M. Sangermano, I. Roppolo and M. Messori*

|  |  |  |
|---|---|---|
| 7.1 | Introduction | 121 |
| 7.2 | Scratch-Resistant UV-Cured Coatings and Multifunctional Coatings | 122 |
| 7.3 | Conductive UV-Cured Coatings | 125 |
| 7.4 | Photoluminescent UV-Cured Coatings | 127 |
| 7.5 | UV-Cured Coatings in Photocatalysis | 129 |
| 7.6 | Conclusions | 130 |
| References | | 131 |

**Chapter 8   Photoreactive Polymers For Microarray Chips** — 134
*Di Zhou, Ponnurengam Malliappan Sivakumar, Tae Il Son and Yoshihiro Ito*

|  |  |  |  |
|---|---|---|---|
| 8.1 | Introduction | | 134 |
| 8.2 | Photoreactive Polymers for Microarrays | | 136 |
| | 8.2.1 Photoreactive Amphiphilic Polymers | | 137 |
| | 8.2.2 Photoreactive Zwitterionic Polymers | | 139 |
| 8.3 | Applications of Photoimmobilized Microarrays | | 139 |
| | 8.3.1 Proteins | | 140 |
| | 8.3.2 Allergens | | 142 |
| | 8.3.3 Autoantigens | | 142 |

Contents   xi

|  |  | 8.3.4 | Viruses | 144 |
|---|---|---|---|---|
|  |  | 8.3.5 | Cells | 145 |
|  | 8.4 | Conclusion | | 147 |
|  | References | | | 147 |

**Chapter 9  Boron/Phosphorus-Containing Flame-Retardant Photocurable Coatings**     **150**
*Emrah Çakmakçı and Memet Vezir Kahraman*

|  | 9.1 | Flame Retardancy | 150 |
|---|---|---|---|
|  | 9.2 | Phosphorus-Containing Flame-Retardant Photocurable Coatings | 152 |
|  | 9.3 | Boron-Containing Flame-Retardant Photocurable Coatings | 178 |
|  | 9.4 | Conclusions | 183 |
|  | References | | 185 |

**Chapter 10  Lamellar and Circular Constructs Containing Self-Aligned Liquid Crystals**     **188**
*Luciano De Sio, Nelson Tabiryan and Timothy Bunning*

|  | 10.1 | Introduction | 188 |
|---|---|---|---|
|  | 10.2 | Polymeric Template | 189 |
|  |  | 10.2.1 Lamellar Polymer Templated Gratings | 191 |
|  | 10.3 | Circular Polymer Templated Gratings | 194 |
|  |  | 10.3.1 Electro-Optical Properties | 196 |
|  | 10.4 | Conclusion | 198 |
|  | Acknowledgements | | 198 |
|  | References | | 198 |

**Chapter 11  POLICRYPS: A Multipurpose, Application-Oriented Platform**     **200**
*R. Caputo, M. Infusino, A. Veltri, L. De Sio, A. V. Sukhov and C. P. Umeton*

|  | 11.1 | Introduction | 200 |
|---|---|---|---|
|  | 11.2 | POLICRYPS Photocured Materials | 201 |
|  |  | 11.2.1 POLICRYPS Photocuring Process | 202 |
|  | 11.3 | Tunable Optical Switch | 204 |
|  |  | 11.3.1 Tunable Phase Modulator | 206 |
|  |  | 11.3.2 Microsized Organic Lasing System | 210 |
|  | 11.4 | Photocuring an Arbitrary Morphology POLICRYPS | 214 |
|  | 11.5 | Conclusions | 217 |
|  | Acknowledgements | | 217 |
|  | References | | 217 |

**Chapter 12 SU-8 for Microsystem Fabrication**     **220**
*Yi Chiu and Yu-Ting Cheng*

    12.1   Introduction     220
    12.2   Material Properties and Fabrication Processes     221
           12.2.1   Material Properties     221
           12.2.2   Microfabrication Processes     221
    12.3   SU-8 Structures     223
           12.3.1   Parts and Actuators     223
           12.3.2   Microfluidic Structures and Systems     227
    12.4   SU-8 Applications     230
           12.4.1   SU-8 Sensors     230
           12.4.2   SU-8 Micro-Optical Components     233
           12.4.3   SU-8 for RF Applications     241
    12.5   Conclusion     250
    References     251

**Chapter 13 UV-Based Dual Mechanism for Crosslinking and Stabilization of PAN-Based Carbon-Fiber Precursors**     **256**
*Marlon S. Morales and Amod A. Ogale*

    13.1   Introduction     256
           13.1.1   Polyacrylonitrile Precursors for Carbon Fibers     256
           13.1.2   Stabilization Routes     257
    13.2   Experimental     259
           13.2.1   Materials     259
           13.2.2   UV Treatment     260
           13.2.3   Thermal Treatment     262
           13.2.4   Characterization     263
    13.3   Results and Discussion     265
           13.3.1   Influence of UV Radiation and Photoinitiator     265
           13.3.2   UV-Radiation Time and Temperature     269
           13.3.3   Influence of UV Radiation on Polyacrylonitrile Fibers     275
           13.3.4   Influence of UV Radiation and Photoinitiator on Carbon Fibers     281
           13.3.5   Mechanical Properties     283
    13.4   Conclusions     286
    Acknowledgements     287
    References     287

Contents

## Chapter 14 Analytical Methods for Determining Photoinitiators in Food-Contact Materials — 290
*M. A. Lago, R. Sendón and A. Rodríguez-Bernaldo de Quirós*

- 14.1 Introduction — 290
- 14.2 Photoinitiators — 292
  - 14.2.1 Chromatography — 292
  - 14.2.2 Voltammetry — 312
  - 14.2.3 DART — 313
- 14.3 Nonadditionally Intended Substances (NIAS) — 313
  - 14.3.1 Study of Byproducts — 313
  - 14.3.2 Nontargeted Screening Analysis — 314
- 14.4 Conclusions and Future Overview — 315
- Acknowledgements — 316
- References — 316

## Chapter 15 Methacrylate and Epoxy Resins Photocured by Means of Visible Light-Emitting Diodes (LEDs) — 321
*Claudia I. Vallo and Silvana V. Asmussen*

- 15.1 Introduction — 321
- 15.2 Light-Emitting Diodes — 322
- 15.3 Light-Curable Methacrylate Resins — 326
  - 15.3.1 Methacrylate Monomers — 326
  - 15.3.2 Photopolymerization Reactions of Methacrylate Resins — 327
  - 15.3.3 Photon-Absorption Efficiency — 329
  - 15.3.4 Light Attenuation in Thick Layers — 331
  - 15.3.5 Oxygen Inhibition in Free-Radical Polymerization — 335
  - 15.3.6 Conversion of Methacrylate Groups under Irradiation — 335
- 15.4 Light-Curable Epoxy Resins — 339
  - 15.4.1 Epoxy Monomers — 339
  - 15.4.2 Cationic Photoinitiators — 340
  - 15.4.3 Cationic Photopolymerization of Epoxy Monomers — 341
- 15.5 Conclusion — 345
- References — 345

## Chapter 16 Waste Materials Cured and Modified by Irradiating and their Use in Concrete — 347
*Gonzalo Martínez-Barrera and Osman Gencel*

- 16.1 Introduction — 347

| | | |
|---|---|---|
| 16.2 | Waste Materials Cured and Modified by using Gamma Radiation | 348 |
| 16.3 | Recycled and Waste Materials used in Concrete | 351 |
| 16.4 | Irradiated Concrete Containing Waste and Recycled Materials | 354 |
| 16.5 | Conclusions | 357 |
| Acknowledgements | | 358 |
| References | | 358 |

**Subject Index**     **360**

CHAPTER 1

# Photocured Materials: A General Perspective

ALEXANDER POLYKARPOV*[a] AND ATUL TIWARI*[b]

[a] AkzoNobel Packaging Coatings, Strongsville, OH, USA; [b] Department of Mechanical Engineering, University of Hawaii at Manoa, Honolulu, HI, USA
*Email: alexander.polykarpov@akzonobel.com; tiwari@hawaii.edu

## 1.1 Current Trends and Future Avenues

### 1.1.1 Photocured Applications

Photocured materials are obtained by photoinduced hardening or cross-linking of various monomer-, oligomer-, and polymer-based compositions. Most commonly these materials are cured by UV, visible light, and electron beam (EB). A wide range of inexpensive UV photoinitiators coupled with the compactness, simplicity, and relatively low cost of UV equipment allows UV-cured materials to be used much more often and the interest in photocured materials has continued to grow[1] in recent years.

Photocured materials are commonly used as coatings, inks[2] and adhesives. The global consumption of photocurable coatings, inks and adhesives was 868 million pounds in 2012 worth $4.94 billion, according to a recent study from Kusumgar, Nerlfi & Growney Inc.[3] Photocured materials have also been the mainstay of photolithographic applications playing an important role in the creation of microchips and printed circuit boards. Due to the very rapid and low-temperature cure these single-component systems also found uses in biomedical applications from nail polish and dental restorations to providing scaffolds for tissue and organ regeneration. Increasing demands for higher productivity and lower emissions of volatile organic chemicals (VOC) continue to support the expansion of photocured materials into various areas of human activities.

### 1.1.2 Graphic Arts

As the printing market declined due to reduced use of newspapers, magazines, and paper books the growth in the UV and electron beam (EB) -curable printing inks and coatings continued to be in the packaging applications. Especially attractive has been the use in food packaging where there is a continued need for safer, faster, and cheaper inks, functional coatings, and overprint varnishes.

Recent research in the photocurable inks area has been focused on replacing mercury-vapor-based UV lamps with UV LED cure sources and developing better performing systems with low migration and low odor for use in food packaging. Acrylate oligomers developed for use in UV/EB curable inks have to be fast curing, disperse multiple pigments well, and provide suitable rheology to the ink formulations. Most commonly these are polyester acrylates,[4] though aliphatic epoxy acrylates and polyurethane acrylates are also used. Various special effect inks continue to be developed.[5]

UV inkjet inks are being used to make packaging labels and typically suffer from fewer delamination issues than conventional inks. These inks also allow for easy inline processing such as hot stamping, die cutting, embossing and others.[6] Plastic films are well suited for UV inkjet printing.[7]

EB cure often provides a viable alternative to UV for inks and overprint varnishes.[8] EB cure has been found to be especially effective in food-packaging applications where it leads to higher conversions of photocurable materials, while also removing the need for migrating photoinitiators and cure accelerators. Being an ionizing radiation with higher depth of penetration EB allows for much better through cure of pigmented coatings and inks. EB also enables cure through opaque to UV and visible-light multilayered substrates. EB coatings can be sufficiently durable to replace more expensive film lamination in paper-bag packaging.[9]

### 1.1.3 Adhesives and Sealants

Photocured materials are being used in laminating and pressure-sensitive adhesives and even as structural adhesives.[10] However, they still face a tough competition with the often less expensive on the per weight unit basis solvent, waterborne, and 100% solids two-component systems. Providing a low migration bonding layer for substrates used in food packaging presents an additional level of challenge due to the lower functionality of the oligomers and monomers used in adhesives than in coatings. This is often coupled with the difficulties in curing through opaque bonded substrates.

EB-curable adhesives have recently seen better growth due to superior properties and better process control of EB cure.[11]

Due to the speed of cure and ease of application UV-curable sealants found use in manufacturing of ammunition rounds.[12] In this application UV-curable sealant is applied to the joint between the brass casing and the projectile of the ammunition cartridge after the cartridge has been fully assembled and crimped. The sealant penetrates into the joint by capillary action and is then UV cured to form a waterproof joint. A specially designed LED UV source is used to obtain a narrow strip of UV exposure with well-controlled uniformity.

UV-curable hot melt adhesives continue to be developed and often have superior properties to the conventional systems. UV-cured hot melt adhesives find uses in specialty pressure sensitive tapes, construction, and medical applications.

### 1.1.4 Barrier Coatings

Films and plastic containers with gas barrier coatings found multiple uses in food packaging where protection of flavor and extension of the life of packaged food and beverages is of great importance. Photopolymerizable thiol–ene systems were shown to form films with oxygen permeability that can be dialed from very low to very high depending on the glass-transition temperature of the created polymer network. Modification of these networks with amine functional groups was found to create very good oxygen barrier films.[13] Surface modification of clays with acrylate or thiol-functional groups coupled with the selection of suitable exfoliating surfactants was shown to lead to UV-curable formulations with improved mechanical properties. Such formulations were also shown to result in better gas and water barrier properties.[14]

Various other barrier and sensor applications became possible *via* printing of the functional components using UV- and EB-cure processes.[15] The opportunity in gas and especially oxygen barrier coatings remains to be open for photocured coatings with high flexibility and adhesion to various plastic substrates.

### 1.1.5 Release Coatings

Release coatings are usually applied either on film or paper[16] substrates and used as removable liners for pressure-sensitive adhesive films or as protective films for various displays. Modern UV-curable release coatings are based on acrylated or epoxy functional polysiloxanes.[17]

### 1.1.6 UV Powder Coatings

UV powders are typically based on methacrylates and unsaturated polyesters. These formulations are applied by electrostatic spray to a substrate followed by melting of the powder in an oven and UV cure of the melt. The temperature required for melting is lower than what is required for thermally cured powder coatings and is typically around 120 °C.[18] UV powder coatings are used on a variety of substrates such as metals, medium density fiberboard (MDF), structural foams, plastics, composites and other heat-sensitive materials.[19] Creating cost-effective photocurable polymer powders with better resistance to sintering during storage and shipment remains an industry challenge.

### 1.1.7 Corrosion-Protection Coatings

UV-cure technology capabilities have been extended to the high-performance corrosion protection of steel and aluminum surfaces in industrial applications. Thinner and faster to cure UV coatings were shown to have corrosion protection similar to solvent based two-component urethane and epoxy systems.[20] Adhesion is often a challenge for UV-cured protective coatings on metal and some materials were suggested to address this issue.[21] Very large commercial opportunities for coatings with corrosion protection for metal remain in marine, automotive and aerospace uses, where there is a strong desire to move to rapidly curing, solvent-free and isocyanate-free systems. One of the key challenges remains to be the need for a well-working system that would combine exceptional durability and weatherability of the cured coatings with easy to use and relatively inexpensive curing and application devices. Recent developments in robotic UV systems have been made that allow precision application and cure of UV coatings on surfaces with various curvature and complexity including large vertically positioned surfaces.[22]

UV-curable coatings based on halogenated phenyl acrylates were shown to inhibit biofouling and biocorrosion on plastic panels immersed in waste water.[23]

### 1.1.8 Automotive Refinish

UV coatings are slowly making way into the automotive refinish area.[24] Among the challenges is the need to cure around corners and in cracks. Dual cure and photolatent catalyst approaches have been used to address these issues.[25] Recent advances in UV LED resulted in creation of hand-held cure devices that do not emit in the UVB and UVC range of the spectrum and do not generate ozone, which is expected to accelerate the adoption of the UV-cure technology by the automotive refinish and other applications where the materials are field applied and cured.

### 1.1.9 Aircraft Coatings

UV coatings have found uses in aircraft exterior applications where faster process flow for coatings is often sought.[26] Since the aircraft exterior finishing takes place at the end of the build process the urgency to avoid delivery delays while holding a very expensive inventory pushes the manufacturers to selection of faster curing coatings and UV is one of the options being considered.[27] UV coatings are also being considered as replacement for thermally cured coatings for the exhaust vent areas,[28] where the coatings can be subjected to temperatures as high as 150 °C. Recent developments in UVA LED cure sources and matching photoinitiating systems resulted in increased interest in field repair and stencil coatings for aircraft. These UVA-curable coatings have been shown to have performance that rivals that of the conventional 2K urethane systems.[29]

The idea of creating shark-skin-inspired microscale patterns on the surface of aircraft to improve fuel economy by reducing drag was recently tested using UV-curable materials.[30]

### 1.1.10 Coatings for Plastics

Creation of stable dispersions based on submicrometer scale particles of alumina and silica[31] along with the improvements in polyurethane acrylates led to creation of a new generation of scratch-resistant coatings with exceptional wear resistance. More recent developments in functionalized silicas, hybrid alkoxysilane-epoxy and acrylate systems,[32] and waterborne polyurethane acrylates[33] made UV-curable coatings especially useful for such easy to scratch plastics like polycarbonate, which is widely used in automotive headlamps. Various consumer electronics devices, optical discs, eye glasses, and even vinyl flooring have also benefited from UV-cured scratch- and abrasion-resistant coatings.

Photocured coatings are typically praised for their high gloss. Matte finish or low gloss photocured coatings have been less common due to the challenges in their formulation.[34] Recently, there has been an increased interest in matte UV coatings driven in part by the increasing demand for matte finishes in plastic packaging. There is also a need, particularly for the automotive and consumer electronics markets, for "soft touch" rapidly cured at low-temperature coatings where despite multiple attempts the balance of the soft to touch, velvety texture with high chemical resistance has not yet been attained.

UV-cured coatings are also used as base and top coats for metallization of substrates using physical vapor deposition techniques.[35] The "chrome look" has found uses in automotive, cosmetics and home-appliances markets.

### 1.1.11 Can Coatings

UV coatings found use in protection of beverage can rims from abrasion damage. The coating also allows for easier movement of the cans on conveyors during manufacturing and filling.[36] UV inks are also used in metal container decoration, although to a smaller extent than solvent-based inks.

## 1.1.12  Wood Coatings

Modern coatings for parquet flooring are based on UV-curable formulations. These coatings are multilayered systems consisting of the bottom layer, which is typically a water borne UV-curable formulation, followed by 100% solids UV-curable formulations layers to provide the necessary abrasion and scratch resistance.[37] UV-cured coatings are still used in multiple furniture and kitchen cabinetry applications. 100% solids UV lacquer was found to be the best alternative from an environmental point of view as a surface coating for wood furniture.[38]

## 1.1.13  Concrete Coatings

UV-curable formulations have been developed for coating concrete floors.[39] While these coatings provide almost instantaneous cure and are zero to low VOC they continue to face tough competition with the two-component rapid-cure technologies. Repair of rail seat abrasions on concrete ties has been shown to be very effective when using UV-cured materials.[40]

## 1.1.14  Alkyd Paints

Alkyd paints are based on oxidative drying of the unsaturated fatty acid esters and are used in households to paint doors, trim, and cabinets. Cobalt salts are typically used to accelerate the drying that otherwise can take more than 24 h. Recent studies showed that cobalt salts could be linked to adverse health and environmental effects, which forced the paint industry to seek alternatives to cobalt-based accelerators of the drying. A photochemical system was shown to be a viable candidate for a photoassisted autoxidation drying process in alkyds.[41]

## 1.1.15  Photochromic and Imaging Materials

A full-color imaging system based on photopolymerizable pressure-sensitive microcapsules was invented in 1984 for use in color copying.[42] The system was later developed into a visible light and pressure-sensitive single-sheet full-color printing medium.[43] More recent systems were developed to color irreversibly directly upon the exposure of photopolymerizable materials to UV and even EB without requiring secondary mechanisms such as heat or pressure for color development.[44] These systems found uses as cure indicators in photopolymerizable systems, as sensors and dosimeters for UV and EB radiometry,[45] and presented an opportunity for various photocured formulations to develop color as part of the curing process. Photocurable color imaging systems based on structural color using magnetically tunable photonic crystals are also being developed.[46]

Photocured materials can also be used as parts of the image forming or transferring devices. UV-curable formulations containing carbon nanotubes found uses in electrostatographic devices as transfer belts.[47] Such formulations were shown to have better control over the uniformity of electrical resistivity and superior mechanical properties to solvent-based thermoplastic and thermosetting systems.

### 1.1.16 Photoresists

Photocure can be used to prepare negative-tone photoresists, where photopolymerization or crosslinking can be used to render the material insoluble in the developer. Positive-tone photoresists, where the polymer is rendered soluble in the developer after the exposure, are typically not prepared by photopolymerization and can be classified as photosensitive, rather than photocurable materials.

Relatively simple photopolymerizable systems can be used to make low-resolution stencils and masks for etching glass and ceramics. Flexography is based on the use of the printing plates prepared from photocurable materials image wise exposed by UV.[48] Stencils for screen printing can also be prepared using photoresists.

High-resolution nanolithography and patterning is obtained using immersion 193 nm systems in making microchips.[49] EUV (extreme UV), EB, X-ray, and other imaging technologies are under different stages of evaluation and development for fabricating even smaller nanometer scale patterns for next-generation microchips.

### 1.1.17 Stereolithography

UV-curable formulations find use in stereolithography – a process of printing 3D objects using a computer-controlled laser layer-by-layer using photocurable materials. The photopolymerization takes place using a 355 nm laser and features as small as 40 µm in the lateral direction can be obtained.[50] Two-photon absorption-induced photopolymerization allowed to further expand the capabilities in creating 3D microstructures using photocured materials – features as small as 120 nm can be created.[51]

### 1.1.18 Optics and Electronics

UV-curable coatings for optical fiber have very stringent requirements for resistance to humidity and temperature variations. Acrylate monomers used in such formulations also have very high purity requirements to help minimize potential data transmission losses. As optical fiber began to be used in subterranean exploration for natural resources additional requirements for the coatings were added such as resistance to high pressure and corrosive environment of the drilling wells.[52]

Radiation-curable materials are used in integrated optical and optoelectronic applications (optical discs, diffraction gratings, antireflective coatings, imaging sensors and photonic devices).[53] UV-curable conductive inks are very suitable for making printed electronics for uses in RFID tags and OLED displays.[54]

UV-curable formulations found use in liquid-crystal displays.[55] Anti-reflective coatings for displays can also be obtained by using UV-cured coatings and films.[56]

### 1.1.19 Photovoltaics

UV-curable materials were shown to be capable of meeting the performance requirements for use in photovoltaic structures as adhesives.[57] UV-curable hard coatings found use in silvered films used in production of lead-free mirrors, where they greatly reduced the weight and cost of the system.[58]

### 1.1.20 EB Crosslinked Polymers and Composites

Low-energy EB (120 keV to 300 keV) is typically used to cure inks, coatings and adhesives. Medium-energy EB (300 keV to 5 MeV) is used to crosslink wire and cable insulation and to modify and cure fiber and composites.[59] High-energy (10 MeV) EB crosslinking of high density polyethylene was shown to result in superior dielectric properties *vs.* thermal crosslinking using peroxides.[60] Although, low-energy EB has also been found to be useful in crosslinking low-density polyethylene.[61] EB crosslinked wood–plastic composites are being developed to present a sustainable and nontoxic alternative to pressure-treated lumber for building materials.[62]

### 1.1.21 Nail Polish

UV-curable coatings have been a common alternative to solvent-based formulations in nail-polish applications. Polyurethane methacrylates are typically the basis of these formulations.[63] Renewable UV-curable coating materials are sought for this use.[64]

### 1.1.22 Dental Work

Photocured materials have been used in restorative dentistry since 1969.[65] The cure is now done using visible light, typically the blue region of the electromagnetic spectrum and often blue 450 nm LEDs are used.[66] Camphorquinone-based photoinitiating systems are most common, though other photoinitiating systems are constantly being developed in order to obtain faster and better curing composite systems with minimum shrinkage.[67]

### 1.1.23 Biomaterials

Photocured materials can be used in stereolithography based construction of artificial joints[68] or as scaffolds supporting tissue repair. Hydrogels are used as scaffolds for tissue engineering and can be obtained from

Photocured Materials: A General Perspective      9

photocured materials.[69] The use of UV-cured materials for creation of nerve-implanting surfaces with high spatial resolution may lead to creation of prosthetic materials with better outcomes for patients.[70]

## 1.2   An Overview of Chapter Contributions

Section 1.1 described several interesting applications of photocured materials. It was difficult to cover all the mentioned areas due to scarce of available literature and proprietary nature of technologies. This section briefly details the technical articles included in this book.

The area of UV light-cured coatings has been well established, and the coating industry was probably the beginner to adopt the use of photocuring. Soucek *et al.*, in Chapter 2 describes the fundamentals, the chemical components and their use in the development of photocured coatings. The authors also detail the mechanism of cationic photopolymerization and free-radical photopolymerization along with the advantages and disadvantages of both the methods. Novel synthesis of photocrosslinkable polymers containing pendant chalcone moieties with different substituents has been described by Kumar *et al.*, in Chapter 3. The authors utilized the free-radical polymerization method for the development and claimed that polymers containing pendant chalcone moiety exhibits liquid-crystalline behavior.

Torgersen *et al.*, in Chapter 4 briefly describe laser-assisted two-photon polymerization. Their precise fabrication technique enabled the authors to create three-dimensional micro- and nanostructures. The chapter provides insight into optimization of photoinitiators to absorb two photons for the radical formation leading to photopolymerization. Formation of polymer liquid-crystalline composite material by a photoinduced inhomogeneous curing pattern has been shown by Veltri *et al.* in Chapter 5. The authors found that two parameters related to diffusion and curing intensity primarily governed the structures of the final product.

It is worth mentioning that photosensitive materials are now playing a vital role in precise fabrication of micro/nanocomponents and devices. Chen, in Chapter 6, comprehensively reviews the fundamental principles and interdisciplinary approach behind the development of miniaturized objects. The author also describes various schemes of microfabrication in photolithography, light stereolithography, soft lithography and inkjet printing. The developments of UV-curable scratch-resistant, functional coatings are detailed by Sangermano *et al.* in Chapter 7. Several variations of coating containing series of nanoparticles were achieved by photocuring technique. Zhou *et al.* in Chapter 8 demonstrate a photoimmobilization technology that uses photoreactive and nonbiofouling polymers for the preparation of microarray biochips. The authors expect that use of this technology will assist in understanding the fundamental aspects of biological interactions, along with other useful applications in clinical analysis. Cakmakci *et al.*, study the flame-retardant photocurable coating containing boron and phosphorous in Chapter 9. The effectiveness of various monomers and additives in flame retardancy is discussed in detail. Similar to Veltri *et al.*, in Chapter 10, Sio *et al.*, demonstrated the fabrication of curved periodic microstructures containing self-aligned liquid crystals. The holography and lithography techniques assisted the authors in developing periodic nano-/microcomposite gratings. The diversified nature of polymer liquid-crystal polymer slices (POLYCRIPS) technology is mentioned by Sio *et al.* as well as in Chapter 11 by Caputo *et al.* The author's study suggested that POLYCRIPS technology is valuable for switchable diffraction phase grating, switchable optical phase modulator, arrays of mirrorless optical microresonators for tunable lasing effects and a one-step fabrication of fork gratings.

The use of photocuring in fabrication of thick high aspect ratio structures is shown by Chiu *et al.*, in Chapter 12. The authors have selected a negative-tone photoresist to fabricate complex three-dimensional structures. The basic material properties and fundamentals of the fabrication process related to the selected negative photoresists are summarized. In Chapter 13, Morales *et al.* adopted a unique approach for rapid crosslinking and stabilization of carbon fibers. The authors noticed that with the use of short UV treatment, precursor fibers could be thermo-oxidatively stabilized and successfully carbonized rapidly. The resulted fibers retained a higher extent of molecular orientation and displayed higher tensile modulus.

Although photocured technologies have demonstrated extremely high potential in various commercial arenas, they show limitations when used in food contact. The photoinitiators from packaging materials tend to leach out into food products that poses serious human-health issues. Lago *et al.*, in Chapter 14 review the methods available for the detection and determination of photoinitiators in food packaging and foodstuffs. The authors comprehensively summarize the state-of-the-art systems for the determination of nonintentionally added substances derived from the photoinitiators.

Apart from conventional UV light lamps for photocuring and photopolymerization, LED light curing and polymerization show significant promises. Vallo *et al.*, in Chapter 15 suggest that since spectral output of LED sources is concentrated in a comparatively narrow wavelength range, more efficient curing is possible, resulting in reduced curing time and increased depth of cure compared to conventional light sources. Moreover, with LED curing there will be less heat transfer to the substrate without the possibility of harmful UV rays. Finally, in Chapter 16, Barrera *et al.*, discusses the importance of gamma-radiation technology for the structural and physicochemical modification of waste materials such as PET bottles, TetraPak® packing containers and tire rubbers. Additionally, the utilization of such modified waste as fillers and reinforcement in concrete is emphasized.

# References

1. Y. Yagci, S. Jockush and N. Turro, Photoinitiated Polymerisation: Advances, Challenges, and Opportunities, *Macromolecules*, 2010, **43**, 6245.

2. P. Glockner, T Jung, S. Struck, K. Studer, *Radiation Curing Coatings and Printing Inks*, Vincentz Network, Hannover, 2008, vol. 5, p. 126.
3. Kusumgar, Nerlfi & Growney Inc., *The Global Radiation-Cured Products Industry 2012–2017 Coatings, Inks, Adhesives*, May 2013. A snap shot of the study can be found at: http://www.kusumgar-nerlfi-growney.com.
4. S. Yin and M. O'Brien, New Polyester Polyols for Energy-Curable Ink Applications, *Radtech Rep.*, 2013, **1**, 18.
5. G. Iftime, G. Allen, M. Birau, C. Wagner, D. Vanbesien, J. Wosnick and P. Odell, Radiation Curable Ink Containing Fluorescent Nanoparticles, US8222313, 2012.
6. J. Stines, UV Technology Places Hub Labels at Center of Innovation, *Radtech Rep.*, 2012, **2**, 17.
7. J. Samuel and J. Renner, UV Inkjet Label Printing Getting it Right on the Customer's Substrate, *Radtech Rep.*, 2011, **Summer**, 11.
8. M. Laksin, S. Evans, K. Fontaine and S. Chatterjee, EB Gravure: Novel Printing Concept for Sustainable Packaging, *Radtech Rep.*, 2011, **Summer**, 22.
9. I. Rangwalla, *Electron Beam Curing: An Important Option for Sustainable Packaging, Radtech UV&EB 2012*, conference proceedings.
10. C. Bachmann and S. Cantor, UV Structural Adhesives and Sealants, *Adhes. Age*, 1999, April, 24.
11. S. Lapin, EB Technology for Pressure-Sensitive Adhesive Applications, *Radtech Rep.*, 2011, **Fall**, 9.
12. P. Dehkordi and J. Lennox, An Explosion Proof UV Curable Sealant System, *Adhes. Seal. Indus.*, 2011, **May**, 35.
13. J. Goetz, L Kwisnek and S. Nazarenko, From Gas Barriers to High Gas Flux Membranes: UV-Cured Thiol-ene Networks for Transport Applications, *Radtech Rep.*, 2012, **4**, 27.
14. S. Kim and C. Guymon, Photopolymer-Clay Nanocomposite Performance Utilizing Different Polymerizable Organoclays, *Radtech Rep.*, 2013, **1**, 33.
15. G. Sadler, *Activating Surfaces with Antimicrobial, Chemical, and Biochemical Functionalities Using Energy-Curable Polymers, Radtech UV&EB 2012*, conference proceedings.
16. S. Irifune and A. Itagaki, Low Viscosity Ultraviolet Curable Silicone Composition for Release Paper, US8211547, 2012.
17. I. Khudyakov, M. Hawkins, S. Barth and L. Winckler, *Release Coatings: Defining Performance, Radtech UV&EB 2012*, conference proceedings.
18. V. Maurin, X. Allonas, C. Croutxe-Barghorn, J. Bessieres and A. Merlin, UV Powders for Wood Coatings, *Radtech Rep.*, 2012, **2**, 12; Also see: L. Moens, N. Knoops and D. Maetens, Radiation Curable Low Gloss Powder Coating Compositions, US7816420, 2010.
19. R. Schwarb, M. Knoblauch and R. Rutherford, Innovation and Sustainability Drive UV-Cured Powder Coating Developments, *Radtech Rep.*, 2012, **3**, 17.

20. B. Curatolo, UV Technology for High-Performance Industrial Applications, *Radtech Rep.*, 2011, **Fall**, 24.
21. E. Spyrou and M. Ebbing-Ewald, Method for Producing Radiation Curable Formulations with Increased Corrosion Protection on Metal Substrates and Formulations of this Type, US8222312, 2012.
22. P. Mills, *Robotic UV Curing for 3D Parts, Radtech UV&EB 2012*, conference proceedings.
23. B. Cavitt, C. Holt and J. Lowry, UV-Curable Coatings as Inhibitors to Biofouling and Biocorrosion, *Radtech Rep.*, 2012, **March**, 19.
24. B. Bonham, Status of UV in the Automotive Refinish Market, *Radtech Rep.*, 2009, April/May/June, 16; M. Dvorchak, *1 K UVA Automotive Refinish: Clear Coats and Primers, Radtech UV&EB 2014*, conference proceedings.
25. M. He, X. Huang, Z. Zeng and J. Yang, Phototriggered Base Proliferation: A Highly Efficient Domino Reaction for Creating Functionally Photo-Screened Materials, *Macromolecules*, 2013, **46**, 6402; For a review on photobases see: K. Dietliker, K. Misteli, A. Carroy, E. Peregi, B. Blickenstorfer and E. Sitzmann, *Basics and Applications of Photopolymerization Reactions*, ed. J. Fouassier and X. Allonas, Research Signpost, 2010, vol. 1; For an example of photoacid generation see: J. Crivello and J. Lam, Diaryliodonium Salts. A New Class of Photoinitiators for Cationic Polymerization, *Macromolecules*, 1977, **10**, 1307.
26. M. Bowman, UV-Curable Aerospace Primer and Topcoat, *Radtech Rep.*, 2011, **Spring**, 9.
27. R. Baird, UV-Curable Paints for Commercial Aircraft Exteriors, *Radtech Rep.*, 2011, **Fall**, 40.
28. R. Baird, Progress Toward a Heat-Resistant UV-Curable Clearcoat for Aircraft Exteriors, *Radtech Rep.*, 2013, **3**, 9.
29. C. Williams, M. Dvorchak and C. Gambino, Development of UV-A Curable Coatings for Military Aircraft Topcoats, *Radtech Rep.*, 2011, **Spring**, 13.
30. http://www.lufthansa-technik.com/en/multifunctional-coating, accessed on 25th June, 2012.
31. R. Cayton, *Enhanced Wear Protection of UV-Cured Clear Coats using Sub-Micron Aluminum Oxide Additives, Radtech UV&EB 2012*, conference proceedings.
32. C. Belon, A. Chemtob, C. Croutxe-Barghorn, S. Rigolet, V. Le Houerou and C. Gauthier, *J. Polym. Sci., Part A: Polym. Chem.*, 2010, **48**, 4150.
33. P. De Groote, M. Tielemans, J. Zhang and M. Hutchins, High Solids Solvent-Free and Sprayable Waterborne UV Hardcoat Resins, *Eur. Coat. J.*, 2013, **10**, 20.
34. R. Behl, R. Mertsch, K. Schottenloher and M. Nargiello, *Novel Matting Agent for Low Gloss UV Coatings, Radtech UV & EB 2012*, conference proceedings.
35. J. T. Eich, E. Weber, P. Abell, K. Wagner and C. Mack, Opportunities for Functional Design Using Physical Vapor Deposition and UV-Curable Coatings, *Radtech Rep.*, 2014, **1**, 30.

36. S. Colville, UV-Cured Coatings Speed up the Beverage Can Production Process, *Radtech Rep.*, 2011, **Fall**, 28.
37. K. Menzel and K. Sass, Formulating UV Parquet Coatings to Improve Adhesive, Abrasion and Scratch Resistance, *Radtech Rep.*, 2009, **January/February/March**, 17.
38. L. Gustafsson and P. Borjesson, Life Cycle Assessment in Green Chemistry, *Int. J. Life Cycle Assess.*, 2007, **12**, 151.
39. P. Weissman, Field-Applied, UV-Curable Coatings for Concrete Flooring, *Radtech Rep.*, 2009, **January/February/March**, 23; J. Arceneaux, UV-Curable Concrete Coatings, *Radtech Rep.*, 2009, **January/February/March**, 33.
40. D. Delmonico and F. Weber, Methods for Repair and Preventive Maintenance of Railroad Ties Using UV Curable Polymers, US20110206867, 2011.
41. F. Courtecuisse, C. Ley, G. Ye, C. Croutxe-Barghorn, X. Allonas, P. Taylor and G. Bescond, Photoassisted Drying of Alkyd Resins: Towards Environmentally Friendly Paints, *Prog. Org. Coat.*, 2012, **73**(4), 366.
42. F. Sanders, G. Hillenbrandt, J. Arney and R. Wright, Photocopy Sheet Employing Encapsulated Radiation Sensitive Composition and Imaging Process, US4440846, 1984.
43. J. Camillus, M. Johnson, J. Taylor, D. Terry, W. Lippke and T. Brammer, Self-Containing Imaging Assembly, US 5783353, 1998.
44. A. Mejiritski, Color Change Technology: Applications for Radiation Curing and the Consumer, *Radtech Rep.*, 2009, **April/May/June**, 24.
45. A. Mejiritski, O. Grinevich, M. Swain and I. Rangwalla, *Quick and Easy Way to Characterize Low Voltage (80-125 kV) EB Accelerators Using Fast Check Strips, Radtech UV & EB 2010*, conference proceedings.
46. H. Kim, J. Ge, J. Kim, S. Choi, H. Lee, H. Lee, W. Park, Y. Yin and S. Kwon, Structural Colour Printing Using a Magnetically Tunable and Lithographically Fixable Photonic Crystal, *Nat. Photonics*, 2009, **3**, 534.
47. G. Foley, J. Wu and S. Mishra, UV Cured Heterogeneous Intermediate Transfer Belts (ITB), US8541072, 2013.
48. P. Richter, A. Wigger, G. Fahrbach, E. Seiler and H. Barzynski, Laminates for the Manufacture of Flexographic Printing Plates Using Block Copolymers, US4162919, 1979.
49. D. Sanders, Advances in Patterning Materials for 193 nm Immersion Lithography, *Chem. Rev.*, 2010, **110**, 321.
50. B. Rundlett, UV Curing in a 3-Dimensional World, *Radtech Rep.*, 2013, **3**, 21.
51. K. Belfield, C. Yanez, Z. Huang and A Novel, Two-Photon Absorbing PAG and Its Application in 3-D Microfabrication, *Radtech Rep.*, 2007, **September/October**, 42.
52. D. Wilson, M. Ellison and X. Zhang, Designing UV-Curable Materials for High Temperature Optical Fiber, *Radtech Rep.*, 2007, **September/October**, 24.
53. J. Wang, M. Hutchins, K. Woo, T. Konish and M. Idacavage, Halogen-Free, Radiation-Curable, High-Refractive Index Materials, *Radtech Rep.*, 2010, **September/October**, 11.
54. K. Eiroma, L. Hakola, J. Hast, A. Maaninen and J. Petaja, UV Curing for Printed Electronics, *Radtech Rep.*, 2007, **September/October**, 31.

55. J. Kim, T. Song, Y. Nam and Y. Cho, UV Curable Liquid Prepolymer and Liquid Crystal Display Device Using the Same and Manufacturing Method Thereof, US8105501, 2012.
56. J. Kang, M. Han, Y. Lee, Y. Hong and Y. Chang, UV Curable Antireflective Coating Composition, Antireflective Film Using the Same, and its Manufacturing Method, US8110249, 2012.
57. J. Shaw and J. Arceneaux, *Suitability of Energy Curable Oligomers and Monomers as Encapsulants in Photovoltaic Modules, Radtech UV&EB 2012*, conference proceedings.
58. G. Jorgensen, M. GiGrazia, K. Wagner and P. Peach, UV-Curable Hardcoats Enter Solar Energy Market, *Radtech Rep.*, 2010, **September/October**, 27.
59. A. Chmielewski, M. Al-Sheikhly, A. Berejka, M. Cleland and M. Antoniak, Recent Developments in the Application of Electron Accelerators for Polymer Processing, *Radiat. Phys. Chem.*, 2014, **94**, 147.
60. N. Goodarzian and M. Shamekhi, Chemical Crosslinking versus High Energy Electron Beam Crosslinking of HDPE: Electrical Properties Study, *J. Iran. Chem. Res.*, 2009, **2**, 189.
61. S. Norasetthekul, R. Reddy and S. McCarthy, Material Effects of Low-Energy EB Treatment of Polypropylene and Low-Density Polyethylene, *Radtech Rep.*, 2010, **April/May/June**, 42.
62. A. Palm, J. Smith, M. Driscoll, L. Smith and S. Larsen, *Flexural Strength Properties of an Electron Beam Radiation Crosslinked Wood-Plastic Composite, Radtech UV&EB 2014*, conference proceedings.
63. T. Barclift, *Chemistry & Formulation of UV Curable Nail Enhancements, Radtech UV&EB 2012*, conference proceedings.
64. L. Steffier, UV Curable Nail Coating Formulations Based on Renewable Polyols, US8574558, 2013.
65. H. Strassler, The Physics of Light Curing and Its Clinical Implications, *Compend. Contin. Educ. Dent.*, 2011, **32**(6), 70.
66. N. Malhotra and K. Mala, Light-Curing Considerations for Resin-Based Composite Materials: a Review, Part I, *Compend. Contin. Educ. Dent.*, 2010, **31**(7), 498.
67. C. Gorsche, M. Griesser, C. Hametner, B. Ganster, N. Moszner, D. Neshchadin, A. Rosspeinter, G. Gescheidt, R. Saf and R. Liska, *New Generation of Highly Efficient Long-Wavelength Photoinitiators for Dental Restoratives: From Research to Applications, Radtech UV&EB 2014*, conference proceedings.
68. T. Vanasse, G. Gupta and J. Meridew, Metallic Structures Having Porous Regions from Imaged Bone at Predefined Anatomic Locations, US20140025181, 2014.
69. J. Sun, X. Zhao, W. Illeperuma, O. Chaudhuri, K. Oh, D. Mooney, J. Vlassak and Z. Suo, Highly Stretchable and Tough Hydrogels, *Nature*, 2012, **489**, 133.
70. B. Tuft, L. Xu, A. Hangartner, S. White, M. Hansen, and A. Guymon, *Photopolymerized Patterning and Materials to Enhance Neural Prosthetic Performance, Radtech UV&EB 2014*, conference Proceedings.

CHAPTER 2

# UV-Curable Coating Technologies

MARK D. SOUCEK* AND XIAOFENG REN

Department of Polymer Engineering, University of Akron, OH, USA
*Email: msoucek@uakron.edu

## 2.1 Introduction to UV Curing

It has been acknowledged for decades that the presence of volatile organic compounds (VOCs) in the atmosphere can not only contribute to serious air-pollution problems due to their toxic or carcinogenic characteristics, but also lead to global warming indirectly.[1] Due to concern over these problems, regulations regarding emission of VOCs into the atmosphere were established. Most solvents have been classified by the US Environmental Protection Agency (EPA) as photochemically reactive, and their usage has been regulated since the 1970s to reduce air pollution.[2] In 1990, the US Congress listed certain common solvents as hazardous air pollutants (HAPs), limiting their usage and creating pressure on the coatings industry to look for solutions in order to reduce solvent emissions.[3]

Besides the regulation of VOCs the coating industry is facing today, energy is another concern, especially the consumption of fossil-fuel resources since polymer and coatings fields depend mainly on petrochemical products derived from crude oil that will become scarce in the future. As a result, new technologies were invented and many efforts have been directed toward the area of renewable resources. A solution for the VOC restriction and energy problem is to employ a high-technology UV-curable system that is solvent free, and low in energy consumption.[4,5]

UV curing is the process of film transformation from a reactive liquid (except a powder coating) into a solid by radiation in the ultraviolet-energy region rather than by heat. UV cure coatings involve the polymerization and crosslinked polymer network induced by photons. UV cure coatings are one of the most important classes of radiation-cure coatings.

Introduced commercially in the 1970s, the UV-curing industry has enjoyed a steady growth and has become one of the most rapidly developing fields in the entire coatings industry. UV-curable coatings have found application for the surface protection of all kinds of materials such as metals, plastics, glass, paper, wood, *etc.*[6] Today, the use of UV-curable formulations is progressing at a fast rate. The main reason for such a rapid technological growth of UV curing is due to its advantages, which could account for its popularity.

### 2.1.1 Advantages

(i) Fast cure. UV-curing technology uses the energy of photons from a radiation source to form reactive species, such as free radicals or cations, which can initiate a fast chain-growth polymerization of monomers and oligomers. UV-induced curing allows fast transformation of a liquid resin into a solid material by polymerization of a solvent-free formulation at ambient temperature.[7] The curing time of UV-curable coatings is of the order of magnitude of seconds or minutes rather than hours in thermal curing, which means a significant reduction in cycle time.

(ii) Ambient curing temperatures. Heating in UV curing is not required, thus the overall energy consumption is lower than the thermally initiated curing processes.[7] This contributes to not only lower thermal stresses, but less expensive tooling. Ambient-temperature curing is imperative for heat-sensitive substrates, including paper, some plastics, and wood.

(iii) Space and energy efficiency. The UV-curing requires smaller curing units as compared with ovens for baking in thermal cure, which saves building space.

(iv) Solvent-free formulations. Reduced solvent emission is an important motivation for use of UV-curable coatings. Generally, the formulations contain no solvent and VOC emissions are negligible.

(v) Improved coating properties. Functionalized monomers and oligomers can be easily formulated to meet a variety of applications. Due to the wide range of viscosities inherent in the reactive oligomers used in UV-curable components, coatings can be readily formulated to meet demanding viscosity required by certain coatings applications.

All of the aspects listed above combine together to make UV curing a unique coating means that can be tailored for any application normally met by thermally cured coatings, but with less complication. UV-radiation curing has become a well-accepted technology that has found a large variety of industrial applications because of those distinct advantages. UV polymerization is commonly used for curing of relatively thin polymer films in applications such as fast drying of varnishes, paints, printing inks and adhesives, as well as in the production of printing plates, microcircuits, and optical disks.

### 2.1.2 Disadvantages

There are also some drawbacks for UV-curing technology as follows:

The radiant power of the UV lamp decreases with the square of the distance, which limits the UV-curing technology to three-dimensional products that lead to no curing in the shadow areas. As a result, UV curing cannot be used to cure complex-shaped parts.

UV radiation is most limited in terms of penetration into matter, thus it has difficulty with pigmented coatings. Rapid shrinkage during curing can create adhesion problems. In composites processing, the use of UV curing has so far been limited to open-mold processes and relatively thin laminates.

UV radiation is best known for its deleterious effects on organic compounds, particularly upon prolonged sunlight exposure. By breaking chemical bonds, UV radiation causes severe changes in the mechanical and optical properties of polymer materials, thereby reducing their service life in outdoor applications.

## 2.2 UV Energy and its Properties

Ultraviolet radiation, part of the electromagnetic spectrum, shows wavelength in the range from 40 to 400 nm with wavelengths below those of visible blue light (see Figure 2.1). It is divided into the following four regions:

1. vacuum UV: 40–200 nm;
2. UV C: 200–280 nm;
3. UV B: 280–315 nm;
4. UV A: 315–400 nm.[7]

Vacuum UV is strongly absorbed by the oxygen in air and most common materials such as quartz. Vacuum UV has a small penetration depth. Thus, vacuum UV is not suitable for the usual radiation curing.

Because the energy of radiation increases with increasing frequency or decreasing wavelength, radiation with short waves contains a large amount of energy. This energy is capable of bringing about certain chemical reactions in a system that is sensitive to light and the absorbed energy can then generate species that are capable of initiating polymerization or crosslink reactions.

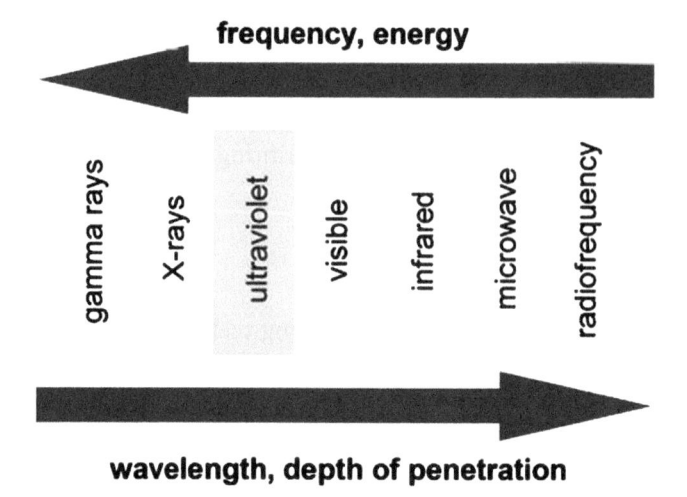

**Figure 2.1** Electromagnetic spectrum.[7]

One of the requirements for UV curing is a UV source that produces high-intensity UV radiation without generating excessive infrared radiation. Major radiation sources in commercial use are medium-pressure mercury vapor lamps. For initiation *via* any lamp source, there must be absorption of radiation by the photoinitiator or substance that leads to the generation of an initiator.

The fraction of radiation absorbed within a coating film, $I_A/I_0$, is related to the molar absorptivity $\varepsilon$, concentration of photoinitiator $C$, and optical path length of radiation in the film $X$. Assuming that no other absorber is present and neglecting surface reflection, the fraction of radiation of a given wavelength absorbed is expressed as $I_A/I_0 = 1 - 10^{\varepsilon CX}$.

When a photoinitiator absorbs a photon, it is raised to an excited state, which leads to generation of an initiating species.[8] The efficiency of generation of initiating species is an important factor in the selection of a photoinitiator.

## 2.3 Equipment

The UV-curing equipment consists essentially of three components: the lamp, lamp housing, and the power supply. The electrical energy supplied to the bulb is converted into UV energy inside the lamp. The housing is designed to direct and deliver to the substrate or the part to be irradiated. The lamp housing reflects and focuses the UV energy generated by the lamp. The power supply delivers the energy needed to operate the UV lamp.[9]

A typical UV-curing unit might house one or more lamps. In most situations, the materials to be cured are passed by or under one or more lamps *via* a moving belt. The speed determines how long the surface is exposed to the light. The light generated by the lamp is reflected by a reflector that can either focus or defocus it depending on the process.[9]

# UV-Curable Coating Technologies

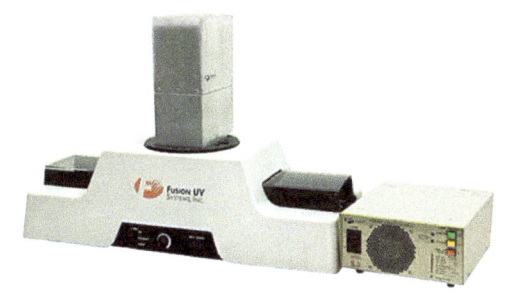

**Figure 2.2** FUSION LC-6B Benchtop Conveyor UV Curing System (http://www.caeonline.com/listing/product/182588/axcelis-fusion-lc-6b).

A FUSION LC-6B Curing System is shown in Figure 2.2. The LC6 Benchtop Conveyor is a production UV-curing unit that is suitable for laboratory and R&D applications. It can also be used in testing of adhesives, inks and coatings for qualification, cure response testing, or performance evaluation. It can handle a variety of substrates up to 7.5 inches wide with an effective curing width up to 6 inches. The lamp and housing are adjustable to accommodate parts up to 3 inches high.

## 2.3.1 Light Sources

UV-curable coatings require a specific type of energy with a wavelength of approximately 400 nm in the UV spectrum to initiate chemical crosslinking of the coating components. One of the most important factors in the success of a UV-curing operation is in the correct choice of the lamp system.

### 2.3.1.1 Mercury Lamp

By far the most popular source for UV curing has traditionally been the medium-pressure mercury lamp with output at specific lines from the deep UV to the visible, superimposed on a low-level continuous band. The standard mercury lamp is used in most of the existing applications. Such electrode lamps are tubes with different length and outputs. The radiation has continuous wavelength distribution with different major peaks (see Figure 2.3), and the main spectral output is in the wavelength range below 300 nm and at 365 nm (H-bulb). In transparent clearcoat application, this spectrum can well be used to tune the spectral output with photoinitiator absorption. By adding iron to the mercury lamp, the lamp spectrum contains additional bands in the range of 350 to 400 nm, thus in the UV-A range (D-bulb). Such types of lamps are used when additives in the coating have strong absorption in the UV-C range, like UV absorbers, and the best overlap with the photoinitiators has to be tuned to longer-wavelength ranges. A further shift to longer-wavelength absorption can be done with gallium and indium addition. Here, intense output is added in the wavelength range of 400–450 nm (V-bulb). Such lamp systems are mainly used for pigmented coatings (see Figure 2.3).[10]

**(a) H Bulb (13 mm)**

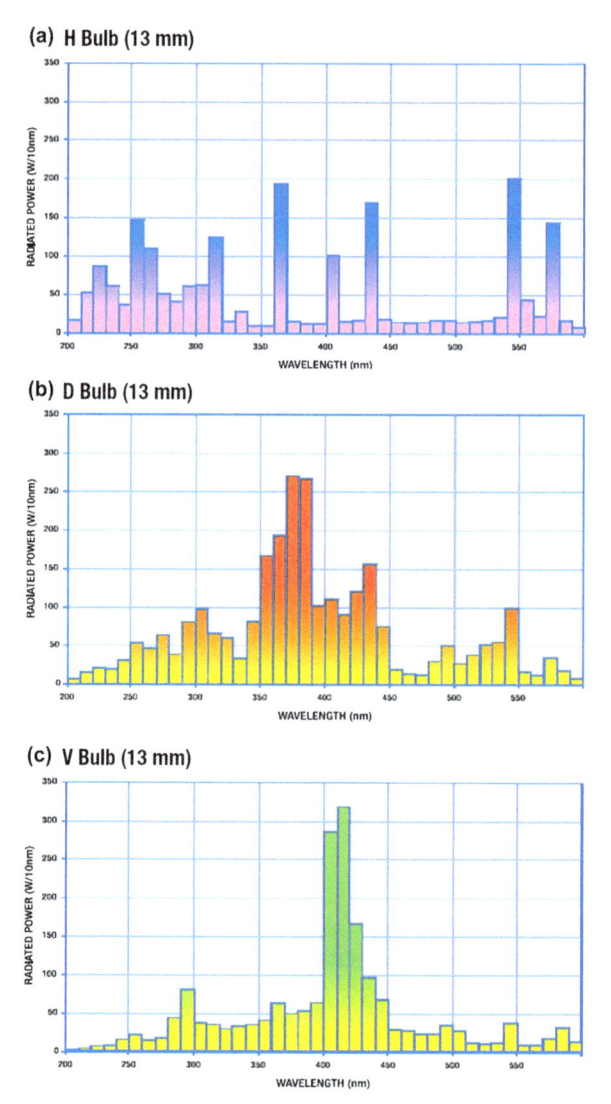

**(b) D Bulb (13 mm)**

**(c) V Bulb (13 mm)**

**Figure 2.3** Spectral power of different mercury bulbs.

A limitation of UV curing is that the distance between the lamp and the coating on various parts of the object being coated must be fairly uniform. Hence, UV curing is most easily applicable to coating flat sheets or webs that can be moved under the UV lamps or cylindrical objects that can be rotated under or in front of the lamps.

## 2.3.1.2 *Electrodeless Lamps*

In recent years, a number of UV operations have turned to the use of the electrodeless lamp systems. Electrodeless lamps are powered by microwaves and are more suitable for doping, since the lifetime of lamp electrodes is generally reduced by dopants. Electrodeless lamps have the further advantage of essentially instantaneous start-up and restart. However, electrodeless lamps are more expensive.[9]

### 2.3.2 Coating Methods

A variety of methods are used to apply liquid systems to different surfaces in UV-curable coatings.

1. Roll coating. Roll coating is widely used as an efficient way to coat uniform, generally flat or cylindrical, surfaces of rigid or flexible substrates. Direct roll coating and reverse coating are two of the most commonly used types of roll coating systems. The typical UV formulations are in a viscosity range (<4000 mPa s) to be used preferably in roll and curtain coating applications.[9,10]
2. Curtain coating. Curtain coating is widely used in coating of flat sheets of substrate, such as wood panels. A coating is pumped through a slot in the coating head so that it flows as a continuous curtain of liquid. The material to be coated is moved under the curtain by a conveyor belt. The film thickness is controlled by the width of the slot and the speed of the substrate being coated.
3. Airless or conventional spray guns, used for three-dimensional or shaped objects.
4. Vacuum coaters.
5. Electrostatic application.

Spay application; dip coating, flow coating, spin coating, and rod coating are some other application methods.[9,10]

## 2.4 Components

UV-curable coatings are very different from conventional solvent or water-based coatings. They are up to 100% solids, containing little or no solvent or water carrier. The main components of such formulations are: photoinitiators, reactive oligomer, reactive diluent, and additives.

### 2.4.1 Photoinitiator

A photoinitiator is a compound-generating reactive species that will initiate polymerization or crosslinking. Photoinitiators are the basic link in the UV-curable formulation between the lamp source and the resin system. The function of a photoinitiator is:

1. absorbing the incident UV radiation;
2. generation of reactive species (free radical or ions);
3. initiation of photopolymerization.[11]

The basic principal of photochemistry indicates that in order to have a photochemical process the first necessary step is to absorb the incoming radiation. In order for a photoinitiator to be effective, it must have a relatively high extinction coefficient in the region of the electromagnetic spectrum matching the output of the lamp source. Besides a high extinction coefficient, the photoinitiator must also be capable of efficiently producing reactive species such as radicals (free-radical polymerization) or acids (cationic). Additionally, the species generated must be capable of rapidly initiating the curing process while participating minimally in the termination step.

The main steps of the UV curing can be illustrated as below:

When a photoinitiator absorbs a photon, it is transformed into an excited state and from there on to initiating radicals or cations. The rate of initiation and penetration of the incident light depend on the type, absorption wavelength and efficiency of the photoinitiator. During this process the excited state may be quenched by a component of the resin formulation or atmospheric oxygen. Different types of deactivation of the excited photoinitiator molecule are demonstrated by the modified Jablonsky diagram in Figure 2.4.[12] The triplet states shown are formed through intersystem crossing from the excited singlet state. In the triplet state electrons have the same rotation. The formation of the triplet state by direct absorption of a photon by the photoinitiator ground state is a forbidden transition.

Some common types of photoinitiators fall into the classes of:

1. aromatic ketones and synergistic amines;
2. alkyl benzoin ethers;
3. thioxathones and derivatives;
4. benzil ketals;
5. acylphosphine oxide;
6. ketoxime ester or acyloxime esters;
7. cationic curing quaternary ammonium salts; and
8. acetophenone derivatives.[11]

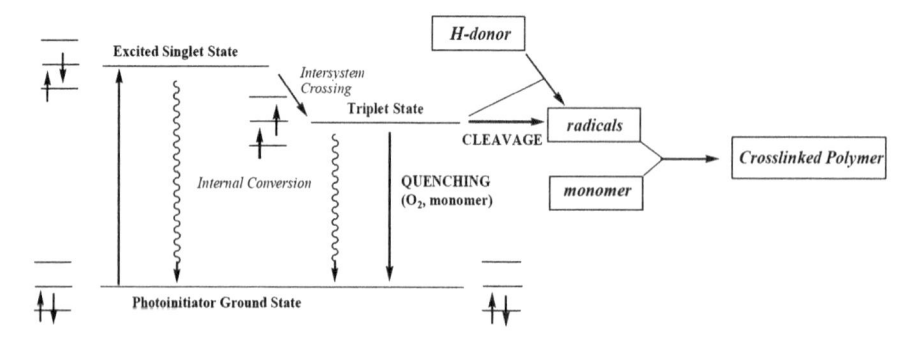

**Figure 2.4** Deactivation pathways of the excited photoinitiator.

## 2.4.1.1 Photoinitiator Selection

It is imperative to select the correct photoinitiator to achieve the desired speed and balance of cure-film properties. A photoinitiator absorbs light of specific wavelengths. During formulation, the absorbance characteristics of the photoinitiator are matched to the radiation characteristics of the lamp output. A suitable photoinitiator system must first have a high absorbance in the emission range of the light source. In addition, the excited states thus formed must have both a short lifetime to avoid quenching by oxygen or oligomer, and split into reactive radicals or ionic species with the highest possible quantum yield.[11] Besides this, selection of the best individual photoinitiator is dependent on a number of factors such as the chemistry of the resin system, the UV lamp type and orientation, the cure speed required, coating property requirements, and the substrate.

## 2.4.1.2 Optimize Concentration

The rate of polymerization reactions is related to the concentration of initiating radicals or ions. As the concentrations are increased from a lower amount to a somewhat higher amount, the rate of cure increases. However, there is an optimum concentration.

The optimum photoinitiator concentration is dependent on film thickness: the greater the film thickness, the lower the optimum concentration. Photoinitiators absorb the UV light, therefore prohibits the penetration of the light to lower parts. Below the optimum photoinitiator concentration, through cure will be poor. Too much photoinitiator can prove just as problematic as not enough because of the "filter effect" caused by high levels of surface crosslinking that does not allow the UV light to effectively penetrate the coating to the lower layers required for through curing.

Both surface and through cure can be achieved by using photoinitiator mixtures that have two distinct absorption maximums near the emission bands of the UV source. The emission band that is highly absorbed by the photoinitiator is absorbed more strongly near the surface, and less UV is available for absorption in the lower layers. This band is most important for counteracting oxygen inhibition but does not contribute substantially to through cure. The weaker absorption of a second emission band is more uniform throughout the film to provide through cure.[13]

### 2.4.2 Oligomers

Functionalized oligomers provide the UV-curable coating with its basic viscoelastic properties. Acrylated oligomers are most frequently used for UV free-radical curing. This is due to higher reactivity and lower volatility.[6] Acrylated resins are preferred over methacrylated ones due to higher cure rates at room temperature and lower oxygen inhibition.

**Table 2.1** Acrylated reactive oligomers for UV-curable coatings.

| Oligomer type | Merits | Drawbacks |
|---|---|---|
| Epoxy | Adhesion, hydrolytic stability, hardness, acid resistance | Poor UV-resistance (BPA) High viscosity |
| Polyether | Fairly good adhesion, flexibility, elasticity, UV-resistance | Weatherability Low Tg |
| Polyester | Fairly good adhesion, fairly good versatility, fairly good hardness | Poor UV-resistance Hydrolytic stability |
| Polyurethane | Versatility, best combination of hardness and elasticity, adhesion, weatherability, adequate H$_2$O resistance | Aromatic yellow, aliphatics costly, aliphatics toxic, plasticized by H$_2$O |

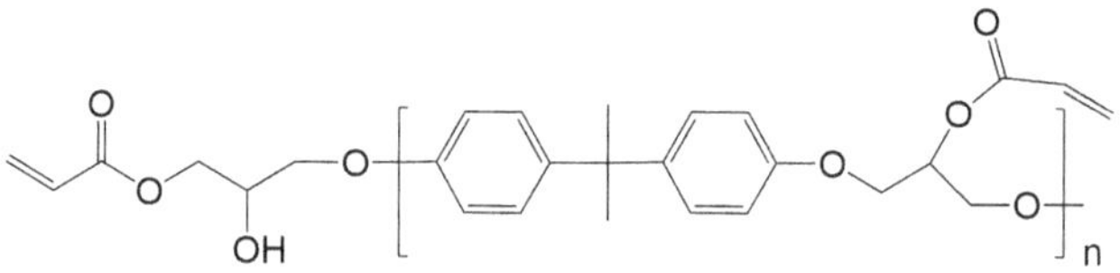

**Figure 2.5** Acrylated BPA epoxy.

Reactive oligomers used in UV curing mainly include acrylated epoxy, acrylated polyether, acrylated polyester, and acrylated polyurethane. Their advantages and disadvantages are shown in Table 2.1.

### 2.4.2.1 Acrylated Epoxies

The reaction of an epoxy group with acrylic or methacrylic acids gives rise to an epoxy acrylate (or methacrylate) (see Figure 2.5). Both aromatic and aliphatic acrylated epoxies are widely used as reactive oligomers for UV-curable coatings. Acrylated epoxies are highly reactive and produce hard and chemically resistant films. Coatings formulated with acrylated epoxies promote good adhesion, chemical resistance, and flexibility. They are especially useful as coatings on metallic substrates where high gloss, hard coatings that cure rapidly are required. Epoxy acrylates are extensively used in UV-curable inks and varnishes, roller coating varnishes for paper and board, printed circuit board, wood and plastic coatings.

### 2.4.2.2 Acrylated Polyethers

The reaction of ethylene or propylene oxide with a polyol in the presence of basic or acidic catalysts such as BF$_3$ or NaOH will give a polyether. Acrylated polyethers have lower viscosity compared to acrylated polyester and are relatively inexpensive to prepare (see Figure 2.6).

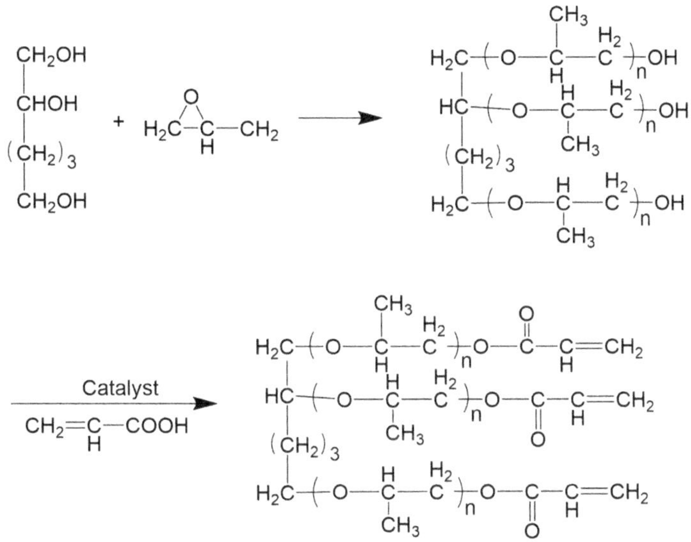

# UV-Curable Coating Technologies

**Figure 2.6** Acrylation of polyethers.

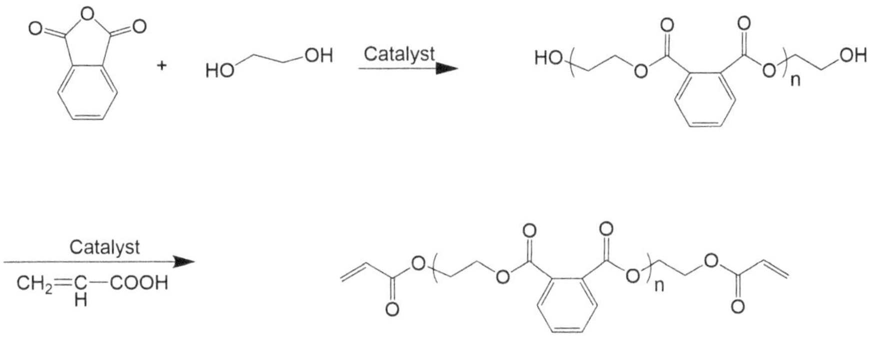

**Figure 2.7** Synthesis and acrylation of polyesters.

### 2.4.2.3 Acrylated Polyesters

Acrylated polyesters are prepared by reacting the OH group of polyesters with acrylic acid or hydroxyl acrylate with acid group of the polyester structure (see Figure 2.7).

Acrylated polyesters are relatively cheap compared with other acrylated prepolymers. One of the major advantages offered by acrylated polyesters over the other prepolymers is their low viscosity, which makes the oligomers require little or no reactive diluent. By varying the molecular weight of polyester, it is possible to obtain acrylated polyesters with different viscosities at ambient temperature.

The compatibility of acrylated polyesters with other prepolymer is good. Thus, acrylated polyesters can be used in many formulations. Acrylated polyesters are mainly used in UV roller coat varnishes for paper and board and in UV wood coatings.[11]

One of the major problems associated with the use of the low molecular weight acrylated polyester is the reduction in reactivity and the increased surface inhibition observed and it is found that this can be compensated for by a number of techniques, *e.g.* incorporation of pendent aromatic group and ether grouping within the polyester backbone. It is possible to prepare acrylated polyesters using transesterification technique with an acrylic monomer such as ethylacrylate.[14]

### 2.4.2.4 Acrylated Polyurethanes

Acrylated polyurethanes are formed by the reaction of isocyanates with hydroxyl-functional acrylate monomers (see Figure 2.8).

Reaction between hydroxyl and isocyanate groups proceeds efficiently at low temperature without the evolution of volatile byproducts. The well-known properties of urethanes, such as hardness, chemical resistance, toughness and light stability can be built into the acrylated prepolymers. Improvement of the flexibility of acrylated polyurethanes may be achieved by chain extension using a long-chain diol to produce a higher molecular weight isocyanate functional prepolymer that is subsequently capped by a hydroxy acrylic monomer.[11]

Acrylated silicone and thiol–ene systems are some other important reactive oligomers that have attracted researchers' attention.

### 2.4.2.5 Acrylated Silicones

Silicones are well known for their release properties, heat and weather resistance. Incorporation of silicone into UV-curable coatings is of interest in a number of areas. The advantages of silicone acrylates in the protection of optical fibers are afforded by their excellent flexibility and extensibility properties, particularly at low operating temperatures.[11]

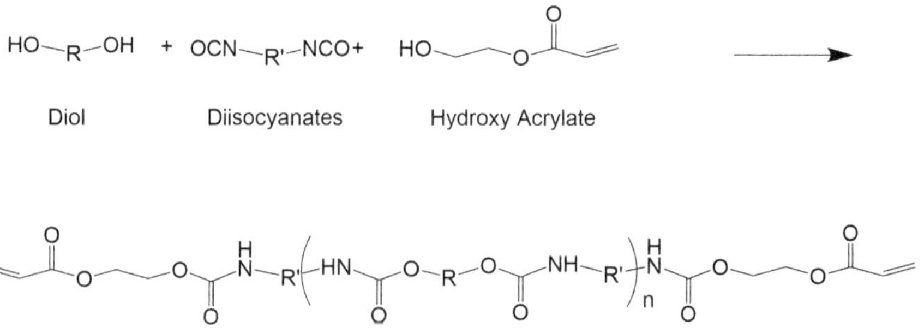

**Figure 2.8** Synthesis and acrylation of polyurethanes.

### 2.4.2.6  Thiol–ene System

The thiol–ene reaction is based on a stoichiometric reaction of multifunctional olefins (*enes*) with *thiols*. The addition reaction can be initiated thermally or photochemically by radical or ionic mechanisms. Thiyl radicals can be generated by the reaction of an excited carbonyl compound with a thiol or *via* radicals. The thiyl radicals added to olefins form the basis of the polymerization process. The addition of a dithiol to a diolefin yields a linear polymer; higher functionality thiols and alkenes form crosslinked systems.[9]

The thiol–ene system has the advantages of no oxygen inhibition. Flexible cured films can be obtained from relatively low-viscosity mixtures without the need to incorporate a diluent. A polyfunctional thiol compound gives very tough abrasive-resistant coatings that are ideal for applications such as flooring coatings. The disadvantage is an unpleasant odor.[11]

## 2.4.3  Reactive Diluents

Most of the currently available oligomers are too viscous to be applied through conventional coating equipment, thus they need be diluted down to application viscosity. Reactive diluents are used in UV-curable systems to provide final-film properties and viscosity control of the resin. They are also important in determining the speed of cure, crosslink density, and final surface properties of the cured film.

A reactive diluent is selected for a system based upon its: viscosity reduction and control; effect on cure speed; effect on film properties; shrinkage during polymerization; cost; shelf life; volatility, odor, and toxicity.

Acrylates are the most widely used reactive diluents in UV-curable systems due to their versatility and reactivity. Not only do they cure at extremely rapid rates compared to other monomer systems, but they are also available in a wide range of structures, which are monofunctional, difunctional, trifunctional, and tetrafunctional.

### 2.4.3.1  Monoacrylates

Some commonly used monoacrylates are shown in Table 2.2.

### 2.4.3.2  Diacrylates

Diacrylates have relatively stronger odor, are skin irritants and carcinogenic.[11] Examples of commonly used diacrylate diluents are listed in Table 2.3.

### 2.4.3.3  Triacrylates

Pentaerythritoltriacrylate (PETA). PETA is widely used in printing inks since it gives rapid cure response. However, PETA is a severe eye irritant and is suspected to be carcinogenic.[11]

**Table 2.2** Commonly used monoacrylates.

| Acrylate type | Structure | Merits | Drawbacks |
|---|---|---|---|
| n-Butyl acrylate (BA) | | Good viscosity reducer / Good flexing action | Slow cure / Poor solvent resistance / Volatile |
| 2-Ethyl hexyl acrylate (EHA) | | Good flexing action | Slow cure / Poor solvent resistance / Volatile |
| Iso decyl acrylate (IDA) | | Good viscosity reducer / Less volatile than EHA / Increase flexibility owing to long aliphatic chains | |
| Iso bornyl acrylate (IBA) | | Fast cure rate / Low shrinkage rate / Low toxicity and low volatility | Strong odor / Imparts a hardness |
| 2-Hydroxy ethyl acrylate (HEA) | | Widely usage | High toxicity |
| 2-Hydroxy propyl acrylate (HPA) | | Good viscosity reducer | High toxicity |

**Table 2.3** Commonly used diacrylates.

| Acrylate type | Structure | Merits | Drawbacks |
|---|---|---|---|
| 1,4-Butanedioldiacrylate (BDDA) | | Widely used in wood coatings | |
| 1,6-Hexanedioldiacrylate (HDDA) | | Relatively good viscosity reducers / Low volatility | Suspected skin irritation sensitisers |
| Neopentylglycoldiacrylate (NPGDA) | | Good viscosity reducer | Low crosslinked density |
| Diethyleneglycoldiacrylate (DEGDA) | | Good viscosity reducer ability | Adverse dermatitic and toxicity properties |

## UV-Curable Coating Technologies

Trimethylolpropanetriacrylate (TMPTA). TMPTA has a low volatility and is widely used in printing inks (Figure 2.9).

### 2.4.3.4 Tetraacrylates

A typical example of tetraacrylates is pentaerythritoltetraacrylate. Pentaerythritoltetraacrylate has low volatility and is used in adhesive and coatings (Figure 2.10).[11]

### 2.4.3.5 Vinyl Ethers

Vinyl ethers have been evaluated as reactive diluents in radical polymerizable acrylate and unsaturated polyester systems, as well as in cationic polymerizable epoxy coatings. Compared to acrylates, vinyl ethers have been shown to have the advantages of high diluting power and low toxicity (Figure 2.11).

Soucek's group investigated the effect of three different divinyl ethers diethyleneglycol divinyl ether (DEGDE), cyclohexane dimethanol divinyl ether (CHDMDE), and triethylene glycol divinyl ether (TEGDE), on norbornyl epoxidized linseed oil.[15] The curing rate of the resulting formulation was significantly improved when divinyl ether was added into the formulation. The incorporation of divinyl ethers increased the curing rate and overall

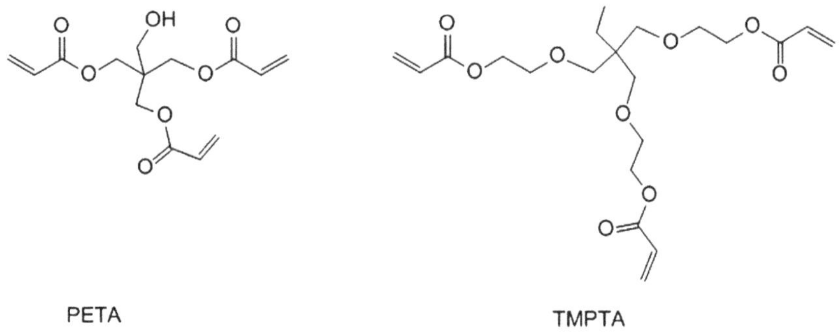

**Figure 2.9** Structures of PETA and TMPTA.

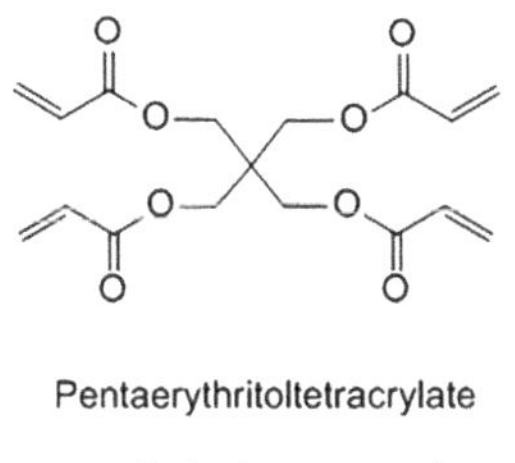

**Figure 2.10** Structures of pentaerythritoltetraacrylate.

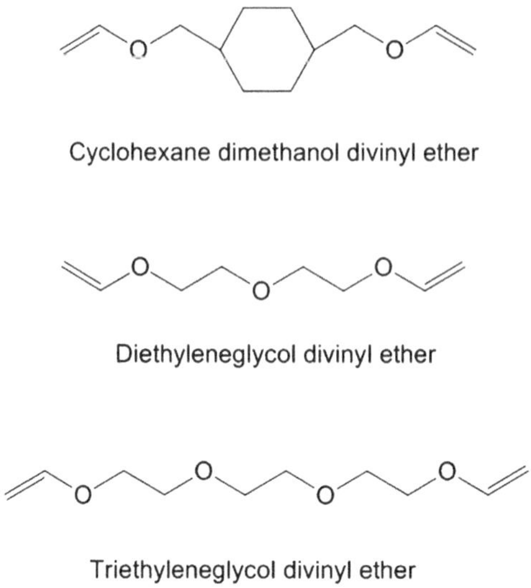

Cyclohexane dimethanol divinyl ether

Diethyleneglycol divinyl ether

Triethyleneglycol divinyl ether

**Figure 2.11** Structures of different divinyl ethers.

conversion of the epoxide groups. The extent of the improvement was dependent on the structure of the divinyl ether; the more flexible the divinyl ether, the higher the curing rate. Of the three divinyl ethers used, coating with TEGDE showed the highest curing rate and coating with CHDMDE showed the lowest curing rate.[15]

### 2.4.3.6 Styrene

Styrene is the most widely used reactive diluent for unsaturated polyester resin systems. It works well and is relatively low cost. The low cost of styrene and its generally beneficial properties are the basis for its widespread acceptance and use as the most common reactive dilute.[16] However, styrene is a VOC and HAP. The US Department of Health and Human Services added eight substances including styrene to its Report on Carcinogens (ROC), a science-based document that identifies chemicals and biological agents that may put people at increased risk for cancer on June 14, 2011. The National Toxicological Program's (NTP) 12th Report on Carcinogens classifies styrene as "reasonably anticipated to be a human carcinogen".[17] Therefore, a means of developing a styrene-free UPE resin system is highly desired.

### 2.4.3.7 N-Vinyl Pyrrolidone (Vinyl Pyrol)

N-Vinyl pyrrolidone is highly flexible when cured, thus it is reported to improve flexibility of the resulting film. It has low toxicity, and is often used with acrylates. N-Vinyl pyrrolidone has no acrylate functionality but copolymerizes with acrylates when used at the optimum molar ratio (Figure 2.12).[11]

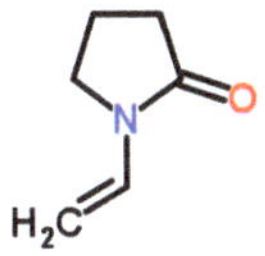

**Figure 2.12** Structures of *N*-vinyl pyrrolidone.

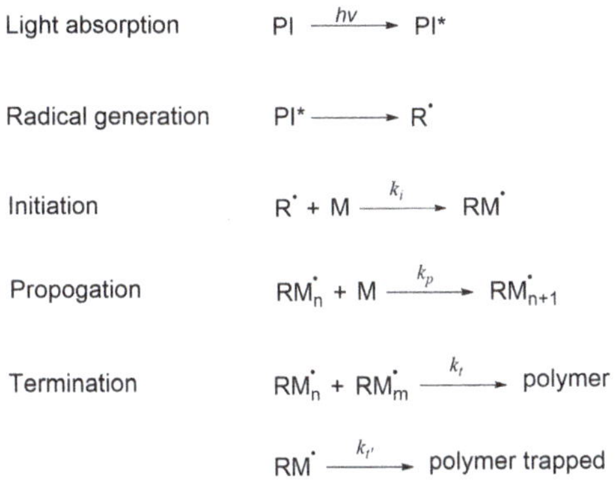

| Step | Reaction |
|---|---|
| Light absorption | PI $\xrightarrow{h\nu}$ PI* |
| Radical generation | PI* $\longrightarrow$ R• |
| Initiation | R• + M $\xrightarrow{k_i}$ RM• |
| Propogation | RM$_n^•$ + M $\xrightarrow{k_p}$ RM$_{n+1}^•$ |
| Termination | RM$_n^•$ + RM$_m^•$ $\xrightarrow{k_t}$ polymer |
|  | RM• $\xrightarrow{k_{t'}}$ polymer trapped |

**Scheme 2.1** Free-radical photopolymerization process initiated by UV radiation.

## 2.4.4 Additives

As in any coatings system, it is often the additives that give the coatings their most important properties. Additives often included in UV-curable coatings are substrate adhesion surfactants, wetting and dispersing additives, surface-improvement additives, viscosity reducers, and stabilizers.

## 2.5 Free-Radical Photopolymerization

The two types of UV-induced polymerization are free-radical and cationic polymerizations that are distinguished by their polymerization mechanism and the reactive species.

A typical UV-induced free-radical photopolymerization process includes light absorption, radical generation, initiation, propagation and termination, as depicted in Scheme 2.1. First, the energy of the light ($h\nu$) is absorbed by the photoinitiator (PI) to produce excited species (PI*) that will decompose to radicals. The radicals then initiate polymerization. This step is followed by successive addition of monomer units to the growing polymer chains in the propagation step. Since photoinduced polymerization takes place in a short time, usually several seconds for acrylated resins, it is possible that the growing polymer chain may not be terminated and is trapped in the crosslinked network after radiation.

### 2.5.1 Free-Radical Photoinitiator

For free-radical photopolymerization, there are two classes of photoinitiators: those that form free radicals with unimolecular photocleavage and those that generate radical through bimolecular hydrogen abstraction.

### 2.5.1.1 Unimolecular Photoinitiator

Unimolecular photoinitiators are compounds that undergo a cleavage reaction and generate two free radicals. Benzoin ethers were the first commercially used class of unimolecular photoinitiators and they cleave into benzoyl radical and benzyl ether radical by UV absorption (Figure 2.13).

However, the package stability of UV-cure coatings containing benzoin ethers tends to be limited due to the ease of hydrogen abstraction on the benzyl ether carbon. The ketal, 2, 2-dimethoxy-2-phenylacetophenone, is an effective photoinitiator with good package stability. Photocleavage produces benzoyl and dimethoxybenzyl radicals. The dimethoxybenzyl can undergo further cleavage to the highly reactive methyl radical (see Figure 2.14). The extent of this cleavage increases with increasing temperature.

### 2.5.1.2 Bimolecular Photoinitiator

Bimolecular photoinitiators are a class of compounds that can get excited after absorbing the photon and abstract the hydrogen or electron from a suitable donor molecule. Both free radicals generated in this process can initiate the polymerization, meaning this type of photoinitiator usually generated more radicals. (both radical can initiate in Norrish I initiators as well) In bimolecular initiation, the photoexcited initiators such as benzophenone (see Figure 2.15), thioxanthone or other related dialkyl ketones do not cleave to give free radicals but can abstract hydrogen from hydrogen donors to yield excited complexes that initiate polymerization.

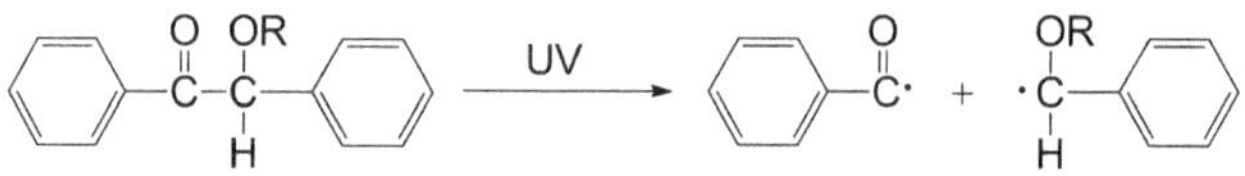

**Figure 2.13** Formation of free radicals from benzoin ethers.[18]

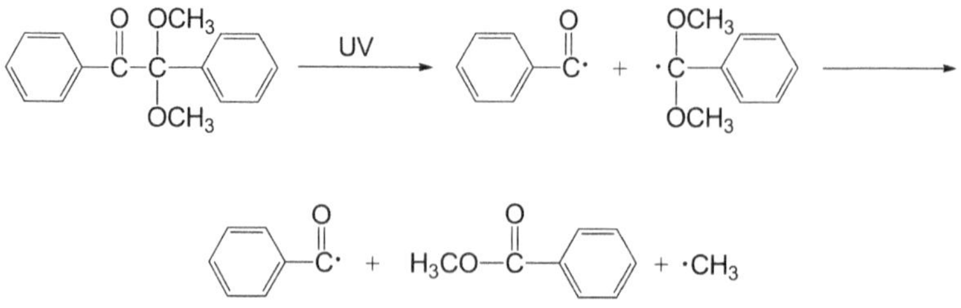

**Figure 2.14** Formation of free radicals from 2,2-dimethoxy-2-phenylacetophenone.[13]

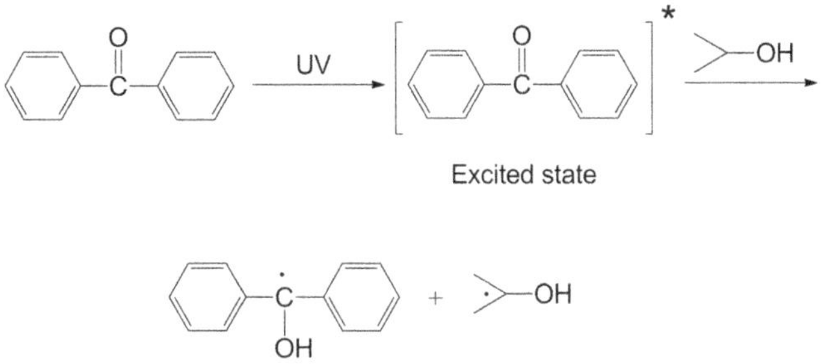

**Figure 2.15** Formation of free radicals from benzophenone.[19]

The most common hydrogen donors are aliphatic amines such as triethylamine, methyl diethanol amine, or dimethyl ethanol amine.[20] The abstraction process is highly dependent on the hydrogen donor. An advantage of bimolecular initiators with hydrogen donors is reduced oxygen inhibition. A disadvantage is that excited states of these initiators are generally longer lived than those of unimolecular photoinitiators and therefore more readily quenched by oxygen as well as by vinyl monomers with lower triplet energies.[13]

### 2.5.1.3 Oxygen Inhibition

One of the disadvantages of free-radical photopolymerization is oxygen inhibition. Oxygen inhibition causes trouble due to the fact that coatings have a high ratio of surface area to total volume that contributed to high oxygen exposure. Oxygen reacts with the terminal free radical on a propagating molecule to form a peroxy free radical. The peroxy free radical does not readily add to another monomer molecule that terminated the growth of the chain. The excited states of certain photoinitiators are quenched by oxygen, thus reducing the efficiency of generation of free radicals.

Several approaches are available to minimize this problem.

(i) Perform the UV exposure under inert atmosphere such as nitrogen.[21] This is the most efficient method but it is relatively expensive. The most economical inert atmosphere approach is to use a $CO_2$ atmosphere. $CO_2$ is relatively inexpensive and its high density makes it easier to contain. It permits reduction in photoinitiator concentration and use of slower curing reactions.

(ii) Increase the reactivity of the radical cure system to shorten the UV exposure time and decrease the time that oxygen diffuses into the film. This can be done by increasing the amount of photoinitiator or by intensifying the UV light.[22] High-intensity UV sources minimize but do not eliminate the problem with fast-curing systems. In effect, free radicals can be generated so rapidly that their high concentration can deplete the oxygen at the film surface, permitting other radicals

to carry on the polymerization before more oxygen diffuses to the surface.
(iii) Add a chain-transfer agent such as amines, alkyl ether, and thiols into the system. The peroxy radicals will abstract H atoms from the chain-transfer agent and produce new radicals. Free radicals on carbon atoms alpha to amines and ethers also react rapidly with oxygen. Thus, amines and ethers not only act as chain-transfer agents but also serve to deplete oxygen levels. However, the presence of amines has some detrimental effects to the coating such as odor, plasticizing effects and yellowing of the film.[23]
(iv) Perform the polymerization under water[24] or prevent the diffusion of oxygen into the uncured film using a barrier such as wax. As the coating is applied and cured, a layer of wax comes to the surface of the coating, shielding the surface from oxygen. This is an effective method, but the disadvantage is that the residual wax detracts from the appearance of the film and makes recoating difficult.

## 2.6 Cationic Photopolymerization

UV-induced free-radical polymerization has enjoyed commercial success over cationic polymerization. However, the UV-induced cationic systems are finding increasing application in many specialized fields. Cationic photopolymerization is an especially convenient technique for carrying out the polymerization of highly reactive monomers such as vinyl ether and epoxide. The advantages of UV-induced cationic polymerization are rapid cure, and no oxygen inhibition compared to free radicals. UV-induced cationic polymerization is capable of delivering polymer materials with a broad diversity of chemical and mechanical properties. This versatility makes it possible to use UV-induced cationic polymers for a wide range of applications as coatings, adhesives, printing inks, microelectronic photoresists, printing plates, and composites.

### 2.6.1 Cationic Photoinitiators

Photoinitiators for cationic polymerization are typically onium salts of very strong acids. The most commonly used onium salts are iodonium and sulfonium salts of hexafluoroantimonic and hexafluorophosphoric acids. Diaryliodonium and triarylsulfonium salt photoinitiators are highly photosensitive, and have unique and excellent thermal stability in the absence of light. Thus, these photoinitiators are incorporated into very highly reactive oligomers/diluents and contribute to fast cure. Another advantage of onium salt photoinitiators is the ability to modify their structures in several straightforward ways to achieve specific desired UV absorption characteristics and to tailor their reactivity and compatibility with a variety of reactive oligomers/diluents.[25] The structures of onium salt photoinitiators can be modified for some purposes. For example, the solubility of some iodonium

## UV-Curable Coating Technologies

and sulfonium salts is often poor in nonpolar monomers. However, it was found that the simple attachment of an alkoxy group leads to markedly better solubility characteristics. This attachment was shown not only to improve solubility in both polar and nonpolar monomers, but to reduce the orders of oral, eye and skin toxicity when the alkoxy chain length reached eight carbons or longer.[25]

The structures of photoinitiators UVI-6974 {10% diphenyl-4-thiophenyl-phenyl sulfonium hexafluoroantimonate and 90% bis[4-(diphenylsulfonio) phenyl]sulfide bishexafluoroantimonate in propylene carbonate} and UV-9385C {bis(4-dodecylphenyl) iodonium hexafluoroantimonate} are shown in Figure 2.16. UVI-6974 has limited solubility and is toxic. UV-9385C has both higher solubility due to the long alkoxy chain and lower toxicity.

The general mechanism for UV-induced cationic polymerization is given in Scheme 2.2[26] using diaryliodonium salt as photoinitiator. Absorption of UV light results in both homolytic and heterolytic cleavage of a carbon–iodine bond of the diaryliodonium salt (eqn (2.1)).

The products of the photolysis reaction are radical and radical cation species that further react to form superacids (eqn (2.2)). The strength of the acid that is produced is highly dependent on the anion present in the

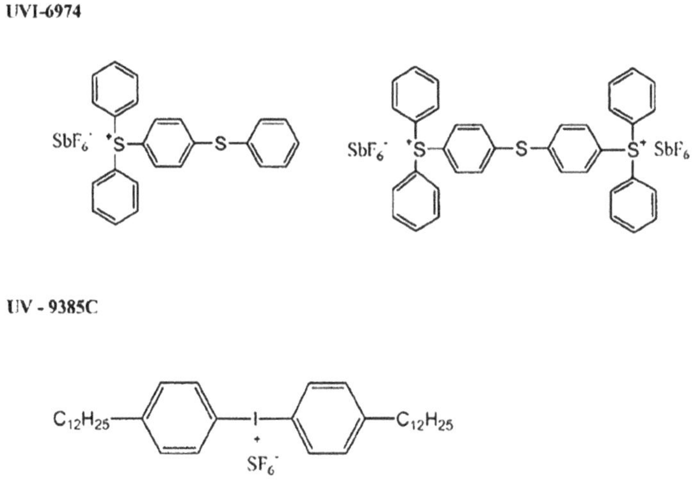

Figure 2.16  Structures of two cationic photoinitiators.

$$Ar_2I^+X^- \xrightarrow{UV} Ar^{\bullet} + Ar-I^{\bullet+}X^- \quad \text{Eq. (1)}$$

$$Ar-I^{\bullet+}X^- \xrightarrow{S-H} Ar-I + S^{\bullet} + HX \quad \text{Eq. (2)}$$

$$M \xrightarrow{H-X} H-M^+X^- \quad \text{Eq. (3)}$$

$$H-M^+X^- + nM \longrightarrow H-(M)_nM^+X^- \quad \text{Eq. (4)}$$

Scheme 2.2  General mechanism of UV-induced cationic polymerization.

starting onium salt. Photoinitiators bearing anions such as $BF_4^-$, $PF_6^-$, $AsF_6^-$, and $SbF_6^-$ generate superacids. These strong acids are the primary species responsible for the initiation of cationic polymerization (eqn (2.3)) by the direct protonation of the monomer. Both the initiation (eqn (2.3)) and the propagation steps (eqn (2.4)) are dark reactions (dark cure) and proceed spontaneously in the absence of light. Consequently, the polymerizations could continue in the absence of light once initiated.[25]

The rate of any UV-induced polymerization depends on several factors that are determined by both the photoinitiator and oligomers. The rate of consumption of the oligomer is directly related to the number of initiating species generated, which is determined by the concentration of the photoinitiator, its absorption and wavelength characteristics, the quantum yield of photolysis and the intensity of the light. In a cationic polymerization, the rates of the initiation and propagation process (eqn (2.3) and (2.4)) are highly dependent upon the oligomer structure.

### 2.6.2 Cationic Oligomers

Cationic polymerization is the most versatile polymerization in terms of the variety of oligomer types that can be employed. Virtually all cationically polymerizable oligomers can be photochemically polymerized in the presence of onium salt photoinitiators. Oligomers that are inactive towards free radicals, such as epoxides, or vinyl ethers, undergo rapid polymerization upon UV irradiation in the presence of aryl sulfonium or iodonium salts. These compounds were shown to generate superacids upon their photolysis in the presence of a hydrogen donor molecule.[25]

Scheme 2.3[25] depicts some very representative oligomer types. The oligomers mainly comprise two basic types: those that undergo vinyl-type

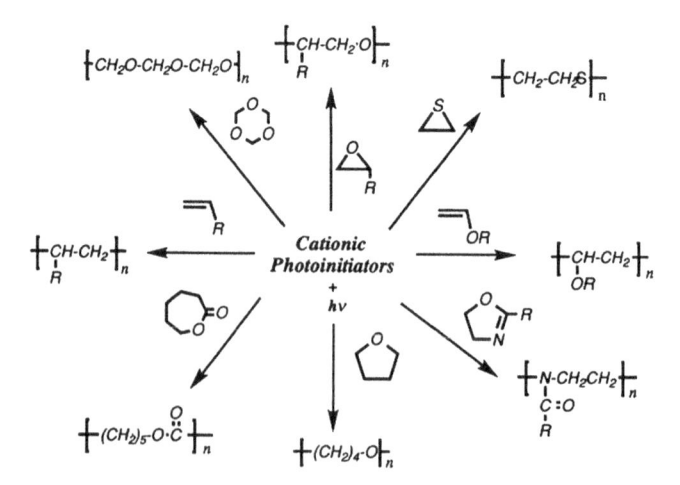

**Scheme 2.3** Oligomer types for cationic polymerization.[25]

polymerization, and those that polymerize by a ring-opening mechanism. Vinyl ethers polymerize much faster than epoxides because of the electron rich C=C bond and the stabilization by resonance of the carbocation. Vinyl ethers are useful as a highly reactive diluent in epoxy-resin coatings.

### 2.6.2.1 UV-Curable Epoxide Oligomers

Polymers obtained from multifunctional epoxide oligomers typically possess excellent adhesion, low shrinkage, and superior chemical resistance as compared to their acrylate counterparts. Due to this reason, they are presently finding wide applications in high-performance protective and decorative coatings for metals, plastics, and wood and in pressure-sensitive adhesives. Such applications require rapid polymerization rates. Extensive studies showed that the UV-induced cationic ring-opening polymerization of epoxy oligomers is highly dependent on the structure of the oligomer.

There are several factors that affect the reactivity of epoxides in UV-induced cationic polymerization. One factor that may contribute to the reactivity of an epoxy oligomer is the ring strain. It has been shown that glycidyl esters are less reactive than glycidyl ethers, which, in turn, are less reactive than epoxidized α-olefins. The most reactive oligomers are cycloaliphatic epoxy oligomers bearing the epoxycyclohexane moiety. The differences in reactivity might be due to the substantial difference in ring strain in the cycloaliphatic epoxides as compared to their linear aliphatic or glycidyl ether counterparts.[15,25,27,28]

To further figure out this issue, Soucek's group prepared epoxynorbornane linseed oils (ENLOs) (see Figure 2.17). The cationic photopolymerization of ENLO and epoxidized linseed oil (ELO) was investigated with real-time infrared (IR) spectroscopy and photodifferential scanning calorimetry (photoDSC). It was found that the curing rate of ENLO was higher than ELO. The incorporation of reactive diluent divinyl ethers increased the curing rate and overall conversion of the epoxide groups.[15] The ring strain in the norbornene ring system is considerably higher than that in the corresponding linear system, thus contributing to the faster curing of epoxynorbornene linseed oils.

Crivello's group investigated the cationic polymerization of several cycloaliphatic epoxies with ring sizes from 5 to 12 carbon atoms (see Figure 2.18). It was found that the reaction rates of epoxies 5–7 are much higher than epoxy 8. Epoxy 8 is more reactive than epoxy 12.[25] It is concluded that the most reactive epoxies are those with the highest ring strain and smallest cycloaliphatic rings. Reaction rates drop off substantially with larger ring sizes. This research confirmed that, together with ring strain, ring size also plays a certain role in the UV-induced polymerization of epoxies.

There are some other additional factors that determine the rate of a cationic polymerization besides ring strain and ring size. One of those factors is the steric environment around the epoxide bond. The ring size not only influences the ring strain in cycloaliphatic epoxides, but also significantly

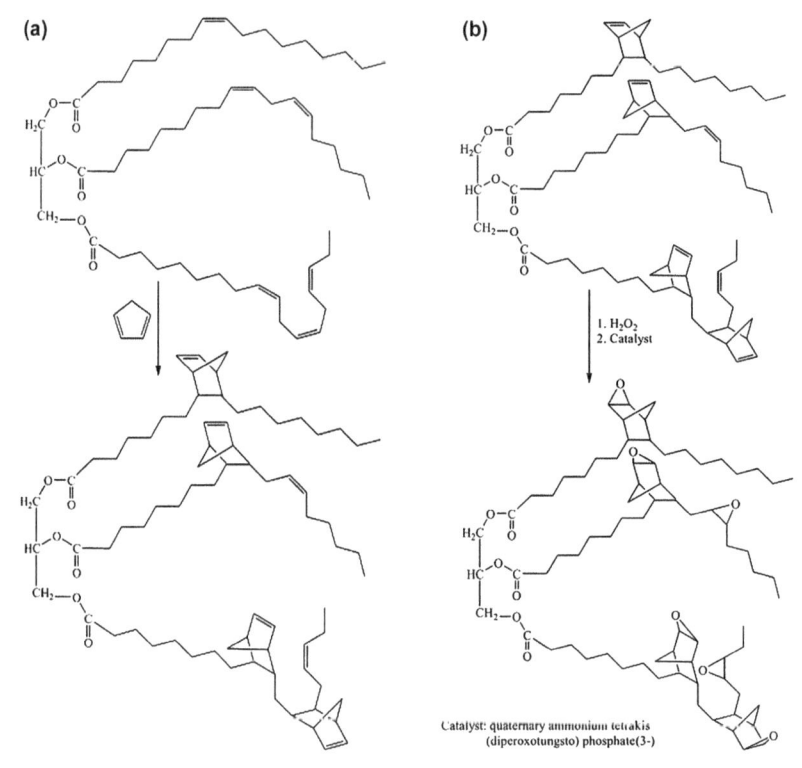

**Figure 2.17** Preparations of (a) NLO and (b) ENLO.[15]

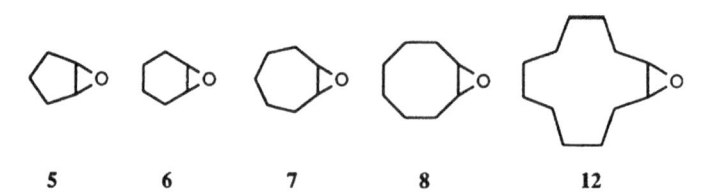

**Figure 2.18** Cycloaliphatic epoxies with ring sizes from 5 to 12 carbon atoms.[25]

influences the rate of polymerization due to the steric requirements. In general, the steric hindrance about an epoxide group increases with larger rings due to the greater bulkiness of the ring substituents and to their greater conformational freedom. This may be the reason for the slow rate of polymerization for monomer 12.[25]

To obtain more information about the influence of structural parameters, cycloaliphatic epoxides with different degrees of steric hindrance about the epoxide group were polymerized by Crivello's group (see Figure 2.19).[25]

It was found that the polymerization rates decrease with an increase in the steric hindrance about the epoxide bond: 6 > 6ME > 6PI. This can be explained by an increase in the steric barrier provided by both the methyl

*UV-Curable Coating Technologies* 39

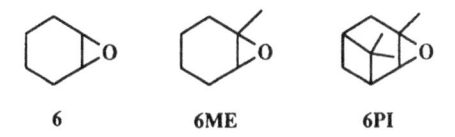

**Figure 2.19** Cycloaliphatic epoxies with different steric environment.[25]

group on the epoxide ring of 6ME and the two geminal methyl groups on the bridge of the bicyclic ring system of 6PI.[25]

### 2.6.3 Comparison of Cationic with Free Radicals

#### 2.6.3.1 Oxygen Inhibition/Water Effect

Unlike radical polymerization, cationic polymerization is not inhibited by atmospheric oxygen. However, the addition of proton donors such as water has an effect on the photopolymerization rate. It has long been anticipated that water would inhibit the polymerization reaction because of proton-transfer mechanisms. However, Soucek's group research tells a different story and indicates that a certain amount of water would promote the photopolymerization.[29]

In their research, the effect of relative humidity on the UV-induced curing kinetics of cyclohexyl epoxide and epoxide/polyol coating formulations was investigated. The formulations were exposed to a maximum of seven different relative humidities (6, 16, 20, 30, 51, 62, 75 RH). It was found that the lack of water had a deleterious effect on the UV-initiated homopolymerization of cycloaliphatic epoxides. The curing speed increased with increasing water concentration until a maximum was reached, after which the curing speed was again retarded. It was found that water also has a synergistic effect with polyols on the UV-curing kinetics of cycloaliphatic epoxide coating formulation. It is proposed that both water and polyol participate in the proton transport, stabilize the reactive intermediate (AM or ACE), and mediate a transition between chain growth and step-growth polymerization.[29]

#### 2.6.3.2 Dark Cure

In the absence of nucleophilic anions, cation-initiated polymerization can continue after exposure to the radiation source until the reactive cations become immobilized.

#### 2.6.3.3 Shrinkage

Compared to free-radical polymerization, the shrinkage of the polymers during curing is lower for cationic polymerization, typically between 4% and 6%, which results in low internal stresses.[7] The cured material shows, in general, high mechanical properties and good adhesion to various substrates.

## 2.7 Evaluation of Cure Extent

### 2.7.1 Real-Time Infrared (RT-IR) Spectroscopy

Since the UV-induced polymerization typically happens in seconds or minutes, it is a tough challenge to study the kinetics of photopolymerization quantitatively and qualitatively. The instrument setup for the RT-IR is shown in Figure 2.20.

For cationic UV curing, the relative humidity is controlled by a humidity chamber with the aid of a salt solution. The formulation is uniformly coated on one KBr crystal or sandwiched between two KBr crystals. The UV light is delivered through a flexible light guide. The end of the light guide is positioned at an angle of 30–45° and at a distance of 3–5 cm from the KBr plate to ensure full KBr plate exposure to UV light. IR illumination is at normal incidence to the sample. During the UV radiation, IR spectra are collected at a certain rate. The transmission mode spectral series collection coupled with data processing generate the conversion *vs.* time profiles. The continuous change of a specific absorbance reflects the reaction extent.

RT-IR, with its milliseconds time resolution, has been successfully used to monitor photopolymerizations occurring within a fraction of a second under intense UV irradiation.[30,31] From the recorded profile the final degree of conversion can be evaluated, *i.e.* the amount of unreacted functional groups that remains in the UV-cured polymer.

Soucek's group determined the cure rate of cycloaliphatic epoxy following the course of epoxy bonding during UV polymerization. With these methods, the reaction conversion could be followed in the seconds range by the decreasing absorption of the epoxy bond at 850 cm$^{-1}$, while simultaneously

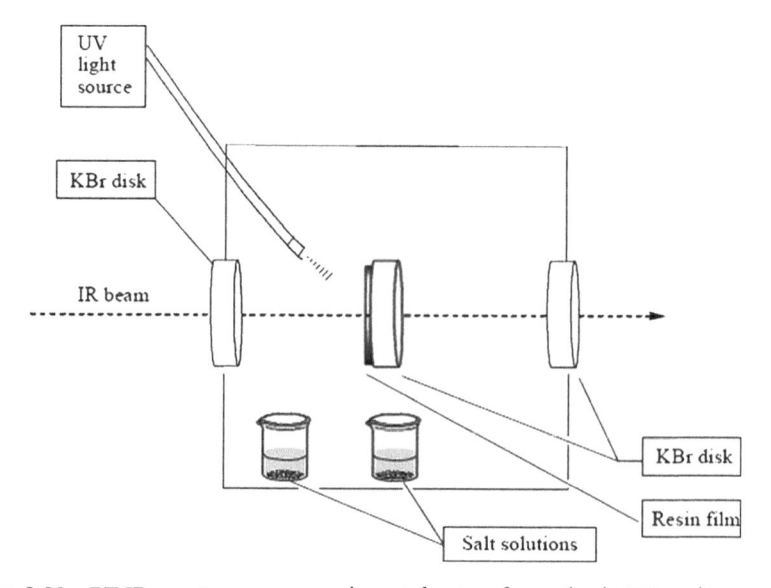

**Figure 2.20** RT-IR spectroscopy experimental setup for cationic UV curing.

curing with UV light.[27] Important information on the polymerization rate, content of the unreacted epoxy bonds, function of water, and so on can be gained with this method. The decreasing absorption, for example of the 850 cm$^{-1}$ band, can be directly converted to conversion of epoxy bands as a function of exposure time. The rate of polymerization ($R_p$) can easily be obtained from the slope of the conversion *versus* exposure time curve and the initial monomer concentration:

$$R_p = [M_0]\mathrm{d}x/\mathrm{d}t$$

Furthermore, the final conversion can be extracted from the conversion-time curve and the residual remaining unreacted epoxy calculated.[27]

### 2.7.2 Confocal Raman Microscopy

The RT-IR measurement gives a good correlation with the total bond conversion of the coating layer, but no information about the cure extent throughout the film thickness can be deduced. Thus, another technique has been developed, which allows the measurement of chemical bonds throughout the film thickness, confocal Raman microscopy.

Confocal Raman microscopy combines the chemical information from vibrational spectroscopy with the spatial resolution of confocal microscopy. The spatial resolution of this technique is in the range of 1 μm$^3$. This allows a mapping of the surface as well as a depth profiling of the remaining chemical bonds.[9]

Real-time IR spectroscopy and confocal Raman microscopy are two complementary nondestructive techniques useful for the characterization of UV-curable coatings. The RT-IR gives information about the conversion development during curing and the residual double bonds over the whole film thickness, confocal Raman spectroscopy is able to give a detailed picture about the spatial distribution of residual chemical bonds within the coating depth as well as a surface mapping. Thus, IR and confocal Raman spectroscopy are complementary techniques applied to gain information about conversion, residual monomer content, oxygen inhibition, water function effects and so on.

### 2.7.3 Photo-DSC

Differential scanning calorimetry (DSC) is the most widely used technique to study photocuring reactions. It is very well suited for the determination of kinetics parameters, like enthalpy, degree of conversion, rate constants, Arrhenius parameters, *etc.* (Figure 2.21).

Photo-DSC has the main limitation of relatively long response time (about 2 s), which makes it impossible to monitor accurately polymerization reactions occurring within less than 10 s, thus requiring operation with low-intensity UV radiation.[6]

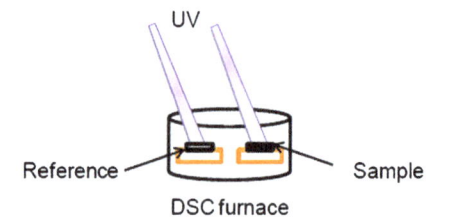

**Figure 2.21** Instrument setup of photo-DSC.

## 2.8 Development and Prospect

### 2.8.1 Oil-Based Coatings

In coating applications, volatile organic compounds (VOCs) are a major environmental and toxic issue worldwide.[32] To alleviate problems associated with VOCs, Reactive oligomers and reactive diluents have been studied with a particular focus on UV curing.[33,34] With increasing environmental regulations in recent years, many natural products have been modified to become UV-curable materials.[35–37] Vegetable oil and its derivatives have traditionally been major ingredients in commercial coatings and inks. In fact, vegetable oil derivatives such as alkyds, uralkyds, and epoxy esters are widely used as low-cost coatings. To reduce the VOC emission, and to reduce the dependence on petroleum as a chemical feedstock for coatings, there has been a push to derive new classes of coating resins and reactive diluents from renewable resources.

Epoxidized vegetable oils can be photopolymerized and are used in UV-curable coatings and inks. Although they are cost effective, commercial epoxidized oils have suffered from low glass-transition temperatures and sluggish reaction rates. As a result, it would be very interesting to design new derivatives of epoxy vegetable oils with enhanced polymerization rates and more stiffness built into the polymer backbone. A study of Soucek and coworkers demonstrated the synthesis and subsequent photopolymerization of an epoxide-modified linseed oil. The obtained epoxynorbornane linseed oils (ENLOs) showed higher reactivity than the conventional epoxidized linseed oils.[15]

UV-curable tung oil and UV-curable tung-based alkyd were prepared by reacting trimethylolpropane trimethacrylate (TMPTMA) onto the α-eleosterate of a tung oil and tung oil alkyd molecule *via* a Diels–Alder reaction by Soucek's group (see Figure 2.22). The results indicated that the formula of UV-curable tung-based alkyd had a faster curing speed than the formula of UV-curable tung oil.[37]

### 2.8.2 Water-Reducible UV Coatings

Water-reducible UV coatings have the advantages of excellent gloss and viscosity control, which enable the resulting coatings to be applied easily with minimal or no process issues with VOC/HAP.

# UV-Curable Coating Technologies

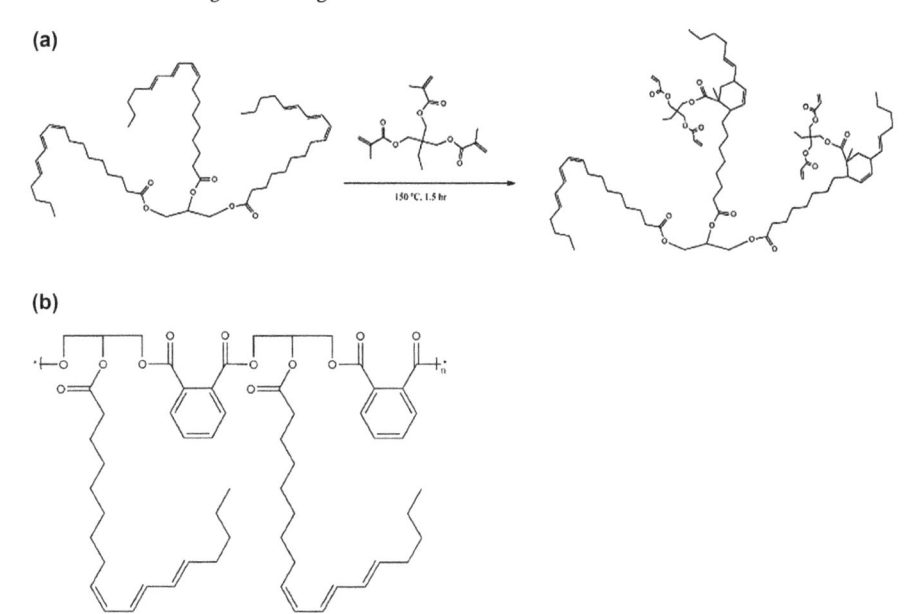

**(a)**

**(b)**

**Figure 2.22** (a) Reaction scheme of UV-curable tung oil (UVTO); (b) tung-based alkyd.[37]

The demand for water-reducible UV-curable coatings is predominant in the high-volume door and panel finishing industry. These surfaces typically have a low build appearance comparable to that created with traditional or conventional solvent borne coatings. The coatings are easily applied, dried, and UV cured to provide excellent finish appearance, durability, and resistance.[9,10]

In addition, pigmented versions are available and the use of many universal type colorants can be used to offer the capability of custom color matching.

More importantly, any method of application can be used for water-reducible UV coatings, although it is most common to see spray and vacuum coating lines in operation.

However, finish defects will occur if the coating is not fully dried prior to UV cure. A common defect observed is the presence of white spots in the finish especially where the coating thickness is slightly higher. These white spots are created when suspended UV-sensitive material undergoes cure before it has a chance to dry and form a continuous film. The cured, nonfilm-formed mass of material scatters reflected light and it will appear as a white spot or area. Other finish defects that are observed include blisters and bubbles present in the cured finish. These are formed by the combination of heat from the UV lamps and from the exothermic reaction of the dry UV-sensitive material during cure. Water and other volatiles will rapidly flash out of the coating causing these defects.[9,10,13]

### 2.8.3 UV Powder Coatings

Conventional powder coatings gained popularity with the coating industrial due to their advantages such as low VOC content, reduction in overspray, and good recyclability. Powder coatings are now starting to be used for high-performance applications such as automotive top coats. Those coatings are mainly used on metal substrates, since they need high curing temperatures above 140 °C and curing times of up to 30 min. The temperatures have to be increased up to 300 °C if the curing times need to be shortened to the range of minutes.[10,13]

The disadvantages of powder coatings, however, have prevented the usage of powders in some markets. The conventional powder coatings are useful only on heat-tolerant substrates, eliminating their use on heat-sensitive substrates such as plastics or wood. They also require long bake cycles at high temperatures, and thus are very energy intensive. Conventional powder coating creates problem in terms of the smoothness of the final coating surface, which is due to the flow characteristics.[10]

To deal with the problems associated with conventional powder coatings, UV-curable powder coatings have been developed. The UV powder coating formulations developed nowadays are mostly based on radically polymerizable systems essentially consisting of UV-curable powder resins, photoinitiators, additives and optionally pigments. To be suitable for UV-curable powder coatings, materials must contain unsaturated double bonds. The chemistry used in UV powder coating resins is mainly based on polyesters, either unsaturated or (meth) acrylate functionalized, or other acrylate functionalized polymers.[9,13]

UV powder coatings provide excellent coating properties and combine the advantages of powder coating technology with the advantages of UV curing. Combination of the benefits of UV-curing technology with those of powder coatings gives the coating formulator superior, UV-curable powder coating systems that have a variety of advantages. The fundamental benefit of UV powder coatings is that they can be processed at considerably lower temperatures than traditional thermoset powders. In contrast to the conventional powder coating, the melt and cure processes of UV-curable powders are decoupled, so that, while heat is used to flow the material, it does not cure the powder and the crosslinking occurs only after the exposure to UV light. Because the temperature required for melting the UV-curable powders is in the range of 80 °C to 120 °C, much less heat is required, and is needed only long enough to melt the powder, not to cure it.[9] Crosslinking by the UV light is nearly instantaneous, thus, the process is finished in a fraction of the time required for conventional powders. Of the wood-based substrates the most interesting for coating with UV powders is medium-density fiberboard (MDF), because of its uniform density, controllable humidity, dimensional stability and stability at processing temperatures up to 140 °C.[10]

The formulation of the powder has to be done in an extrusion process, which should be carried out at temperatures above $T_g$, but below $T_m$, usually in the range of 50–70 °C.[10] The application of the UV powders to the substrate is preferably done in an electrostatic application process, which is essentially the same in the conventional powder application process. The powder is pumped by air through a gun that generates an electric field and imparts a static charge to the powder particles. These charged particles are attracted uniformly to a grounded part. The part has to be conductive or coated by a conductive primer coat so that it can be rounded. The electrostatic spray gun has several functions: to shape and direct the flow of powder, to control the pattern size, shape and density of powder, to impart the electrostatic charge to the powder, to control the deposition of powder on the part.[9,10,13]

The coated substrate is then heated in an oven or with IR lamps until the powder forms a smooth surface film that usually takes several minutes and the hot film is passed under UV lamps to be cured completely within several seconds. The whole process may be run continuously and takes minutes at most.[9]

### 2.8.4 UV-Moisture Dual-Curable Coatings

UV-moisture dual-curing inorganic–organic hybrids are very complex systems. In the literature, network-structure formation of the UV-curing hybrid coatings was not effectively studied. Soucek's group held that inorganic–organic hybrids can be a good alternative to unsaturated polyester (UPE)/styrene systems. Inorganic silicate oligomers may function as a reactive diluent and not only help to reduce viscosity of the resin but also may increase the mechanical properties of the films.[38]

UV-moisture dual-curable inorganic–organic hybrid coatings based on UPE binders and silicate-clusters were successfully prepared. *In situ* formed nano- to microsize inorganic clusters increased the abrasion resistance, adhesion, tensile strength and fracture toughness of the films. A free-radical UV-crosslinking process was used to form the organic phase. An isocyanate functional alkoxysilane coupling agent was used to connect organic and inorganic phases (see Figure 2.23).

### 2.8.5 UV-Thermal Dual-Cure Coatings

The dual-cure coatings combine the fast curing mechanism of the UV-curing reaction with an independent thermal cure, especially necessary in areas where sufficient light is not available, or where special effects, like improved adhesion have to be obtained.

## 2.9 Conclusion

UV-curable coatings enjoyed a steady increase over the last several decades due to their advantages. A wide choice of photoinitiators, reactive oligomers, and reactive diluents even extended the application for UV-curable coatings.

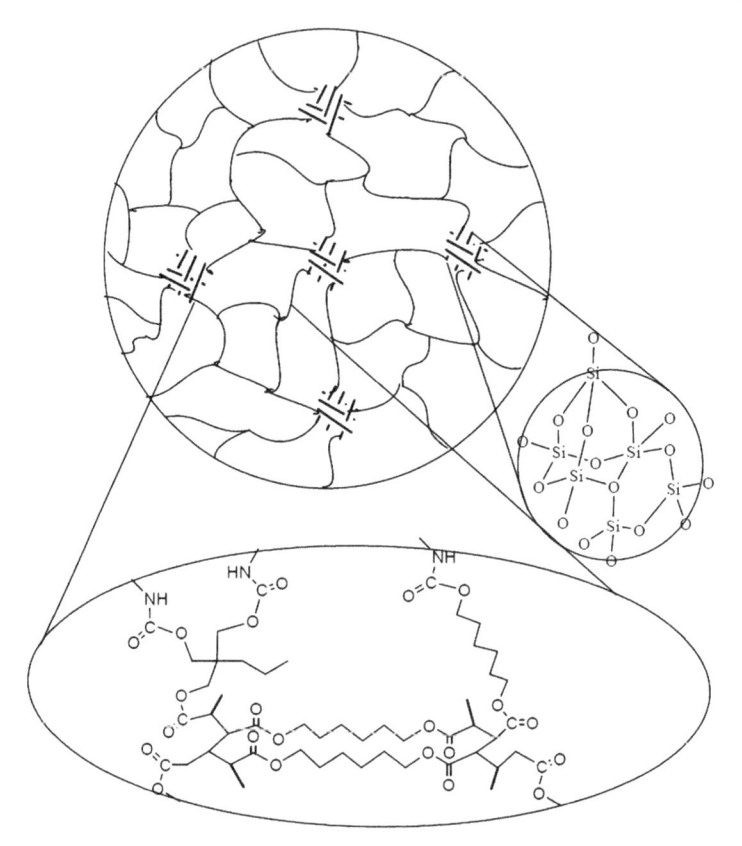

**Figure 2.23** UV-cured polyurethane/polysiloxane organic/inorganic network model.[38]

Free-radical photopolymerizations have had overwhelming commercial success compared to cationic photopolymerization. However, cationic photopolymerizations are finding increasing application due to their advantages such as rapid cure, no oxygen inhibition, and ability for dark cure. The techniques for characterization of UV-curing are developing to meet the requirement of coating research. The real-time FT-IR and photo-DSC have been shown to successfully characterize the curing process. The combination and development of UV-curable technology with water-reducible coatings, powder coatings, moisture and thermal dual coatings are gaining popularity, and may have a bright future for development and application.

# References

1. J. W. Johnson, R. R. Matheson, D. A. White, N. E. Drysdale, P. H. Corcoran and L. A. Lewin, Polymerization of acrylic polymers in reactive diluents, *US Pat.* 20080287622 A1, 2008.
2. Environmental Protection Agency, "Clean Air Act: History of the Clean Air Act".

3. A. Wojciechowski, J. Lewis and T. Braswell, Development of a low VOC chemical agent resistant coating for use with supercritical $CO_2$, *Fed. Facil. Environ. J.*, 1998, **9**(3), 2.
4. R. Golden, Low-emission technologies: A path to greener industry, *RadTech Rep.*, 2005, **19**(3), 14.
5. R. Mehnert, A. Pincus, I. Janorsky, R. Stowe, A. Berejka, *UV & EB Curing Technology & Equipment*, John Wiley and Sons Inc., New York, 1998.
6. C. Decker, The use of UV Irradiation in Polymerization, *Polym. Int.*, 1998, **45**, 133.
7. A. Endruweit, M. S. Johnson and A. C. Long, Curing of Composite Components by Ultraviolet Radiation: A Review, *Polym. Compos.*, 2006, 119.
8. P. Swaraj., *Surface Coatings: Science and Technology*, John Wiley and Sons Inc., New York, 2nd edn, 1996.
9. J. Drobny, *Radiation Technology for Polymers*, CRC Press, Florida, 2003.
10. R. Schwalm, *UV Coatings: Basic, Recent Developments and New Applications*, Elsevier, Netherlands, 2007.
11. V. Shukla, M. Bajpai, D. K. Singh, M. Singh and R. Shukla, Review of basic chemistry of UV-curing technology, *Pigm. Resin Technol.*, 2004, **33**(5), 272.
12. C. Decker, *Acta Polym.*, 1994, **45**, 333.
13. Z. W. Wicks, Jr., F. N. Jones, S. P. Pappas and D. A. Wicks, *Organic Coatings: Science and Technology*, John Wiley and Sons Inc., New York, 3rd edn, 2007.
14. P. Micheli, Pigment wetting characteristics of radiation curing systems, *Surf. Coat. Int.*, 2000, **9**, 455.
15. J. Chen, M. D. Soucek, W. J. Simonsick and R. W. Celikay, Synthesis and photopolymerization of norbornyl epoxidized linseed oil, *Polymer*, 2002, **43**, 5379.
16. A. B. Strong, *Fundamentals of Composites Manufacturing*, Society of Manufacturing Engineers, 2008.
17. Styrene, 12th Report on Carcinogens, National Institute of Environmental Health Science, 2011.
18. R. Mehnert, *et al.*, *UV & EB Curing Technology & Equipment*, John Wiley and Sons Inc., New York, 2011, vol. I.
19. C. Pongsith, Pigmented UV-curable alkyd, A Thesis Presented to The Graduate Faculty of The University of Akron, 2009.
20. C. Decker and A. Jenkins, *Macromolecules*, 1985, **18**, 1241.
21. K. Studer, C. Decker, E. Beck and R. Schwalm, *Prog. Org. Coat.*, 2003, **48**, 92.
22. M. Awokola, W. Lenhard, H. Löffler, C. Flosbach and P. Frese, *Prog. Org. Coat.*, 2002, **44**, 211.
23. R. S. Davidson, The role of amines in UV-curing, in *Radiation Curing in Polymer Science and Technology: Polymerisation Mechanism*, ed. J. P. Fouassier and J. F. Rabek, Elsevier Applied Science, London, 1993, vol. III.

24. E. Selli and I. R. Bellobono, Photopolymerization of multifunctional monomers: kinetic aspects, in *Radiation Curing in Polymer Science and Technology: Polymerisation Mechanism*, ed. J. P. Fouassier and J. F. Rabek, Elsevier Applied Science, London, 1993, vol. III.
25. J. V. Crivello, UV and electron beam-induced cationic polymerization, *Nucl. Instrum. Methods Phys. Res., Sect. B*, 1999, **151**, 8.
26. J. V. Crivello and R. Narayan, Novel Epoxynorbornane Monomers. 2. Cationic Photopolymerization, *Macromolecules*, 1996, **29**, 439.
27. Z. Zong, M. D. Soucek, Y. Liu and J. Hu, Cationic Photopolymerization of Epoxynorbornane Linseed Oils: The Effect of Diluents, *J. Polym. Sci., Part A: Polym. Chem.*, 2003, **41**, 3440.
28. J. V. Crivello and R. Narayan, Novel Epoxynorbornane Monomers. 1. Synthesis and Characterization, *Macromolecules*, 1996, **29**, 433.
29. M. D. Soucek and J. Chen, *J. Coat. Technol.*, 2003, **75**, 936.
30. C. Decker and K. Moussa, Real-Time Kinetic Study of Laser-Induced Polymerization, *Macromolecules*, 1989, **22**, 4455.
31. C. Decker and K. Moussa, A new method for monitoring ultra-fast photopolymerizations by real-time infra-red (RTIR) spectroscopy, *Makromol. Chem.*, 1988, **189**, 2381.
32. EPA, A.US, Hazardous Air Pollutants Strategic Implementation Plan, 1997.
33. R. Mehnert, *et al.*, *UV & EB Curing Technology & Equipment*, John Wiley & Sons, New York, 2011, vol. 1.
34. K. E. J. Barrett and R. Lambourne, Air-drying alkyd paints, The effect of incorporation of nonvolatile monomers on film properties, *J. Oil Colour Chem. Assoc.*, 1966, **49**(6), 443.
35. J. T. Guthrie and J. G. Tait, UV-curable coating from palm oil and derivatives, surface coatings international, *J. Oil Colour Chem. Assoc.*, 2000, **83**(6), 278.
36. P. Phinyocheep and S. Juangthong, Ultraviolet-curable liquid natural rubber, *J. Appl. Polym. Sci.*, 2000, **78**, 1478.
37. N. Thanamongkollit, K. R. Miller and M. D. Soucek, Synthesis of UV-curable tung oil and UV-curable tung oil based alkyd, *Prog. Org. Coat.*, 2012, **73**, 425.
38. A. Nebioglu, G. Teng and M. D. Soucek, Dual-Curable Unsaturated Polyester Inorganic/Organic Hybrid Films, *J. Appl. Polym. Sci.*, 2006, **99**, 115.

CHAPTER 3

# Newly Synthesized Photocrosslinkable Liquid-Crystalline Polymers and their Properties

G. KUMAR AND K. SUBRAMANIAN*

Department of Chemistry, Anna University, Chennai-600 025, India
*Email: kathsubramanian@yahoo.com

## 3.1 Introduction

Photocrosslinkable polymers have recently received considerable attention owing to their potential application in many technologies such as optical switching, data recording and storage, nonlinear optics, and displays. The capability of these polymers to crosslink is due to the carbon–carbon double bonds of the α,β-unsaturated carbonyl groups that undergo $[2\pi + 2\pi]$ cycloaddition reactions under irradiation. Theses photosensitive polymers are regarded as negative-type photoresists.[1,2] The photosensitivity of these materials is based on the π-electron density of the chromophores present in the polymer backbone.[3] The literature reveals that the chalcone groups possess higher sensitivity to UV irradiation than that of the cinnamate groups.[4] Liquid-crystalline polymers (LCPs) have generated considerable interest in recent years and the photocrosslinkable LCPs have driven special attention, if they contain both mesogen and photoactive groups in their structure.[5] Mesogens incorporate liquid-crystalline (LC) properties to the

RSC Smart Materials No. 13
Photocured Materials
Edited by Atul Tiwari and Alexander Polykarpov
© The Royal Society of Chemistry 2015
Published by the Royal Society of Chemistry, www.rsc.org

polymer and photoreactive group facilitates crosslinking of the chain under the influence of UV light radiation. These types of polymers were useful in fabricating anisotropic networks and thin films, information storage devices[6] nonlinear optical devices,[7] aligned membrane for permeation of gases and drugs, *etc.*[8] Among many promising photocrosslinkable groups, the chalcone group has been well recognized and can be incorporated in photocrosslinkable side-chain polymers. The polymers that carry the dual properties of both liquid crystallinity and photocrosslinking are seen to be indispensable material in nonlinear optical (NLO) applications.[9] Many photocrosslinkable LCPs reported so far contain the cinnamate ester group, azogroup, imide as a photoactive group that undergoes photodimerization on irradiation leading to the crosslinking. Thousands of research articles have been published in liquid-crystalline polymers and in photoreactive polymers separately by using the above-mentioned photoreactive group. But research over the last two decades and the present trend find very few articles in photocrosslinkable polymers containing pendant chalcone moiety exhibiting liquid-crystalline property. This chapter discusses the photoresist nature of polymers using UV and fluorescence spectroscopy and the liquid-crystalline nature by DSC and HOPM studies. We believe that the photoreactive nature of photocrosslinkable polymers and a morphological study by SEM analysis, may also give some information about behavior of photocrosslinkable polymers.

## 3.2 Synthesis of Monomers and Polymers

The monomers were synthesized in a two-stage process, the first step was by conversion of chalcones containing carboxylic acid into acid chlorides in the benzene medium and the second step was conversion of acid chloride of chalcones in to monomer by reacting the acid chlorides of chalcones with 2-hydroxy ethyl methacrylate using triethylamine as the reagent in tetrahydrofuran as a solvent medium. The predetermined quantities of monomers were taken with AIBN (5% weight of monomer) as initiator in polymerization tube and dissolved with 20 ml of dry tetrahydrofuran (THF). The above solution was degassed by a $N_2$ gas atmosphere for 10–15 min. Then, the polymerization tube was kept at 70 °C for 48 h, as described in the literature.[10] The following are synthesized polymers (Figure 3.1):

- poly(2-[benzoyl styryloyloxy] ethyl methacrylate) [P1];
- poly(2[4-bromo benzoyl styryloyloxy] ethyl methacrylate) [P2];
- poly(2-[4-methoxy benzoyl styryloyloxy] ethyl methacrylate)[P3];
- poly(2-[3,5-di benzyl oxy benzoyl styryloyloxy] ethyl methacrylate)[P4];
- poly(2-[4-biphenyl oxy benzoyl styryloyloxy] ethyl methacrylate)[P5].

## 3.3 Structural Conformations by Proton and $H^1$-NMR Spectroscopy

The representative $H^1$-NMR spectra of the polymers P1–P5 are shown in Figure 3.2 The aromatic protons of polymers resonate between 6.7–8.3 ppm.

**Figure 3.1** Synthesized polymers.

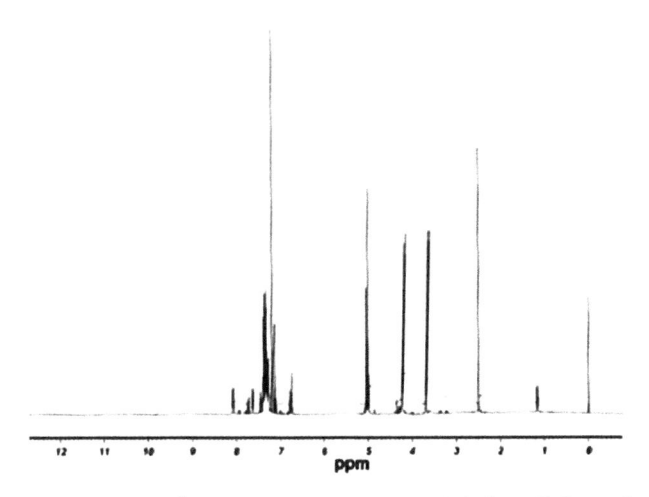

**Figure 3.2** Representative $^1$H-NMR spectrum of poly(2-[3,5-di benzyl oxy benzoyl styryloyloxy] ethyl methacrylate)[P4].

The olefinic protons appear as a doublet in the regions 6.8 and 7.9 ppm. The disappearance of two singlet protons in the regions 5.8 and 6.2 ppm confirms the polymerization of monomers. The ethyleneoxy protons show their resonance as multiplet in the region between 4.0–4.8 ppm. The resonating signals appear between 0.9–2.5 are attributed to spacer protons and other aliphatic protons.

## 3.4 Structural Conformations by Proton and $^{13}$C-NMR Spectroscopy

The C$^{13}$-NMR spectra of the polymer P4 is shown in Figure 3.3. In all the polymers, aliphatic spacers and other aliphatic carbons resonate between 20–64 ppm. The resonance occurs in the region between 163–178 ppm due to the presence of an ester carbon. The methoxy carbon resonates at 52.6 ppm. The olefinic double-bonded carbons appear in the region between 121–145.5 ppm. The resonance signal appears around 190 ppm due to a ketone carbonyl carbon. The benzyl oxy ether carbon has its resonance at 70.5 ppm. All the aromatic carbons in the polymers appear between 113–162 ppm.

## 3.5 Molecular Weight of Polymers

The number average ($\bar{M}_n$) and weight average ($\bar{M}_w$) molecular weight of polymers P1–P5 were determined by a Waters PL-GPC650 model using THF

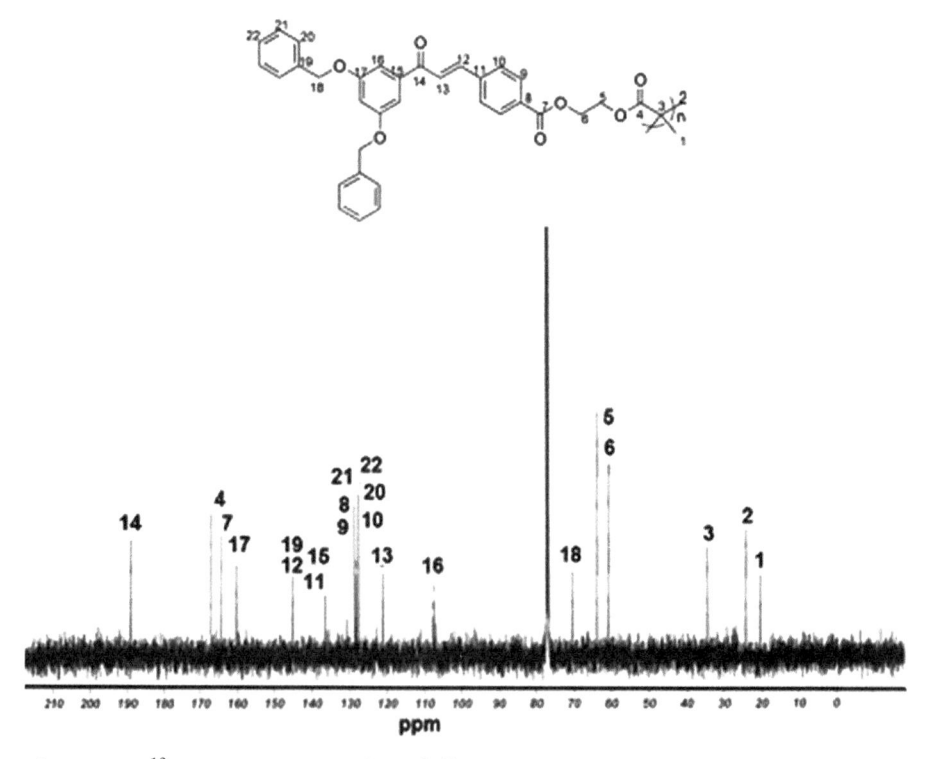

**Figure 3.3** $^{13}$C-NMR spectrum of poly(2-[3,5-di benzyl oxy benzoyl styryloyloxy] ethyl methacrylate) [P4].

as eluent. Polystyrene standards with PDI of less than 1.2 are typically used to calibrate the GPC. The weight- and number-average molecular weight of the polymers and their polydispersity index (PDI) values are given in Table 3.1. The number ($\bar{M}_n$) and weight ($\bar{M}_w$) average molecular weight of the polymer P1 were $9.33\times10^4$ and $13.95\times10^4$. The molar mass distribution of polymer is given by polydispersity index (PDI) value 1.49. For the polymer P2, the $\bar{M}_n$ and $\bar{M}_w$ were $8.58\times10^4$ and $12.67\times10^4$, respectively, and the polydispersity index (PDI) value was 1.47. The theoretical values of PDI for polymer produced *via* radical combination and disproportionation are 1.5 and 2.0, respectively.[11] The PDI value of polymers P1 and P2 were 1.49 and 1.47, respectively. Both polymers P1 and P2 possess the tendency of chain termination by radical combination rather than disproportionation.

The number- ($\bar{M}_n$) and weight-average molecular weight ($\bar{M}_w$) of the polymer P3 were $8.93\times10^4$ and $13.57\times10^4$, respectively. The PDI value of photocrosslinkable liquid-crystalline polymer (PCLCP) P3 was found to be 1.52 from the values of $\bar{M}_n$ and $\bar{M}_w$, which suggests the strong tendency of chain termination by radical combination rather than disproportionation. The weight-average molecular weight ($\bar{M}_w$) of the polymers P4 and P5 are shown in Table 3.1. The polydispersity index (PDI) values of polymers P4 and P5 were 1.56 and 1.55 respectively. All the polymers show PDI values around 1.5, which indicates that the polymers have been produced *via* free-radical recombination.

## 3.6 Thermal Stability

Thermogravimetric analysis (TGA) measures the amount and rate of change in the weight of a material as a function of temperature or time in a controlled atmosphere. Measurements are used primarily to determine the composition of materials and to predict their thermal stability at temperatures up to 1000 °C. The thermogravimetric analysis (TGA) of the prepared polymers was carried out with a heating rate of 20 °C min$^{-1}$ and a gas flow rate of 20 cm$^3$ min$^{-1}$ under a nitrogen atmosphere in the temperature range between 30–800 °C in order to investigate the thermal stability.

The TGA data illustrated in Table 3.2. The data in the table and figure indicates that the homopolymers decompose at higher temperature and they show single-stage decomposition with good thermal stability.

Table 3.1 Molecular weight of polymers (P1–P5)

| S. No. | Polymer | Weight average molecular weight ($\bar{M}_w$) | Number average molecular weight ($\bar{M}_n$) | Polydispersity index (PDI) values |
|---|---|---|---|---|
| P1 | Poly(BSOEMA) | $13.9\times10^4$ | $9.33\times10^4$ | 1.49 |
| P2 | Poly(BBSOEMA) | $12.67\times10^4$ | $8.58\times10^4$ | 1.47 |
| P3 | Poly(MBSOEMA) | $13.57\times10^4$ | $8.93\times10^4$ | 1.52 |
| P4 | Poly(DBOBSOEMA) | $14.25\times10^4$ | $9.1\times10^4$ | 1.56 |
| P5 | Poly(BPOBSOEMA) | $11.50\times10^4$ | $7.41\times10^4$ | 1.55 |

**Table 3.2** Thermogravimetric analysis (TGA) data of polymers P1–P5

| S. No. | Polymer | IDT (°C) | 10 | 25 | 50 | 75 | 90 | Char yield (wt%) at 600 °C |
|---|---|---|---|---|---|---|---|---|
| P1 | Poly(BSOEMA) | 250 | 255 | 270 | 390 | 480 | 720 | 40 |
| P2 | Poly(BBSOEMA) | 265 | 275 | 360 | 400 | 490 | 728 | 42 |
| P3 | Poly(MBSOEMA) | 225 | 300 | 350 | 420 | 460 | — | 16 |
| P4 | Poly(DBOBSOEMA) | 200 | 240 | 321 | 485 | — | — | 44 |
| P5 | Poly(BPOBSOEMA) | 190 | 210 | 280 | 400 | 455 | — | 11 |

Temperature (°C) at weight loss (%)

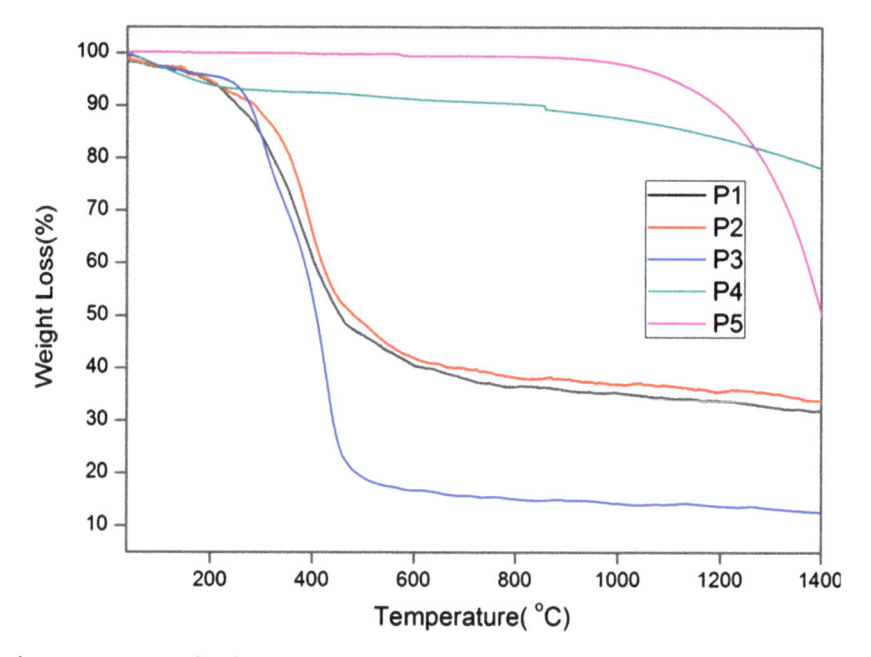

**Figure 3.4** TGA of polymers P1–P5.

The initial decomposition temperatures (IDT) of the polymers P1 and P2 were slightly different from each other due to smaller difference in molecular mass. Both P1 and P2 show 50% weight loss at the temperature 390 and 400 °C, respectively, prove the flame retardancy and stability of polymers. These thermal studies have shown that the polymers possess very good thermal and thermo-oxidative stability required for the negative photoresist polymers.

The TGA of the polymer P3 is shown in the Figure 3.4. The homopolymer poly(MBSOEMA) undergoes single-stage thermal decomposition between 225–475 °C with the weight loss of 90%. The initial decomposition temperature (IDT) of the polymer starts at 225 °C due to the presence of an ethylene spacer in the polymer backbone and a pendant unit of chalcone showed 50% and 75% weight loss at 390 °C and 437 °C, respectively. The thermal decomposition was observed at 450 °C.

TGA of the polymers P4 and P5 show IDT at 200 °C and 190 °C, respectively, based on the aliphatic content present in the polymeric backbone. The polymers P4 and P5 undergo 50% weight loss at the temperatures of 485 °C and 400 °C, respectively. When compared to P4 the polymer P5 has more thermal stability due to the presence of benzyloxy substitution at the 3rd and 5th positions of the polymer. The char yields of all the polymers at 600 °C have been tabulated in Table 3.2 It is observed from the table that all the polymers show char yields from 26–44% except polymers P5. Polymers P1, P2, P3 and P4 have char yields of 40%, 42%, 16% and 44%, respectively, at 600 °C, among these polymers P4 exhibits the higher char yield, since it has a high aromatic content in its polymer backbone. The high char yield of these polymers can be attributed to their high aromatic contents.[12] It is noted from the above results that the homopolymers decompose at higher temperatures in a single stage and possess good thermal and thermo-oxidative stability required for a negative photoresist, in comparison with the homopolymers that contain spacers between the backbone of the polymer and the photoreactive groups.[13] As most of the polymers indicate high char yield they may also be useful as flame-retardant materials.

## 3.7 Photocrosslinking Properties

### 3.7.1 UV Spectral Studies

The photocrosslinking studies were carried out to study the changes occurring in the polymer during UV irradiation to confirm the photoresist nature of the polymer. The polymer solution was prepared in the concentration range of 10–20 mg l$^{-1}$ using chloroform. It was irradiated with UV light of 254 nm and the photocrosslinking ability of the polymer was followed by the rate of disappearance of the C=C bond of the photosensitive group in the UV spectrum. The rate of disappearance of double bonds in the photosensitive group was calculated by:[14]

$$\text{Rate of conversion } (\%) = (A_o - A_t)/(A_o) \times 100 \quad (3.1)$$

where $A_o$, and $A_t$ are absorption intensities due to the $\rangle$C=C$\langle$ group after the irradiation time $t=0$ and $t=T$ respectively.

The UV-spectral changes during photocrosslinking and photoconversions of polymers are shown in Figures 3.6–3.11.

The photochemical changes were followed by UV-visible spectra and fluorescence spectra. The ultraviolet (UV) irradiation of polymers P1 and P2 was carried out by preparing a polymer solution with a concentration range 10.5–100 mg l$^{-1}$, by irradiation with (254 nm) UV light and the effects of UV light with polymers P1 and P2 were observed by measuring the rate of disappearance of olefinic $\rangle$C=C$\langle$ bond of the pendant chalcone unit in UV spectra. When the polymer solutions were irradiated with UV light, the polymers P1 and P2 formed cyclobutane rings during the $2\pi + 2\pi$ cycloaddition reaction, as shown in Figure 3.5.

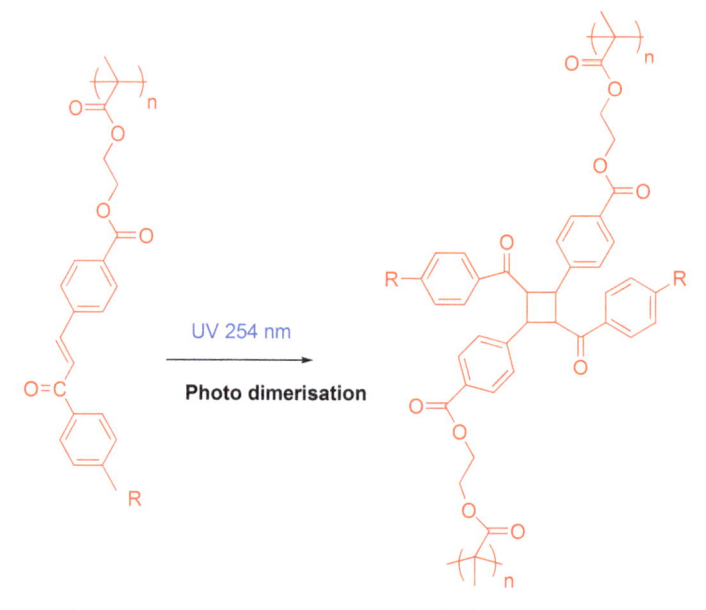

**Figure 3.5** Schematic representations photocrosslinking of polymers (P1–P5).

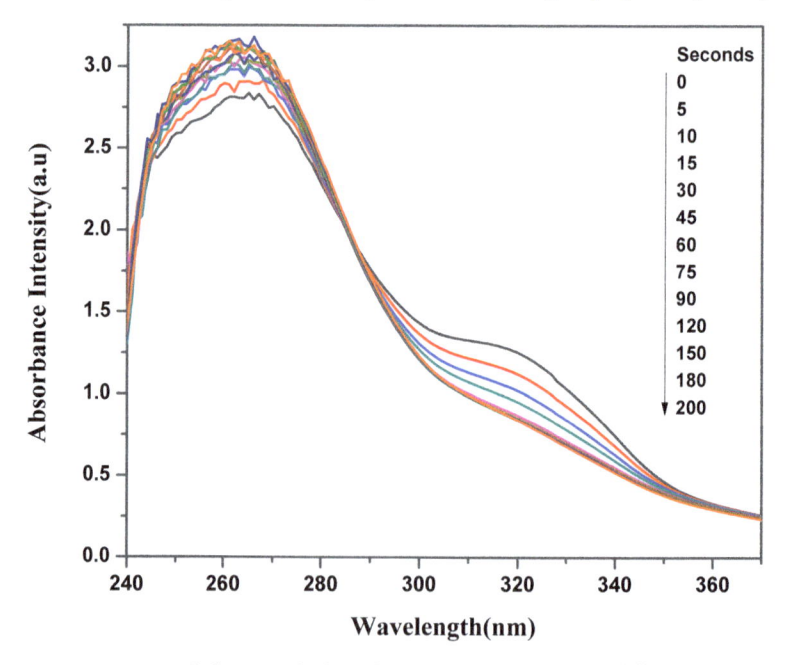

**Figure 3.6** UV-spectral changes during photocrosslinking of poly(2-[benzoylstryloyloxy] ethyl methacrylate) [P1] at various time interval in chloroform solution.

The changes in the UV spectral pattern with respect to unsubstituted and electronegative substituent polymers P1 and P2 are shown in the Figures 3.6 and 3.7. In both the polymers P1 and P2, there is a successive decrease in the

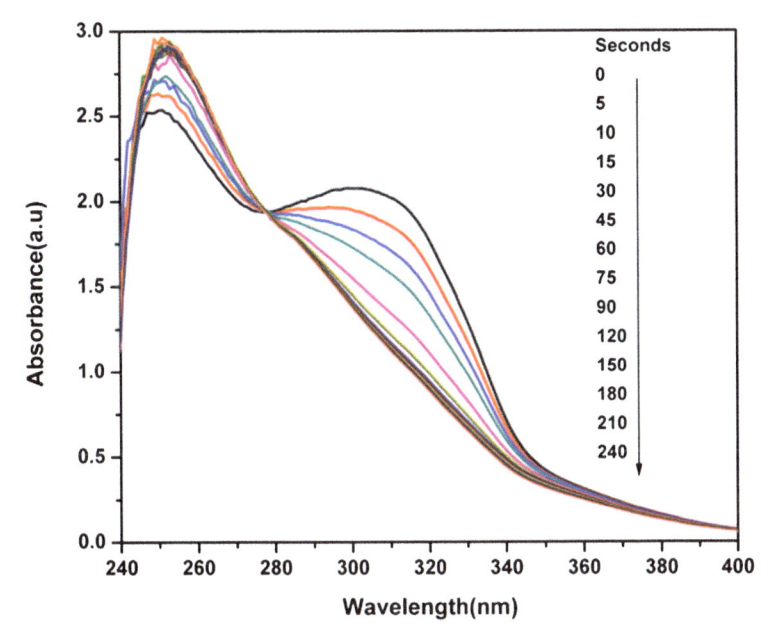

*Newly Synthesized Photocrosslinkable Liquid-Crystalline Polymers and their Properties* 57

**Figure 3.7** UV-spectral changes during photocrosslinking of poly(2[4-bromo benzoyl styryloyl oxy] ethyl methacrylate) [P2] at various time interval in chloroform solution.

absorption intensities observed during irradiation at 254 nm. The absorption maxima at 320 nm and 308 nm are attributed to polymer P1 and P2, respectively. Though the polymer has no substitution in the para position, it has a slightly electron-donating nature *via* the presence of hydrogen in the para position. So, it shows an absorption maxima at longer wavelength. But in the case of polymer P2, it has bromosubstitution in the para position. Due to the electron-withdrawing nature of the bromo group in the para position the polymer P2 shows absorption maxima at the shorter wavelength of 308 nm than P1. The isosbestic points have been observed at 289 and 278 nm for the polymers P1 and P2, respectively. As the $2\pi + 2\pi$ cycloaddition destroys the conjugation in the entire $\pi$-electron system, the UV absorption intensity decreases with irradiation time[15,16] that leads to cyclobutane ring formation. The photoresponsive nature of polymers can be measured in terms of the rate of disappearance of $\rangle C=C\langle$ with irradiation time, which are presented in Figure 3.11.

The irradiation plot of polymer P1 shows a 56.31% photoconversion occurred at 20 s and 73%, 99.6%, 99.8%, 99.9% and 100% of polymer conversion at 15, 75, 150, 180 and 200 s, respectively. But in the polymer P2, it shows photoconversion of 49.9%, 96.6%, 98.5%, 98.7%, 99.5% and 100% occurred at 15, 75, 150, 180, 210 and 240 s, respectively. From the graphical data, the unsubstituted polymer P1 undergoes faster photocrosslinking than the bromosubstituted polymer P2. Because the electron-releasing nature of hydrogen in the para position of P1 facilitates faster photoconversion than the electron-withdrawing bromo group in the para position of polymer P2. The polymers after irradiation were insoluble in solvents in which they were soluble before irradiation and this may be due to crosslinking of polymers chain through $2\pi + 2\pi$ cycloaddition of the olefinic double bond of the pendant chalcone unit.

The change in the UV-absorption spectrum of the polymer P3 is illustrated in Figure 3.8. It can be seen that polymer P3 containing the chalcone moiety shows a maximum absorption at 318 nm. The rapid photocrosslinking may be due to a resonance effect produced by the electron-releasing methoxy group present in the pendant chalcone unit of the polymer P3. The photocrosslinking ability of polymer P3 was accelerated faster and the $\pi$–$\pi$* transition of the pendant methoxy styryl group is evident from the appearance of the isosbestic point at 290 nm. The absorption intensity decreases rapidly with increasing irradiation time and the band disappears almost completely within three minutes of irradiation. The decrease in the UV absorption intensity due to the crosslinking of the polymer through $2\pi + 2\pi$ cyclodimerization of the CH=CH-group of the methoxy styryl group and leads to formation of the cyclobutane ring.[17] The photocrosslinkable liquid-crystalline homopolymer is insoluble in polar aprotic and chlorinated solvents in which it was soluble before irradiation. The rate of photoconversions of polymer P3 with irradiation time is shown in Figure 3.11. Thus, the poly(MBSOEMA) with pendant chalcone moiety has a higher rate of photocrosslinking, even in the absence of a sensitizer, leading to insolubility of the polymer.

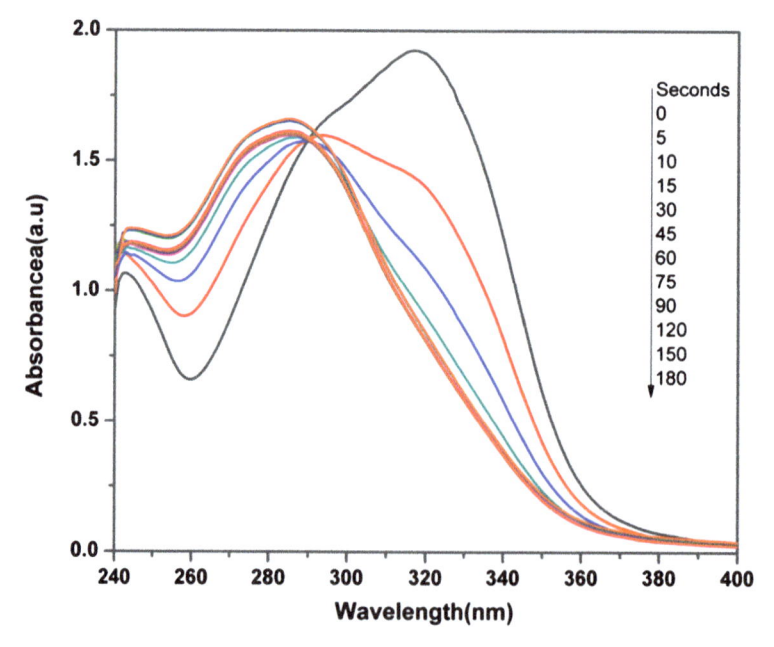

**Figure 3.8** UV-spectral changes during photocrosslinking of poly(MBSOEMA) at various time intervals in chloroform solution (reproduced from ref. 10).

The UV-visible spectral changes during photocrosslinking of polymers P4 and P5 are shown in Figures 3.9 and 3.10. The photocrosslinking study of polymers reveal that polymer P5 shows much faster olefinic C=C decay, *i.e.* photocrosslinking or photodimerization, than polymer P4. Because in the case of biphenyl substituted polymer P5 has biplanarity of the biphenyl ring, Since the two phenyl rings are perpendicular to each other, the close packing of the chains through the higher $\pi-\pi^*$ interactions of two aromatic rings is allowed.[18] Polymer P4, which has dibenzyloxy substitution in the 3,5-positions of the pendant unit undergoes olefin C=C decay to some extent slower than the biphenyl-substituted polymer P5. Generally, the formation of a cyclobutane ring as a result of photocrosslinking leads to bulk structure neighboring the backbone of the polymer chain and leads to a decrease in both the decay rate and photocrosslinking.[19] Thus, P4 undergoes photocrosslinking 450 s slower than P5, although it has donor groups in the 3, 5 positions of the pendant chalcone unit.

In general, the rates of decay in the C=C bond of chalcones on exposure to UV radiation with time indicated that:

(i) the rate was faster in the first few seconds followed by a slow rate
(ii) there was no clear kinetic order that can be determined (iii) the rate of

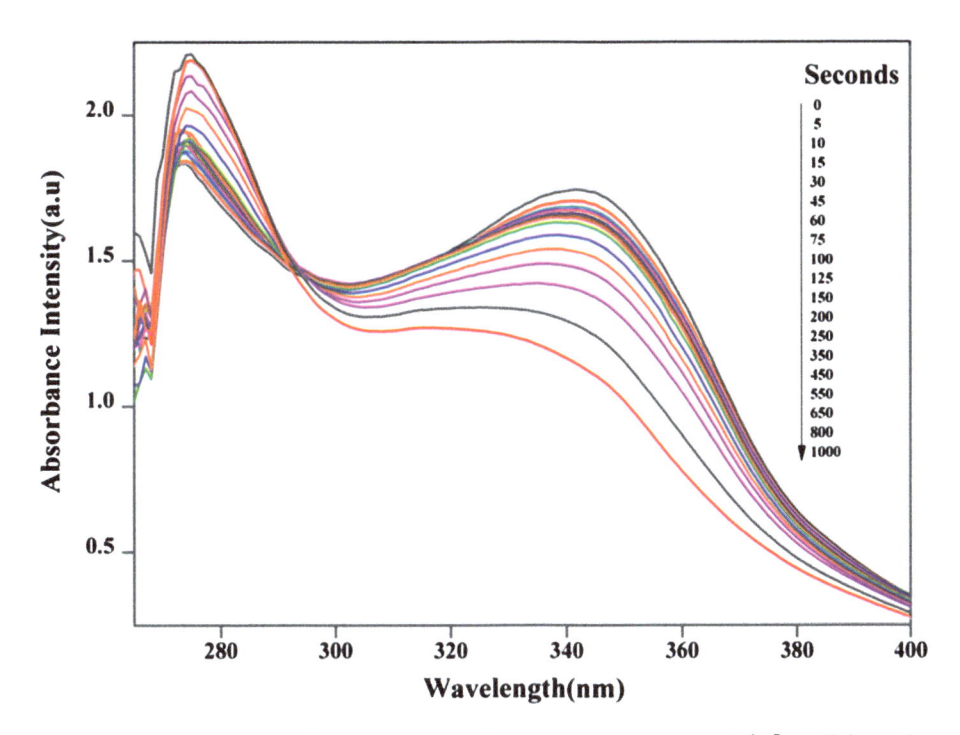

**Figure 3.9** UV-spectral changes during photocrosslinking of poly(2-[3,5-di benzyl oxy benzoyl styryloyloxy] ethyl methacrylate) [P4] at various time intervals in chloroform solution.

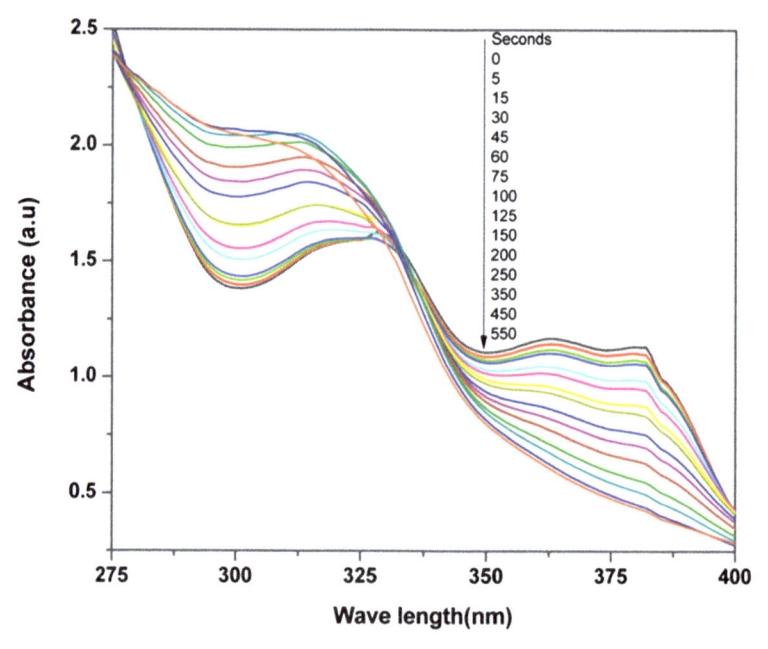

**Figure 3.10** UV-spectral changes during photocrosslinking of poly(2-[4-biphenyloxy benzoyl styryloyloxy] ethyl methacrylate) [P5] at various time intervals in chloroform solution.

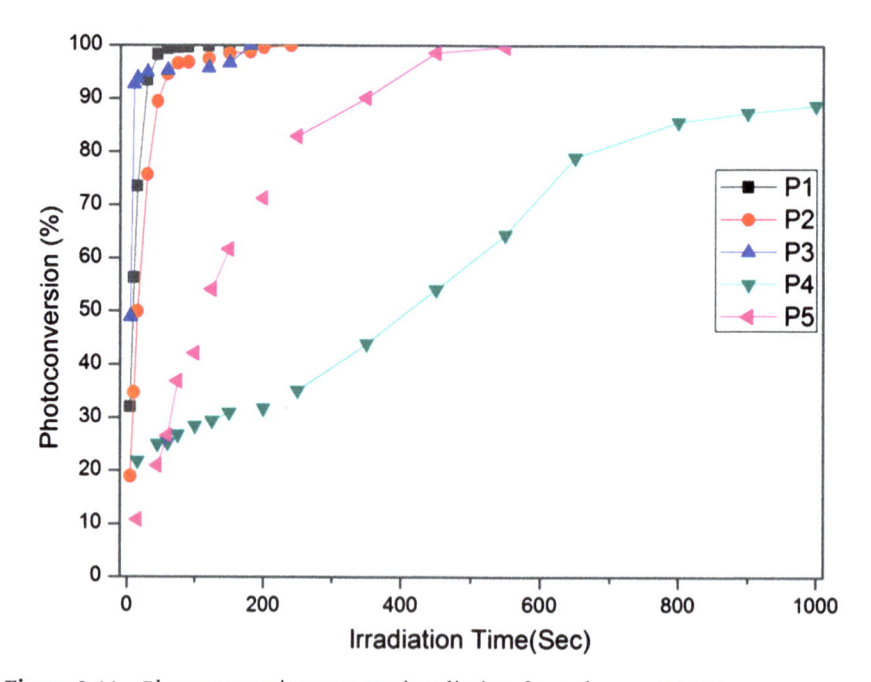

**Figure 3.11** Photoconversions on UV irradiation for polymers P1–P5.

disappearance of the C=C bond depends on several factors, such as spacer units between the sensitive moieties and the backbone of polymer chains, copolymerization, concentration and temperature (iv) the solvent used in the photocrosslinking process has no significant effect on the rate of photo-dimerization of the C=C bond.

## 3.7.2 Fluorescence Studies

Fluorescence occurs when a molecule absorbs light photons from the UV-visible light spectrum, known as excitation, and then rapidly emits light photons as it returns to the ground state. The fluorescence behavior of a molecule or an atom is shown in the Jablonski energy diagram (see Figure 3.12). The examination of the Jablonski diagram reveals that the energy of the emission is typically less than that of absorption. Fluorescence typically occurs at lower energies or longer wavelengths.

This phenomenon was first observed by Sir G. G. Stokes in 1852 at the University of Cambridge. The energy losses between excitation and emission are observed universally for fluorescent molecules in solution. When a system (be it a molecule or atom) absorbs a photon, it gains energy and enters an excited state. One way for the system to relax is to emit a photon, thus losing its energy (another method would be the loss of heat energy). When the emitted photon has less energy than the absorbed photon, this energy difference is the Stokes shift. If the emitted photon has more energy, the energy difference is called an anti-Stokes shift.

The Stokes shift (Figure 3.13) is the difference in wavelength or frequency units between positions of the band maxima of the absorption and emission

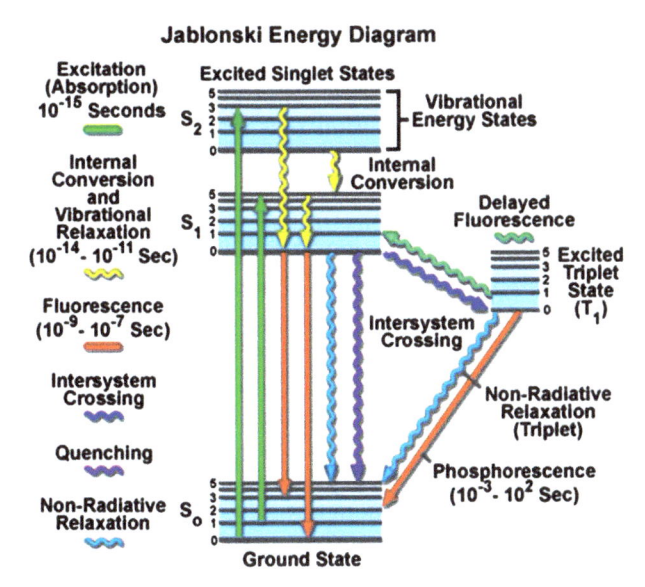

**Figure 3.12** Jablonski energy diagram.

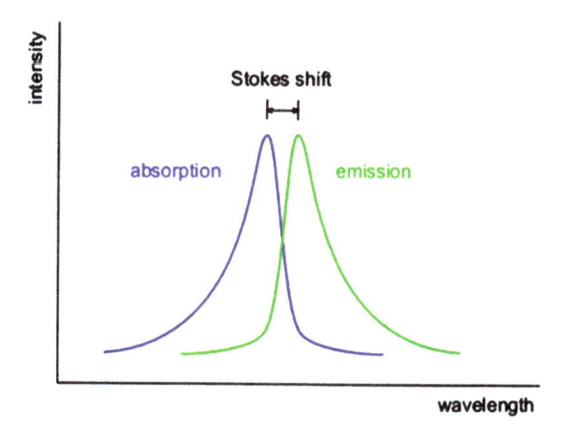

**Figure 3.13** Spectral representation of Stokes shift.

spectra of the same electronic transition. It is named after the Irish physicist George G. Stokes. The existence of photoresponsive behavior of polymers can be evidenced by fluorescence spectra. The fluorescence spectrum of the polymer has been recorded in FluroMax 2.0. Fluorescence-spectral changes during photocrosslinking of polymers at various time intervals in chloroform solution are shown in Figures 3.14–3.18.

The fluorescence spectra of polymers P1 and P2 are shown in Figures 3.14 and 3.15. Both polymers P1 and P2 were irradiated at the excitation wavelengths of 320 nm and 290 nm, respectively. The unsubstituted polymer shows emission at 529 nm, whereas P2 showed emission at 465 nm due to the presence of acceptor in the para position. There is a notable blueshift in the fluorescence wavelength of polymer P2 by the presence of an acceptor bromo substituent in the para position. The fluorescence intensity of a polymer decreases as the time of irradiation increases. The decrease in intensity is due to $2\pi + 2\pi$ cycloaddition that leads to cyclobutane ring formation by destroying $\pi$-electron conjugation. In the figures, the P1 shows a gradual decrease in the intensity, whereas P2 shows a sudden decrease. This is because the presence of an electron-acceptor group in P2 facilitates gradual photocrosslinking and slows down the rate of photoconversion. But in the case of P1, there is a rapid photoconversion due to the absence of an electron acceptor. This is the supporting evidence that indicates photoresponsive behavior of polymers and shows the indispensable uses of polymers in photoresist applications.

The photoresponsive nature of the polymer P3 evidenced by fluorescence spectra is shown in Figure 3.16. The fluorescence study of polymer was carried out by irradiation of the polymer in chloroform solution at 254 nm. The polymer was excited at a wavelength of 300 nm. The polymer P3 shows emission between 320–570 nm. It is seen from Figure 3.16 that the fluorescence intensity shows a drastic decrease after 10 s of irradiation and the decrease in intensity continued gradually. This occurs because the

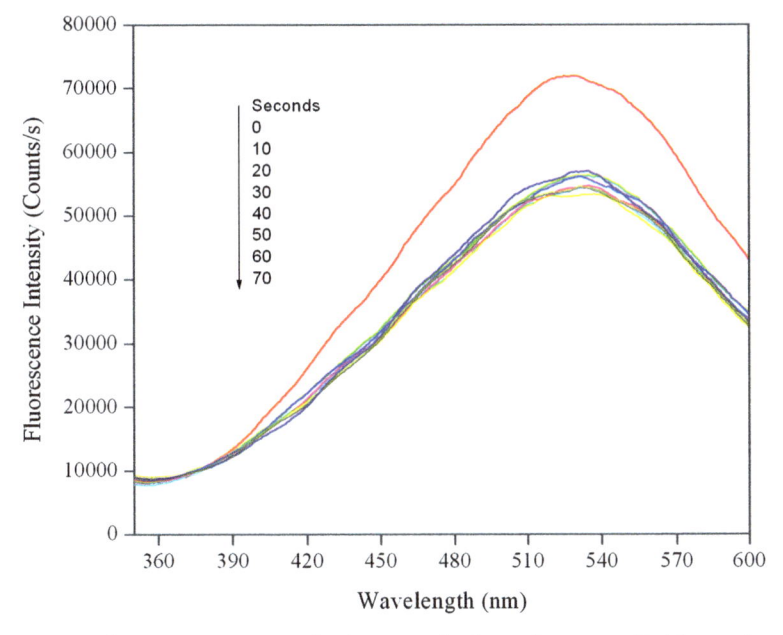

**Figure 3.14** Fluorescence-spectral changes during photocrosslinking of poly(2-[benzoyl stryloyl oxy] ethyl methacrylate) [P1] at various time intervals in chloroform solution.

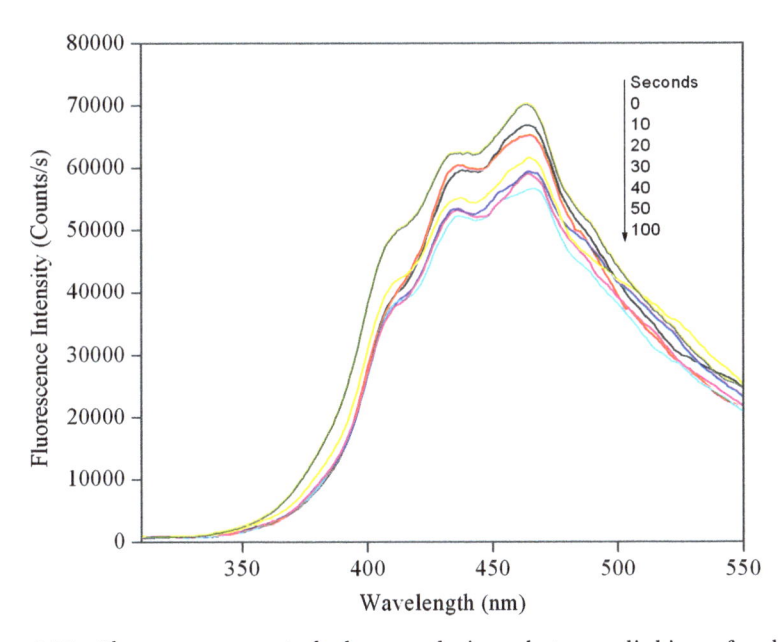

**Figure 3.15** Fluorescence-spectral changes during photocrosslinking of poly(2[4-bromo benzoyl styryloyl oxy] ethylmethacrylate) [P2] at various time intervals in chloroform solution.

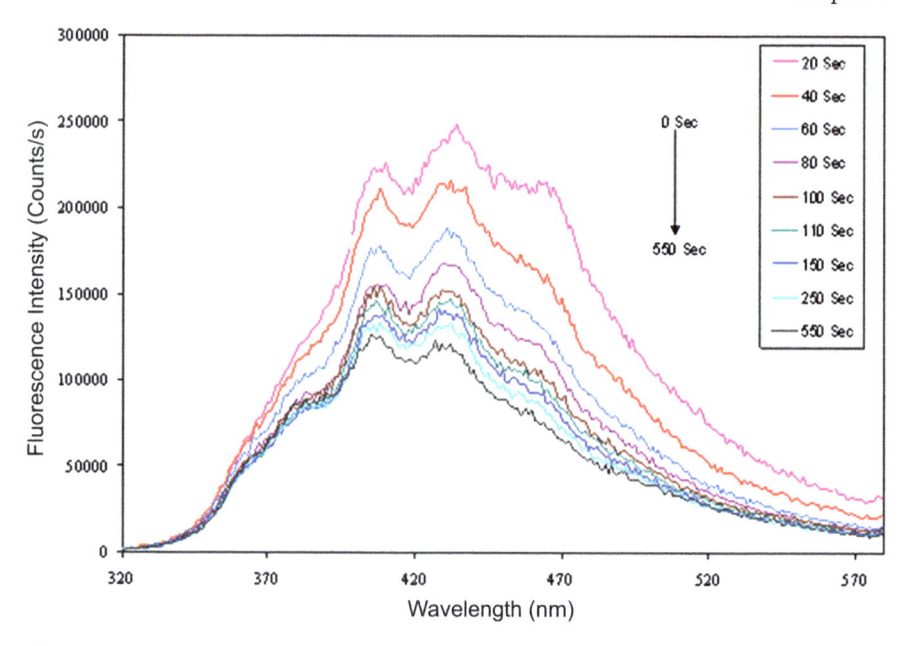

**Figure 3.16** Fluorescence-spectral changes during photocrosslinking of poly-(2[4-methoxy benzoyl styryloyl oxy] ethylmethacrylate) (MBSOEMA) [P3] at various time intervals in chloroform solution (reproduced from ref. 10).

electron-releasing –OCH$_3$ group (donor) facilitates faster photocrosslinking and destroys the π electron conjugation due to 2π + 2π cycloaddition reaction. The decrease in intensity continues until the formation of a cyclobutane ring.[20] The fluorescence study reveals the indispensable use of this polymer P3 in photoresist applications.

The fluorescence spectral changes during photocrosslinking of polymers P4 and P5 shown in Figures 3.14–3.18. The polymers P4, P5 and P6 were excited at the wavelengths of 350, 383 and 325 nm and UV irradiation was carried out as discussed above. Polymers P4 and P5 show emission at 418 and 431 nm, respectively.

All the measurements have been made in the polar solvent, chloroform (CHCl$_3$), The absorption maxima ($\lambda_A$), and emission maxima ($\lambda_F$) are shown in Table 3.3. The Stokes shift ($\nu_s$) calculated for all the photoreactive polymers are given in Table 3.3. The Stokes shifts are expressed in wave number and are calculated from the experimental parameters by taking the difference between absorption and emission maxima expressed in wave number using the equation:[21]

$$\text{Stokes shift, } \nu_S = (\nu_A - \nu_F) \times 10^7 \text{ cm}^{-1}$$

where

$$\nu_A = 1/\lambda_A \text{ (nm)}; \quad \nu_F = 1/\lambda_F \text{ (nm)}$$

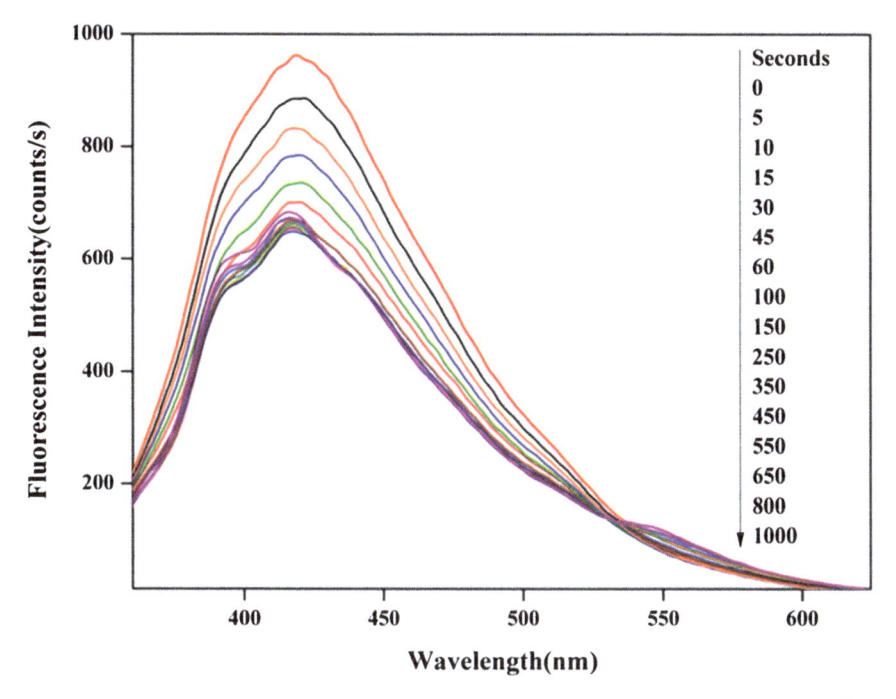

Newly Synthesized Photocrosslinkable Liquid-Crystalline Polymers and their Properties 65

**Figure 3.17** Fluorescence-spectral changes during photocrosslinking of poly(2-[3,5-di benzyl oxy benzoyl styryloyloxy] ethyl methacrylate) at various time intervals in chloroform solution.

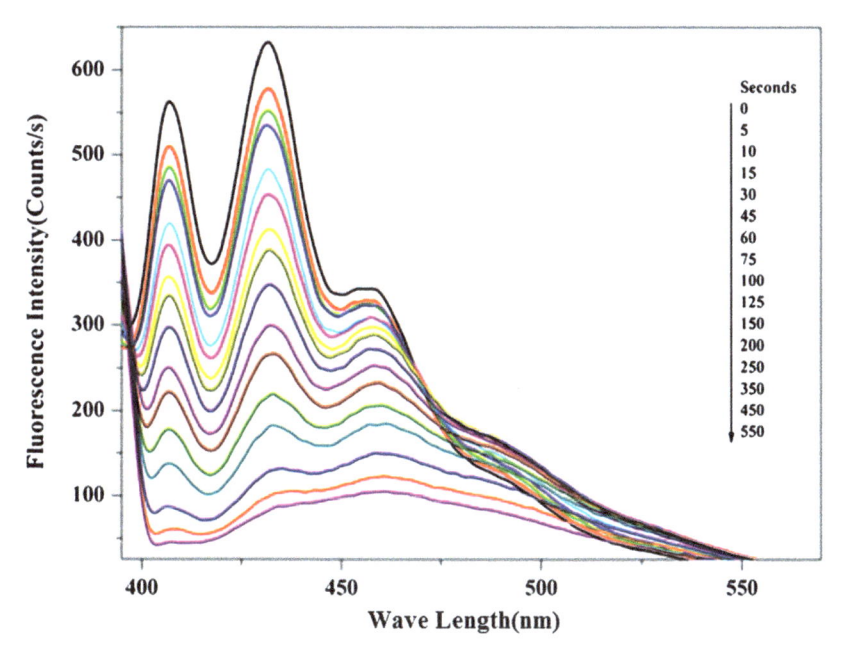

**Figure 3.18** Fluorescence-spectral changes during photocrosslinking of poly(2-[4-biphenyl oxy benzoyl styryloyloxy] ethyl methacrylate) at various time intervals in chloroform solution.

**Table 3.3** Photochemical properties of polymers P1–P5

| S. No. | Polymer | $\lambda_A$ (nm) | $\lambda_F$ (nm) | Stokes shift $\nu_S$ (cm$^{-1}$) |
|---|---|---|---|---|
| P1 | Poly(BSOEMA) | 320 | 529 | 12 350 |
| P2 | Poly(BBSOEMA) | 290 | 465 | 12 977 |
| P3 | Poly(MBSOEMA) | 317 | 435 | 8557 |
| P4 | Poly(DBOBSOEMA) | 350 | 418 | 4648 |
| P5 | Poly(BPOBSOEMA) | 383 | 431 | 2900 |

In all the polymers P1–P5, the most dramatic aspect of fluorescence is its occurrence at wavelengths longer than those at which absorption occurs. This behavior is termed a Stokes shift, and this effect may be due to the following reasons:

- These Stokes shifts, which are most dramatic for polar fluorophore in polar solvents, are due to interaction between fluorophore and its immediate environment. However, spectral shifts also occur due to specific fluorophore–solvent interactions caused by charge separation in the excited state.
- One common cause of a Stokes shift is the rapid decay to the lowest vibrational level of $S_1$. Furthermore, fluorophores generally decay to the higher vibrational level of $S_0$, resulting in further loss of excitation energy by thermalization of the excess vibrational energy.
- In addition to these effects, fluorophores can display a Stokes shift due to the solvent effect, excited state reaction, complex formation, and/or energy transfer.
- Hence, emission from fluorophores generally occurs at wavelengths that are longer than those at which absorption takes place. This loss of energy is also due to various dynamic processes, which occurs following the light absorption.[21]

## 3.8 Scanning Electron Microscopy (SEM)

The SEM technique can give high-resolution images that enables the visualization of morphological information without losing any accuracy during analysis. The virgin polymers (P1–P5) and photocrosslinked polymers (P1–P5) were characterized by a HITACHI scanning electron microscope (SEM) S-3400N model to understand the morphology of both virgin and photocrosslinked polymers. The SEM images of both virgin and photocrosslinked polymers shown in Figures 3.19 and 3.20. As observed from the SEM images, the photocrosslinked polymer sample confirms loosely held dispersion of polymeric materials, while virgin polymer exhibits compact stringent dispersion of polymeric surface. It is inferred from the SEM images that all the virgin polymers from P1–P5 have irregular-shaped flakes in their lattice that packed one over another in a nondirectional manner and they all have hard and crystal-like surfaces. However, SEM images of photocrosslinked polymers from P1–P5 show uniform size of polymer flakes, which are arranged

**Figure 3.19** SEM images of virgin polymers P1 (a), P2 (b) and P3 (c) photocrosslinked polymers P1 (d), P2 (e) and P3 (f) (reproduced from ref. 10).

regularly. It can be clearly observed from the SEM images that their surface after photocrosslinking were smoothed well and all the irregular crystals with rough surfaces have been changed into smooth surfaces. The smoothness in the polymer surface may be due to photodimerization of the polymers. When two polymer molecules undergo cyclobutane ring formation during photocrosslinking, one polymer molecule binds to the other through a cyclobutane ring. This structural interaction may lead to smoothness and regular or ordered arrangement of polymer lattice after UV treatment.

## 3.9 Liquid-Crystalline Properties of Polymers

The development of photosensitive media based on liquid crystalline compounds for data recording, optical storage and reproduction is one of the most rapidly developing areas in the physical chemistry of low molecular mass and polymer liquid crystals.[20] The rigidity of the mesogenic core, the flexible spacer length and terminal units highly influence the melting temperature, mesophase temperature and even molecular arrangement.

The polymers shown in Figure 3.1 are rod-like molecules and contain high polar hindered pendant group revealed to have high interaction leading to formation of liquid-crystalline phases.[22] In some polymers, they take the effect of mesogen and spacer together, a polymer having a rigid mesogen and a shorter spacer should show the higher transition temperature.[23] The phase-transition temperature and mesophase of the polymers are studied by

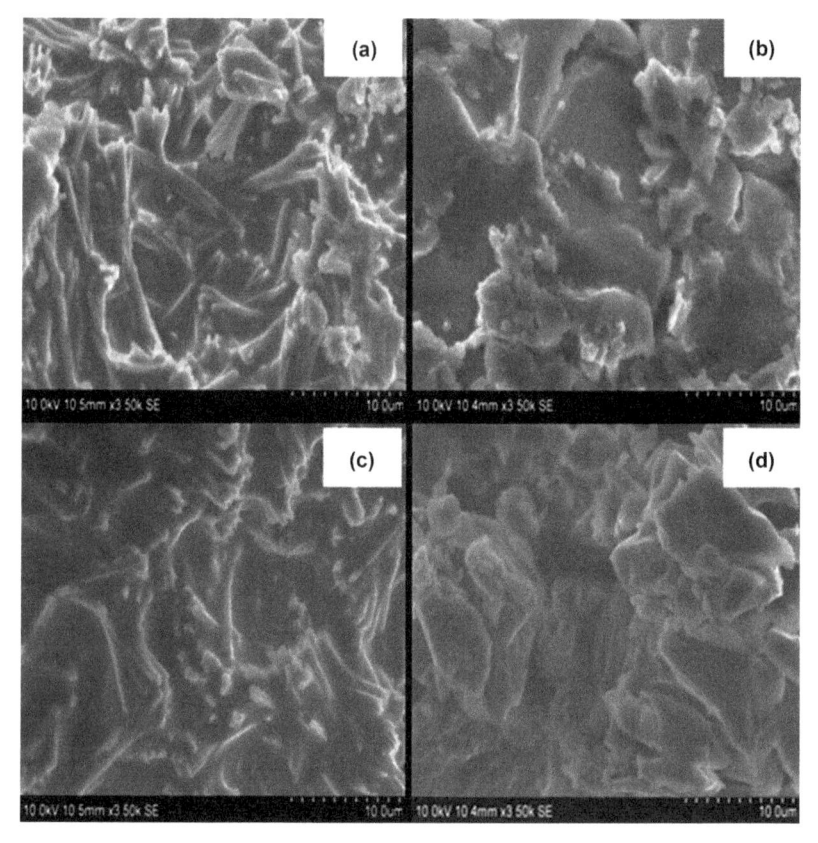

**Figure 3.20** SEM images of virgin polymers P4 (a) and P5 (b) photocrosslinked polymers P4 (c) and P5 (d).

traces of DSC thermogram and HPOM images. Generally in the DSC thermogram, at the highest transition temperature there will be an endotherm corresponding to the transition from LC phase to isotropic phase. The transition in some cases from crystal to liquid crystal is marked by more than one endotherm. When such multiple curves were observed, the one having the highest temperature is attributed to the crystal to mesophase transition.

From the DSC thermogram shown in Figure 3.21 the polymer P1 shows a glass-transition temperature at 60 °C and there are two endothermic curves are observed one at 175 °C and the other at 236 °C. The lower transition at 175 °C is due to a crystalline to nematic mesophase transition and the highest transition at 236 °C is due to nematic to isotropic transition. Both of the transitions during the heating cycle are shown in the HPOM Figure 3.23(a) and (b). Figure 3.23(a) shows a crystalline to liquid-crystalline nematic mesophase and Figure 3.23(b) shows a nematic to isotropic transition. But in the case of polymer P2, there is no noticeable phase transitions

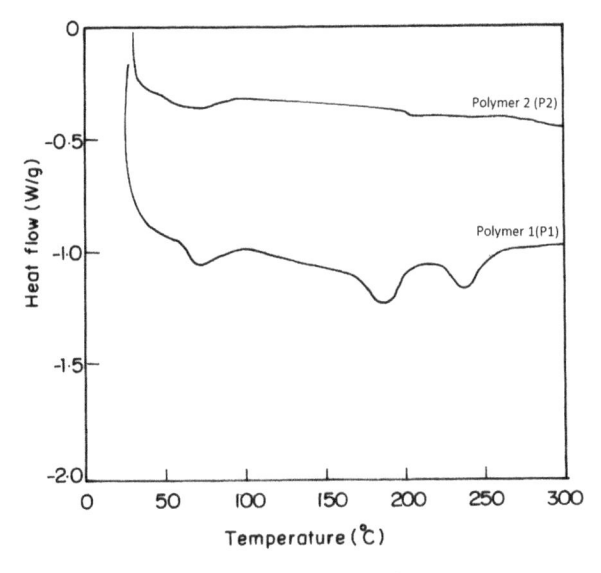

Newly Synthesized Photocrosslinkable Liquid-Crystalline Polymers and their Properties 69

**Figure 3.21** DSC thermogram of polymers P1 and P2.

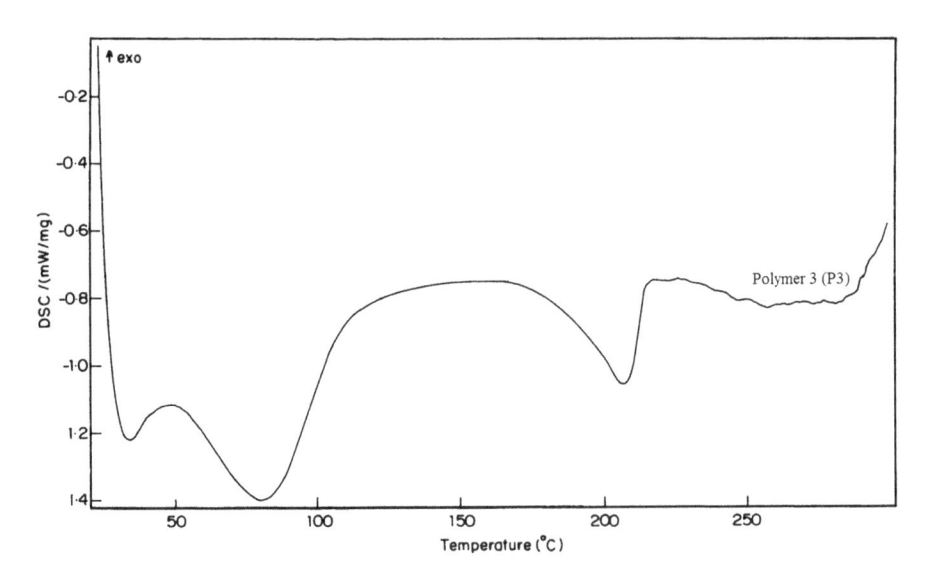

**Figure 3.22** DSC thermogram of polymer P3 (reproduced from ref. 10).

in the DSC thermogram. The DSC shows only melting and isotropic transition of the polymer and there is no $T_g$ occurs in the polymer P2. From the observation of the thermogram there is no liquid-crystalline property for P2. The liquid-crystalline nature of polymer P1 may be due to the presence of rod-like structure containing pendant benzoate mesogenic unit and their interaction may lead to formation of liquid-crystalline phases.

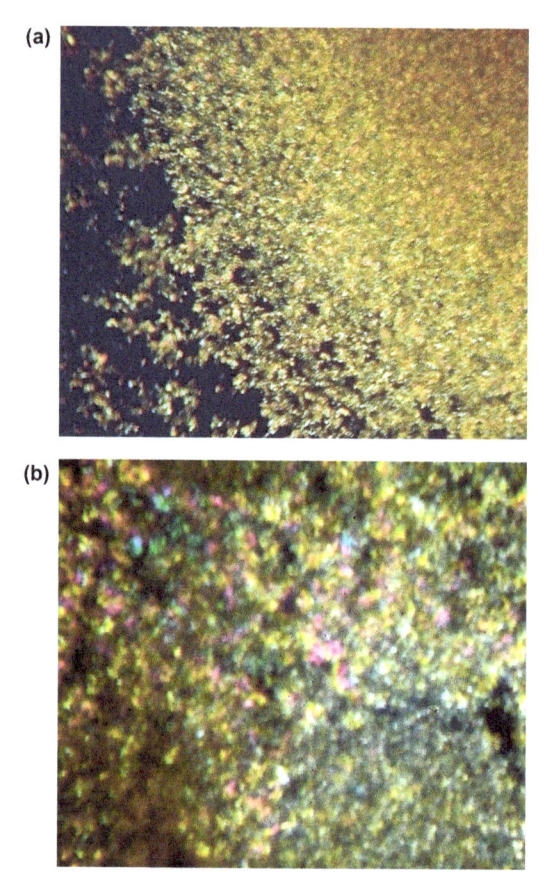

**Figure 3.23** (a) Phase transition from crystalline to nematic mesophase at 175 °C. (b) Nematic mesophase transition of P1 at 236 °C.

**Table 3.4** Liquid-crystalline properties of polymers (P1–P5)

| S. No. | Polymers | $T_g$ (°C) | $T_m$ (°C) | $T_i$ (°C) | $\Delta T = T_i - T_M$ (°C) | Mesophase |
|---|---|---|---|---|---|---|
| P1 | Poly(BSOEMA) | 60 | 175 | 236 | 61 | Nematic |
| P2 | Poly(BBSOEMA) | Nil | 197 | Nil | Nil | No phase transitions |
| P3 | Poly(MBSOEMA) | 52 | 84 | 210 | 126 | Grainy |
| P4 | Poly(DBOBSOEMA) | Nil | Nil | Nil | Nil | No phase transitions |
| P5 | Poly(BPOBSOEMA) | Nil | Nil | Nil | Nil | No phase transitions |

The liquid-crystalline property of polymer (Table 3.4) was identified and confirmed using DSC and characterized from HPOM images. The DSC of the polymer (Figure 3.22) indicates that there are two endothermic transitions that occur on heating the polymer sample. The above recorded thermogram

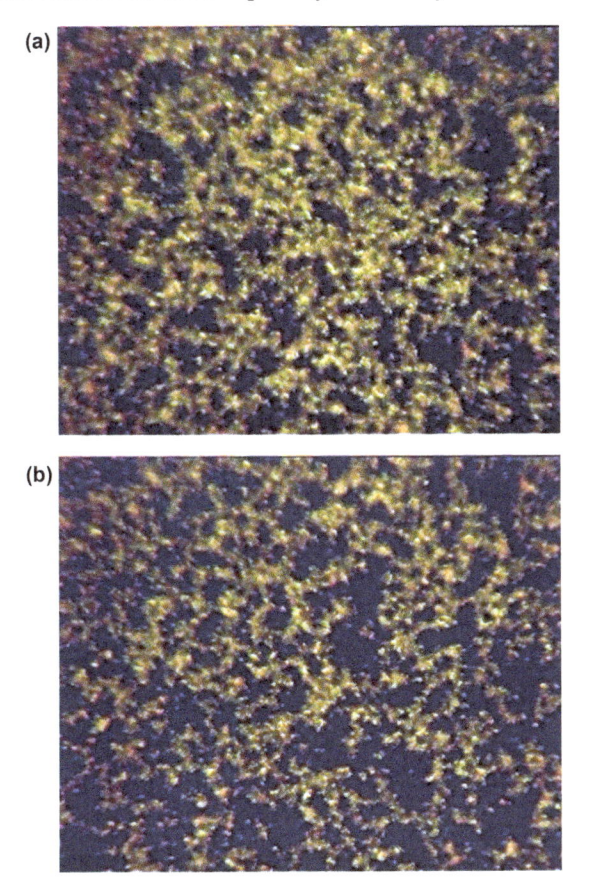

**Figure 3.24** (a) Poly(MBSOEMA) (P3) shows liquid-crystal phase from unidentified phase at 80 °C (reproduced from ref. 10). (b) Poly(MBSOEMA) (P3) at 208 °C shows isotropic transition from mesophase (reproduced from ref. 10).

of polymer was compared with the HPOM image shown in Figure 3.24(a) and (b). In polymer P3, $T_g$ occurs at 52 °C and the lower transition ($T_m$) occurs at 84 °C ($T_m$) that shows unidentified crystalline phase to liquid crystalline phase and a grainy-like mesophase to the isotropic transition occurs at 210 °C as shown in Figure 3.24(a) and (b). Both the DSC and HPOM image showed negligible variations in their liquid-crystalline properties.

The DSC of the polymers P4 and P5 is shown in Figure 3.25. Polymers P4 and P5 do not show any mesophase transition, since polymers have no flexible spacer with methylene groups, a hard/soft ratio above 1.1 that is necessary to generate a mesophase. In order to establish mesophase transition, either the chain geometry or polarity plays the major role. A significant difference of the polarity is expected due to the presence of carbonyl group. Unfortunately, P5 and P4 did not exhibit LC behavior.[24]

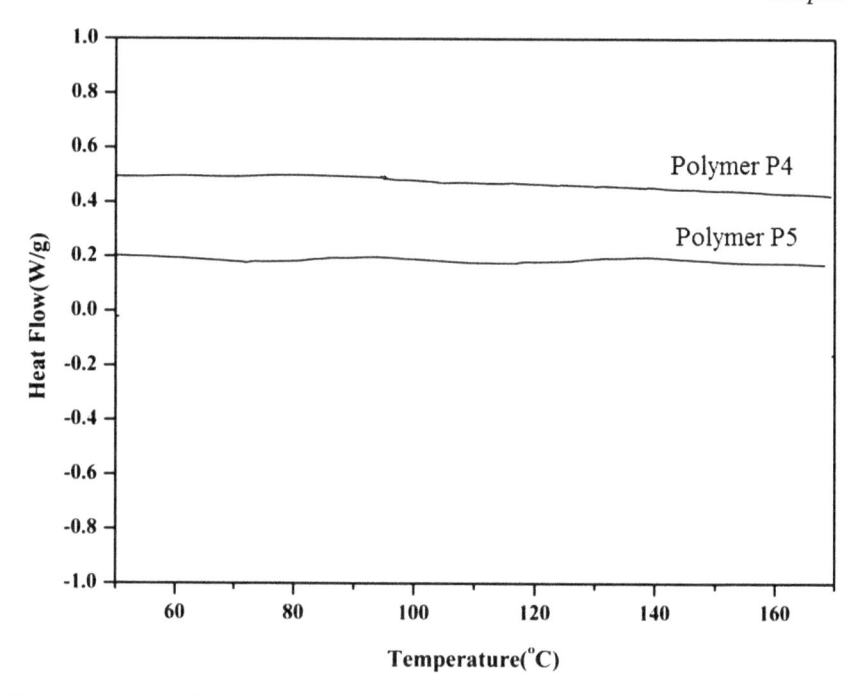

**Figure 3.25** DSC thermogram of polymers P4 and P5.

## 3.10 Conclusion

The present investigation deals with synthesis, characterization of photocrosslinkable polymers containing pendant chalcone moiety exhibiting liquid-crystalline properties. It also includes UV-visible, fluorescence and fluorescence lifetime studies that give detailed information about photocrosslinking nature of polymers. All the five polymers P1–P5 were synthesized by a free-radical polymerization method. The predetermined quantities of monomers were dissolved in THF solvent and AIBN was used as initiator. The polymerization was carried out under nitrogen (inert) atmosphere for 48 h.

The $^1$H-NMR and $^{13}$C-NMR spectral data obtained for polymers coincided with their chemical structure.

All the polymers have $\bar{M}_w$ near to $13 \times 10^4$ and $\bar{M}_n$ near to $8.5 \times 10^4$ and they show polydispersity index (PDI) values around 1.5. This PDI values indicate all the polymers possess the tendency of chain termination by radical combination rather than disproportionation.

Thermogravimetric analysis (TGA) revealed good thermal stability of polymers that decreases with increasing spacer methylene group. The char yield of all the polymers increases at 600 °C with decreasing methylene chain in the polymer backbone.

Liquid-crystalline behavior of the polymers were identified by DSC thermograms and confirmed by hot-stage polarized optical microscope images.

Two out of five polymers exhibited mesophase transitions such as nematic by P1, grainy by P3. They show melting ranges between 97–197 °C, and mesophase transition between 126–236 °C. The liquid-crystalline polymers showed a decrease in their melting point if they have increased spacer methylene chain.

The photocrosslinking behavior of all the polymers was investigated by UV-vis spectroscopy, the crosslinking of all the polymers proceed *via* cyclobutane ring formation during a $2\pi + 2\pi$ cyloaddition reaction on irradiation with UV light at 254 nm.

Bulky substituted polymers such as P4, and P5 undergo faster photocrosslinking as the trend below shows

$$\text{biphenyl (P5)} > \text{dibenzyloxy (P4)}$$

The Stokes shifts of all the polymers were calculated based on the substituent present in the fluorophore unit as they showed different Stokes shift values. The emission of a fluorophore generally occurs at longer wavelength than their absorption wavelength. This energy loss may be due to a dynamic process like absorption of light.

The surface morphology studies of both virgin and photocrosslinked polymers were investigated by scanning electron microscope (SEM) analysis, photocrosslinked polymer sample confirms the loosely held dispersion of polymeric material while the virgin polymers exhibit compact stringent dispersion of the polymeric surface.

The research work presented here concludes the presented polymers that contain photoreactive pendant chalcone unit, mesogen and spacer in its backbone can exhibit dual properties such as photocrosslinking and liquid-crystal behavior.

## References

1. D. M. Haddleton, D. Creed, A. C. Griffin, C. E. Hoyle and K. Venkataram, *Makromol. Chem., Rapid Commun.*, 1989, **10**, 391–396.
2. Y. Nakayama and T. Matsuda, *J. Polym. Sci., Part A: Polym. Chem.*, 1993, **31**, 3299–305.
3. H. R. Allcock and C. G. Cameron, *Macromolecules*, 1994, **27**, 3131–3135.
4. D. H. Choi and S. J. Oh, *Eur. Polym. J.*, 2002, **38**, 1559-1564.
5. T. Ikeda, H. Itakura, H. Lee, F. M. Winnik and S. Tazuke, *Macromolecules.*, 1988, **21**, 3536–3537.
6. C. H. Legge, M. J. Whitcombe, A. Gilbert and G. R. Mitchell, *J. Mater. Chem.*, 1991, **1**, 303–304.
7. S. Marturunkakul, I. J. Chen, L. Li, R. J. Jeng, J. Kumar and S. K. Tripathy, *Chem. Mater.*, 1993, **5**, 592.
8. H. Loth and A. Euschem, *Makromol. Chem., Rapid Commun.*, 1988, **9**, 35.
9. R. Galvan, R. L. Laurence and M. Tirrell, *Polymerization Process Modeling*, S. P. Pappas, Marketing Corp., Norwalk, CT, 1984.

10. G. Kumar and K. Subramanian, *Mol. Cryst. Liq. Cryst.*, 2012, **552**, 158–173.
11. G. Kumar and K. Subramanian, *Int. J. Polym. Mater.*, 2010, **59**, 519–530.
12. S. H. Hsiao, C. P. Yang, C. W. Chen and G. S. Liou, *Eur. Polym. J.*, 2005, **41**, 511–517.
13. D. Madheswarai, K. Subramanian and A. V. Rami Reddy, *Eur. Polym. J.*, 1996, **32**, 417–422.
14. K. Subramanian, V. Krishnasamy, S. Nanjundan and A. V. Rami Reddy, *Eur. Polym. J.*, 2000, **36**, 2343–2350.
15. S. Watanabe, M. Kato and S. Kosakai, *J. Polym. Sci., Polym. Chem. Ed.*, 1984, **22**, 2801–2808.
16. T. Nishikubo, T. Jchijyo and T. Takaodo, *J. Appl. Polym. Sci.*, 1974, **18**, 2009–2013.
17. A. V. Rami Reddy, K. Subramanian, V. Krishnasamy and J. Ravichandran, *Eur. Polym. J.*, 1996, **32**, 919–926.
18. T. Narasimhamurthy, J. C. Benny, K. Pandiarajan and R. S. Rathore, *Acta Crystallogr., Sect. C: Cryst. Struct. Commun.*, 2003, **11**, 620–621.
19. A. Rehab and N. Salahuddin, *Polymer*, 1999, **40**, 2197–2207.
20. A. Bobrovsky and V. Shibaev, *Polymer*, 2006, **47**, 4310–4317.
21. J. R. Lakowicz *Principles of Fluorescence Spectroscopy*, Kluwer Academic/Plenum Publishers, New York, 2nd edn, 1999.
22. P. C. Yang, M. Z. Wu and J. H. Liu, *Polymer*, 2008, **49**, 2845–2856.
23. Gangadhara and K. Kishore, *Macromolecules*, 1995, **28**, 806–815.
24. S. Alazaroaie, V. Toader, V. Carlescu, K. Kazmierski, D. Scutaru, N. Hurduc and C. I. Simionescu, *Eur. Polym. J.*, 2003, **39**, 1333–1339.

CHAPTER 4
# Efficient Photoinitiators for Two-Photon Polymerization

JAN TORGERSEN

Nanoscale Prototyping Laboratory, 440 Escondido Mall, Building 530, Room 226, Stanford, CA 94301, USA
Email: jan.torgersen@stanford.edu

## 4.1 Introduction

Two-photon polymerization (2PP) is a lithography-based additive manufacturing technology (AMT) for the fabrication of complex 3D parts with micro- and nanometer-scale resolution. Similar to two-photon microscopy, photosensitive chromophores inside photopolymerizable formulations are excited in the focal point of a microscope objective. *Via* the interaction with monomers, the formed reactive species trigger two-photon induced chain-reaction polymerization in the focal point. Tracing it through the formulation, 3D polymer lines are created.[1] Any arbitrary 3D shape can be "recorded" into the volume (see Figure 4.1). The basic building unit, where the polymerization takes place (volumetric pixel or voxel) can be regarded as a "3D pen", with which a polymeric line can be created anywhere in the volume of a photosensitive formulation. The resolution can be down to 65 nm,[2] as the nonlinearity of two photon absorption (2PA) provides the possibility to reduce the size of the polymerized volume below the diffraction limit.[1]

Hence, as the first AMT, 2PP offers true 3D polymerization without the need for a layer-by-layer fabrication procedure. All shortcomings related to

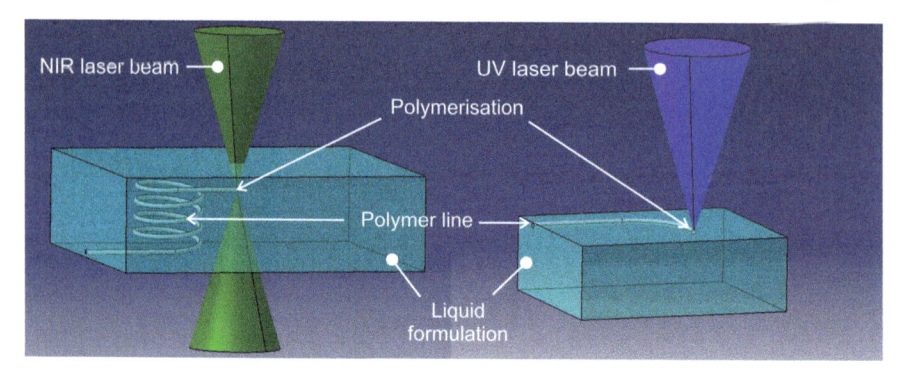

**Figure 4.1** Right image: 2PP is limited to the focal point of a microscope objective; left image: Stereolithography is limited to the surface and requires layer-by-layer manufacturing.

the surface formation such as high viscosities of the formulation (leading to high surface tensions), the necessity of recoating, the need of supporting material and oxygen inhibition, can be discarded.

These prospects may offer 2PP a wide range of application possibilities whenever high resolution, complex geometries and complex three-dimensionality is important. In research, this technique already proved successful in fabricating photonic crystals,[3] microfluidics,[4,5] microelectrical and micromechanical systems[6,7] as well as polymer-based optical waveguides.[8–10] As 2PP allows the exact reconstruction of cell-specific sites in 3D at micro- and nanometer precision,[11–13] the manufacturing of biocompatible structures is another promising application. Furthermore, 2PP can be processed at IR and NIR wavelengths, where biological tissues exhibit a window of transparency. This limits potential compromising photochemical stress for living tissue and cells and thus enables 2PP materials to be formed *in vivo*, providing a dynamic microenvironment.[12,14]

However, 2PP has still one major limitation. It requires long fabrication times, reducing its operating efficiency. This is mostly related to the limited number of available photopolymers optimized for 2PP. Research groups usually use compounds known from one-photon absorption-based lithography, which suffer from limited two-photon absorption (2PA) efficiency. For 2PP being fast and efficiently applied for its numerous potential applications, it is indispensable to create formulations that easily convert into polymers under the absorption of two photons.

But what makes a photopolymerizable formulation efficient for 2PP? To address this question, we will first get to know its basic components.

## 4.2 Photopolymerizable Formulation

A photopolymerizable formulation typically consists of a reactive diluent, a crosslinker and a photoinitiator (PI). However, most formulations contain additional components (Figure 4.2), all influencing its reactivity, viscosity, reaction mechanism and the properties of the resulting polymer:[15,16]

- High molecular weight monomers, *i.e.* crosslinkers with more than one reactive group define the mechanical properties of the resulting polymer.
- Mono- and multifunctional reactive diluents affect the number of reactive groups, decrease the viscosity of the formulation and additionally tune the mechanical properties of the resulting polymer.
- A PI that meets the emission spectra of the used light source efficiently creates radicals upon its activation.
- A solvent swells the polymer network decreasing the stiffness and strength of the obtained structure.
- Filler materials influence the Young's modulus and/or other functional properties of the final polymer structure.
- If the monomers are very reactive, inhibitors can prevent premature polymerization by scavenging formed radicals.

Adding solvents and filler materials helps to decrease the shrinkage during polymerization to obtain better shape accuracy and reduce internal stresses.

In photopolymerization, initiators dissolve into radicals that break the double bonds of the monomers (crosslinkers and reactive diluents) and start the solidification in the radical chain-reaction polymerization. Hence, in the simplest case, the formulation only consists of a radical initiator and a monomer that gets crosslinked upon radical polymerization.

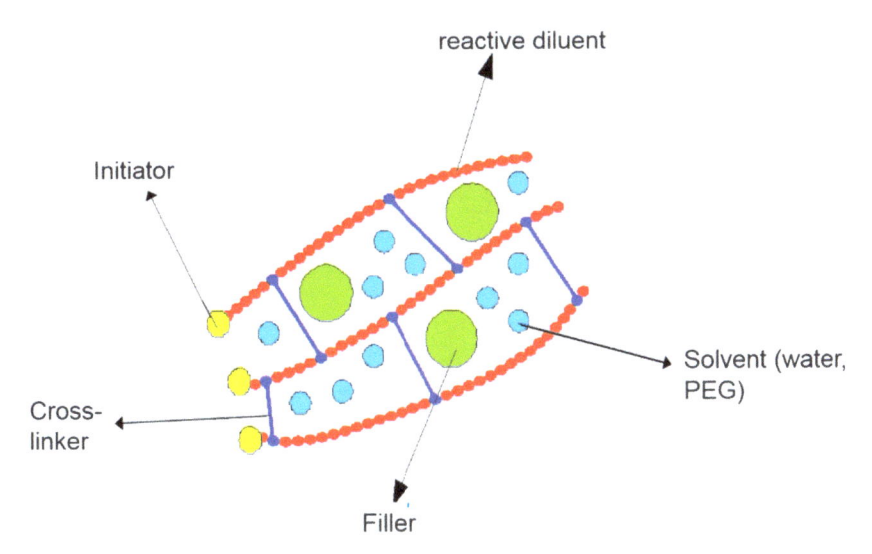

**Figure 4.2** Basic building block of photopolymerizable formulations used for lithography-based AMT.[15]

For 2PP, the initiator's efficient absorption of two photons and its subsequent decay into radicals that trigger polymerization is key for an efficient and fast process. The 2PA cross section ($\delta$) is widely to measure the efficiency of a molecule to absorb two photons. In the next section, we will see how a molecule has to be designed to have a large cross section. We will further explore how this value relates to a PI being efficient for 2PP.

## 4.3 Photoinitiators and Two-Photon Efficiency

PIs are the key substance of photopolymerizable formulations. They are UV or VIS sensitive and convert radiation energy into chemical energy dissolving into radicals[17] (Figure 4.3), molecules that have unpaired electrons on an otherwise open-shell configuration. As they are highly reactive, they can react with another molecule breaking its double bonds.[18] This initiation step starts the free-radical polymerization chain reaction.

In conventional photopolymerization, one absorbed photon elevates the PI molecule from a lower ($S_0$) to a higher and short-lived ($S_1$) vibrational energy level, both of them being singlet states with spin zero. Rather than immediately decaying to the ground state simply emitting fluorescence or converting the energy into internal heat, the PI decays to a long-lived triplet state *via* intersystem crossing. The spin of the molecule is now one. Depending on the molecule, the PI in the excited triplet state can create radicals *via* the monomolecular type-1 or the bimolecular type-2 mechanism of radical formation.

A- or β-cleavage leads to photofragmentation and radical formation of type-I PIs. This usually takes place next to an aromatic carbonyl group and thus results in the formation of one or two benzoyl radicals, which are

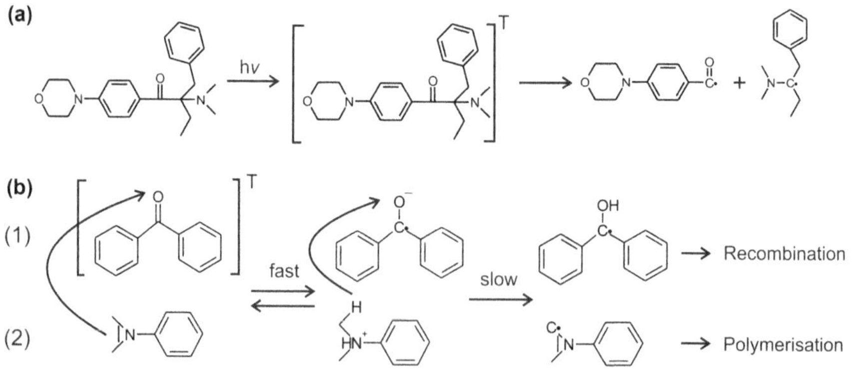

**Figure 4.3** (a) Type-I initiator Irgacure 369 cleaves from the excited triplet state forming two radicals that can start polymerization; (b) an amine (2) quickly transfers an electron to the type II initiator benzophenone (1); proton transfer leads to a reactive amine that can start polymerization. The PI itself recombines.[19]

capable of starting the polymerization. One typical example of a α-cleavage initiator is Irgacure 369 (Figure 4.3(a)).

Chromophores such as benzophenones and analogs as well as donors (coinitiators) such as alcohols, ethers and amines are the basis of bimolecular type-II PIs. They create radicals *via* hydrogen abstraction or electron transfer. In hydrogen abstraction, a benzophenone in the triplet state, for example, abstracts hydrogen from an alcohol, ether or amine. While the formed alcohol, ether or amine radicals start the polymerization, benzophenone radicals recombine to form a nonreactive dimer. Figure 4.3(a) shows the latter type-II reaction. An electron is abstracted from the amine (2) and transferred to the excited ketone (1). Subsequent proton transfer renders reactive amine radicals that can start the polymerization. The benzophenone radicals recombine. Beside amines, ethers or alcohol, also monomers or the formed polymer chains can serve as donors.[19–21]

For one-photon polymerization (1PP), the reaction mechanism of type-I initiators is usually more efficient. It is much simpler and requires shorter excited-state lifetimes not necessitating any interaction with another molecule. For 2PP, however, type-I initiators are rare as shifting their absorption spectra is complicated.

Generally, the efficiency of an initiator to absorb two photons plays a major role when considering a proper design. The two-photon absorption cross section ($\delta$) reflects the efficiency of a molecule for 2PA as an analog to the linear absorption coefficient. It is measured in Göppert–Meyer (GM) units [$10^{-50}$ cm$^4$ s photon$^{-1}$].[22] It has been shown that a high $\delta$ is related to a molecule's charge transfer characteristics.[23] As large charge separation is key for efficient two-photon absorption, efficient two-photon absorbing molecules consist of:

- diaryl or dialkylamino donor (D) and
- heterocycle acceptor (A) end groups coupled to
- a large π-conjugated coplanar chromophore and have
- small 2PA bands.

Now that we have a rough idea of the possible reaction pathways and set initial design parameters of two-photon absorbing PIs, we can think of the chemical structure of such a molecule. As an example, we will take a recently published compound synthesized by our group.[24]

## 4.4 Initiator Design

M$_2$CMK is one of recently published centrosymmetric D–π–A–π–D PIs with different member rings as central acceptor.[24] In these molecules, alkylamino groups act as donors, vinyl as a π-conjugated bridges and carbonyl as a central acceptor. The long conjugation length, its coplanarity and the presence of strong electron donors and central acceptors facilitate high two-photon absorption efficiency. Furthermore, the molecules can be easily

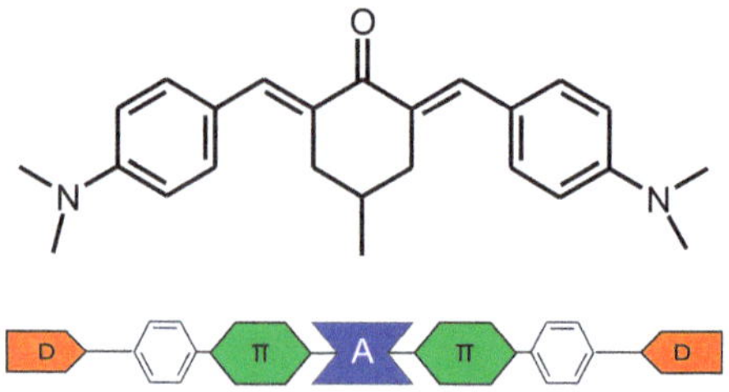

**Figure 4.4** Chemical structure of the centrosymmetric 2PP initiator M$_2$CMK.

synthesized in a one-step synthetic routine and show only low fluorescence quantum yields leading to a high population of active states for polymerization. The molecule is based on a benzylidene ketone with a central cyclohexanone. M$_2$CMK has a two-photon absorption cross section of 191 GM in chloroform when exposed to a 100 fs pulsed, 1 kHz repetition rate laser emitting at 800 nm wavelength (Figure 4.4).

## 4.5 Structuring Results

For evaluating the efficiency of this initiator for 2PP, the compound is mixed to an organo-soluble 1 : 1 mixture of ethoxylated (20/3)-trimethylolpropane-triacrylate and trimethylolpropanetriacrylate (ETA–TTA), an ideally suitable 2PP formulation.[25,26] The concentration was $6.3 \times 10^{-6}$ mol PI per g formulation (0.2 wt%). We performed a screening as described previously,[25] where we structured simple 50×50 µm$^2$ lattice structures. Every structure consists of 40 layers at a layer distance of 5 µm. The distance between the polymer lines was 10 µm resulting in six polymer lines per layer. We fabricated several similar lattice elements in an array, where the markspeed and laser power varied in X and Y, respectively. The writing speeds was varied from 1 to 151 mm s$^{-1}$ in the X-plane and laser powers from 35 to 360 mW in the Y-plane. The laser power was measured before the 20× NA 0.8 objective. The experimental setup is described elsewhere.[24]

After the fabrication process, the specimen was immersed in ethanol for 4 h and observed in the light microscope. When attached properly, the dry structures were investigated in the SEM. The structures were assigned to a classification scheme.[27] The first class included good structures (green) with straight fine lattice bars. Structures of class two (blue) have thicker bars and are slightly wavy. Class 3 (red) includes structures with defects. Yet, their intended appearance can be recognized. These defects include holes or craters usually attributed to overexposure and subsequent bubble formation. Lattices that do not resemble their intended design are assigned to class four (blue).

The whole speed-power array of M$_2$CMK can be seen in Figure 4.5. Almost all structures in the column are pronounced and of good quality. No lattice top left, structured at the lowest speed (1 mm s$^{-1}$) and the highest power

Efficient Photoinitiators for Two-Photon Polymerization 81

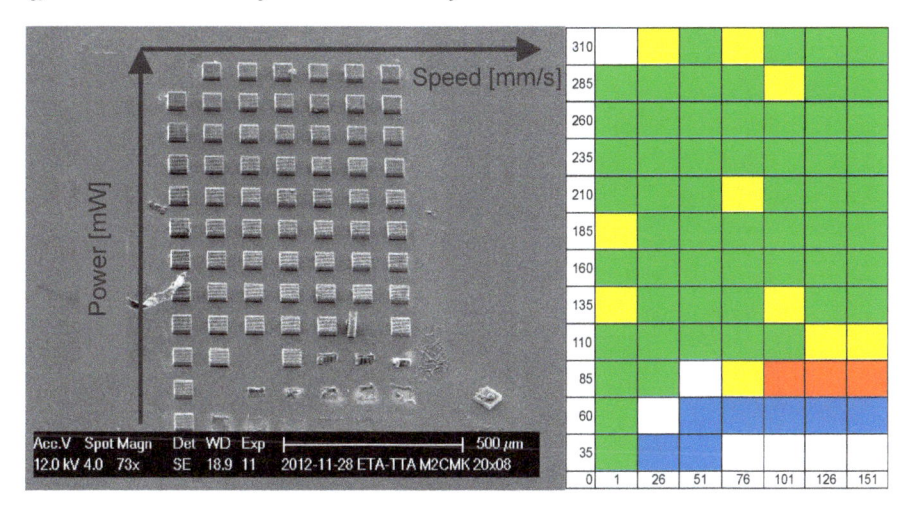

**Figure 4.5** Screening of PI M$_2$CMK in ETA–TTA and qualitative assessment, 20× NA 0.8 objective; left: SEM image, right: qualitative assessment.

(335 mW) can be observed. Presumably, too high intensities caused bubble formation and explosions inside the formulation, which potentially displaced the polymer. The array shows that almost all structures are pronounced and can be classified into the first category (green). Some structures have polymer inclusions and are thus associated to the yellow category. As the yellow structures are randomly distributed within a field of green elements, these inclusions are likely a result of the development process rather than of the structuring parameters. The qualitative assessment shows that defined structures can be produced at laser powers as low as 35 mW at 1 mm s$^{-1}$. Writing speeds of 151 mm s$^{-1}$ are reached at 110 mW. At 135 mW, well-defined lattices are fabricated at this speed.

## 4.6 Discussion

Initiators favoring the fabrication at a wide range of fabrication speeds and laser powers are considered high performing.[25,26] They facilitate a broad process window and low sensibility to laser power fluctuations caused by inertia problems of the experimental setup. M$_2$CMK in ETA–TTA formulation not only facilitates a broad process window but enables precise microfabrication at the highest process speeds ever reported in 2PP.[24] To show the 3D capabilities of the presented formulation at high writing speed, we structured the model of an indy racecar (see Figure 4.6). This structure is 285×130×100 µm$^3$ in dimensions and consists of 100 layers at an average of 200 polymer lines each. The distance between the lines was set to 1 µm. Despite the large number of polymer lines structured in total, the fabrication process only took close to four minutes. The focal point was traced in the formulation at well above 80 mm s$^{-1}$, which is the limitation of the

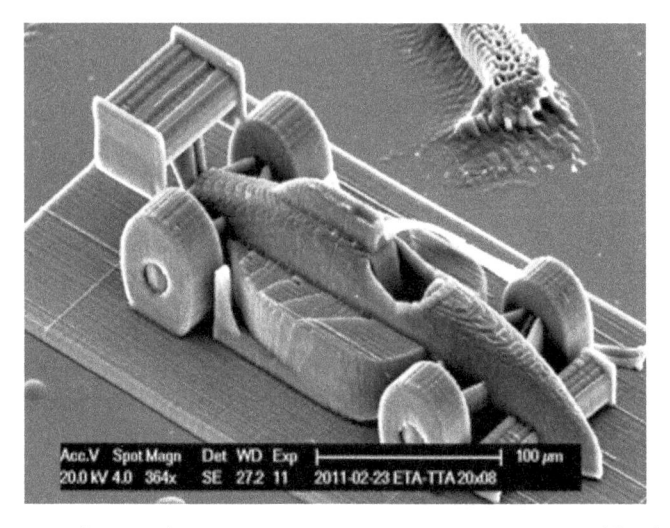

**Figure 4.6** SEM image of racecar, ETA–TTA, M$_2$CMK, 20× NA 08 objective.

experimental setup for this specific CAD structure. ETA–TTA including M$_2$CMK can be processed at even higher writing speeds.

In a recent study,[24] our group compared molecules similar to M$_2$CMK but different central cyclopentanone acceptors. It could be shown that several of the molecule's analogs have substantially higher δ due to the increased planarity and therefore higher conjugation. A similar cyclopentanone molecule has a δ of 466 GM, whereas M2CMK's δ is only 191 GM. Both molecules' two-photon absorption cross sections were measured in chloroform while exposed to a 100 fs pulsed, 1 kHz repetition laser.

These results clearly indicate that the two-photon absorption activity of a chromophore shows the molecules' efficiency to generate excited states. However, it is not the only factor guaranteeing efficient 2PP. From the theoretical description and from previous experiments, it is evident that centrosymmetric D–π–A–π–D and A–π–D–π–A molecules generally show stronger δ than their D–π–A counterparts.[23,28] However, rather than an efficient absorption, it is the energy available in the excited state resulting from a 2PA that is important for subsequent photochemical processes. Increasing the δ with a delocalized π-system and high dipolar moments facilitates an excitation with low energy. But does it always represent a molecule's efficiency for generating radicals for 2PP?

A high δ facilitates the creation of an excited energy state with low energy; however, this means that the excited energy level is also low. It cannot be higher than the sum of the energy levels of two photons of low energy that caused the excitation. This might not be useful for all 2PA-based applications. Moreover, high dipolar moments are a consequence of a resonance with an energy interstate of a centrosymmetric molecule. The moments are higher if the interstate is at approximately half the energy of the excited 2PA state. Yet, the resulting final state can rapidly relax to this existing lower-energy interstate, which causes the photochemical processes to start from this lower energy level instead. The energy of one of the photons converts to heat, whereas in noncentrosymmetric molecules, the energies of both photons are available for subsequent photochemical processes.[29] A process of free-radical polymerization such as 2PP requires a sufficient density of radicals rather than molecules in excited energy levels. $\delta$ is thus an important parameter but does not directly correlate with the 2PP sensitivity, *i.e.* the efficient generation of radicals. Apart from the conversion of one photon's energy to heat in centrosymmetric molecules, other processes might also compromise a PI's efficiency,[30] these include (Figure 4.7):

- energy conversion into fluorescence from the singlet state (F);
- phosphorescence from the triplet state (P); and/or
- molecule deactivation due to monomer (MQ) or radical quenching (RQ).

Hence, although plenty of 2PA active chromophores have been reported, nearly all of them are maybe suitable for fluorescence dyes in bioimaging applications,[31,32] but do not induce photopolymerization upon activation. For an effective 2PP PI, however, low fluorescence quantum yields are preferred, leading to a higher population of active states to initiate subsequent polymerization.

Lu *et al.*, for example, compared a series of D–π–A–π–D chromophores on their 2PP activity. The compound with the lowest $\delta$ was most suitable for polymerizing acrylates.[33] Rhodamine B, an efficient, commercially available sensitizer for two-photon microscopy with a $\delta$ of 200 GM, is not usable for 2PP.[34] Hence, designing an efficient 2PI involves minimizing the instances of Figure 4.7.

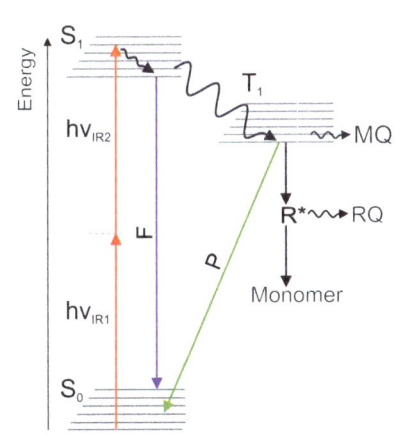

**Figure 4.7** Jablonski diagram for 2PA and initiation efficiency reducing processes including fluorescence (F), phosphorescence (P) and deactivation from monomer (MQ) and radical quenching (RQ); $S_0$, $S_1$: singlet state, $T_1$: triplet state, **hv**: photon energy, R*: radical(s); adapted from ref. 30.

In addition, the solubility of the chromophores and the solvent polarity also determine the 2PA activity and influence the charge-transfer character of excited states. In particular, planar and rigid molecules suffer from limited solubility. A combination of double bonds and benzene rings over anthracene as π-bridges and/or long aliphatic side chains increase the solubility in the formulation or solvent.[21]

## 4.7 Conclusion

Two-photon polymerization (2PP) is a versatile technique that has been used to create microstructures for a variety of applications including optoelectronics, photonics and micromechanical devices.

The low writing speed and the limited number of available formulations so far has prevented 2PP from a broader use where its capabilities would be beneficial. Optimized PIs are the key components for overcoming this bottleneck.

Facilitating high writing speed at low laser power, the presented molecule $M_2CMK$ is a promising candidate. Precise microstructures can be fabricated at speeds above 150 mm s$^{-1}$ and laser powers as low as 110 mW measured before the microscope objective. Processing speed has a lower impact on the quality of the structures than the laser power. This facilitates precise 3D microstructures fabricable at high speed and low power. 150 mm s$^{-1}$ is the current limit of the experimental setup used in this work, it is hence very likely that the presented formulation can even be processed at higher speed.

Although $M_2CMK$ has a lower two-photon activity that its counterparts with different size member rings, this molecule is obviously the most suitable for 2PP. The efficient creation of radicals for subsequent polymerization is challenging. Apart from low two-photon activity, the molecule's conversion into fluorescence from the singlet state, its phosphorescence from the triplet state as well as monomer and radical quenching can compromise its effectiveness. In addition, the molecule's solubility in the solvent plays another important role. Only when optimizing these issues, can 2PP finally be economically used for applications requiring large structures at sub-µm resolution.

## Acknowledgements

This work was financially supported by the Austrian Science Fund (FWF) under the contract J3505-N20. In addition, the author thanks Prof. Fritz Prinz, head of the Nanoscale Prototyping Laboratory for the possibility and time to write this chapter. Furthermore, the work of the former PhD students Dr Zhiquan Li, Dr Xiao-Hua Qin from the Institute of Applied Synthetic Chemistry, Vienna University of Technology; Dr Klaus Cicha, Dr Niklas Pucher and Dr Klaus Stadlmann, from the Institute of Material Science and

Technology, Vienna University of Technology is highly acknowledged. Their work was essential for this report.

The author greatly acknowledges the input and support of Prof. Jürgen Stampfl and Dr Aleksandr Ovsianikov from the Insitute of Material Science and Technology, Vienna University of Technology as well as Prof. Robert Liska from the Institute of Applied Synthetic Chemistry, Vienna University of Technology.

# References

1. S. Passinger, *Two-Photon Polymerization and Application to Surface Plasmon Polaritons*, Cuvillier Verlag, 2008.
2. W. Haske, V. W. Chen, J. M. Hales, W. Dong, S. Barlow, S. R. Marder and J. W. Perry, *Opt. Exp.*, 2007, **15**, 3426.
3. H.-B. Sun and S. Kawata, *NMR • 3D Anal. • Photopolymerization*, Springer, Berlin Heidelberg, 2004, pp. 169–273.
4. C. Schizas, V. Melissinaki, A. Gaidukeviciute, C. Reinhardt, C. Ohrt, V. Dedoussis, B. N. Chichkov, C. Fotakis, M. Farsari and D. Karalekas, *Int. J. Adv. Manuf. Technol.*, 2010, **48**, 435.
5. S. Maruo and H. Inoue, *Appl. Phys. Lett.*, 2006, **89**, 144101.
6. S. Matsuo, S. Kiyama, Y. Shichijo, T. Tomita, S. Hashimoto, Y. Hosokawa and H. Masuhara, *Appl. Phys. Lett.*, 2008, **93**, 051107.
7. S. Maruo and K. Ikuta, *Sens. Actuators, A*, 2002, **100**, 70.
8. K. Stadlmann, *Proc. LPM 2010-11th Int. Symp. Laser Precis. Microfabrication*, K. Sugioka, Stuttgart, 2010, p. 6.
9. J. Stampfl, R. Inführ, K. Stadlmann, N. Pucher, H. Lichtenegger, V. Schmidt, and R. Liska, *Lasers Manuf. 2009*, A. Ostendorf, Munich, Germany, 2009, pp. P1–P4.
10. S. Krivec, N. Matsko, V. Satzinger, N. Pucher, N. Galler, T. Koch, V. Schmidt, W. Grogger, R. Liska and H. C. Lichtenegger, *Adv. Funct. Mater.*, 2010, **20**, 811.
11. J. Torgersen, A. Ovsianikov, V. Mironov, N. Pucher, X. Qin, Z. Li, K. Cicha, T. Machacek, R. Liska, V. Jantsch and J. Stampfl, *J. Biomed. Opt.*, 2012, **17**, 105008.
12. J. Torgersen, X.-H. Qin, Z. Li, A. Ovsianikov, R. Liska and J. Stampfl, *Adv. Funct. Mater.*, 2013, **23**, 4542.
13. B. D. Fairbanks, M. P. Schwartz, A. E. Halevi, C. R. Nuttelman, C. N. Bowman and K. S. Anseth, *Adv. Mater.*, 2009, **21**, 5005.
14. M. W. Tibbitt, A. M. Kloxin, K. U. Dyamenahalli and K. S. Anseth, *Soft Matter*, 2010, **6**, 5100.
15. J. Stampfl, S. Baudis, C. Heller, R. Liska, A. Neumeister, R. Kling, A. Ostendorf and M. Spitzbart, *J. Micromech. Microeng.*, 2008, **18**, 125014.
16. J. Stampfl and R. Liska, in *Stereolithography*, ed. P. J. Bártolo, Springer, US, Boston, MA, 2011, pp. 161–182.

17. C. Heller, *Biocompatible and Biodegradable Photopolymers by Additive Manufacturing: From Synthesis to In-Vivo Studies*, Vienna University of Technology, Vienna, Austria, 2010.
18. C. Decker, *Polym. Int.*, 1998, **45**, 133.
19. A. Ovsianikov, V. Mironov, J. Stampfl and R. Liska, *Expert Rev. Med. Devices*, 2012, **1**, 613–633.
20. P. J. Bártolo, *Stereolithography: Materials, Processes and Applications*, Springer, New York, 2011.
21. N. Pucher, *Synthesis and Evaluation of Novel Initiators for the Two-Photon Induced Photopolymerization Process – a Precise Tool for Real 3D Sub-Micrometer Laser Structuring*, Vienna University of Technology, Vienna, Austria, 2010.
22. M. Sheik-Bahae, A. A. Said, T.-H. Wei, D. J. Hagan and E. W. Van Stryland, *IEEE J. Quantum Electron.*, 1990, **26**, 760.
23. M. Albota, D. Beljonne, J.-L. Brédas, J. E. Ehrlich, J.-Y. Fu, A. A. Heikal, S. E. Hess, T. Kogej, M. D. Levin, S. R. Marder, D. McCord-Maughon, J. W. Perry, H. Röckel, M. Rumi, G. Subramaniam, W. W. Webb, X.-L. Wu and C. Xu, *Science*, 1998, **281**, 1653.
24. Z. Li, N. Pucher, K. Cicha, J. Torgersen, S. C. Ligon, A. Ajami, W. Husinsky, A. Rosspeintner, E. Vauthey, S. Naumov, T. Scherzer, J. Stampfl and R. Liska, *Macromolecules*, 2013, **46**, 352.
25. K. Cicha, Z. Li, K. Stadlmann, A. Ovsianikov, R. Markut-Kohl, R. Liska and J. Stampfl, *J. Appl. Phys.*, 2011, **110**, 064911.
26. K. Cicha, T. Koch, J. Torgersen, Z. Li, R. Liska and J. Stampfl, *J. Appl. Phys.*, 2012, **112**, 094906.
27. K. Cicha, *Struktur-Eigenschaftskorrelation von 3D-Mikrostrukturen, hergestellt mit Hilfe von Zweiphotonenpolymerization*, Vienna University of Technology, 2012.
28. N. Pucher, A. Rosspeintner, V. Satzinger, V. Schmidt, G. Gescheidt, J. Stampfl and R. Liska, *Macromolecules*, 2009, **42**, 6519.
29. M. Pawlicki, H. A. Collins, R. G. Denning and H. L. Anderson, *Angew. Chem.*, 2009, **121**, 3292.
30. H.-B. Sun and S. Kawata, *J. Light Technol.*, 2003, **21**, 624.
31. Y. Liu, X. Dong, J. Sun, C. Zhong, B. Li, X. You, B. Liu and Z. Liu, *The Analyst*, 2012, **137**, 1837.
32. A. Hayek, F. Bolze, J.-F. Nicoud, P. L. Baldeck and Y. Mély, *Photochem. Photobiol. Sci.*, 2006, **5**, 102.
33. Y. Lu, F. Hasegawa, T. Goto, S. Ohkuma, S. Fukuhara, Y. Kawazu, K. Totani, T. Yamashita and T. Watanabe, *J. Lumin.*, 2004, **110**, 1.
34. P. J. Campagnola, D. M. Delguidice, G. A. Epling, K. D. Hoffacker, A. R. Howell, J. D. Pitts and S. L. Goodman, *Macromolecules*, 2000, **33**, 1511.

CHAPTER 5

# Inhomogeneous Photopolymerization in Multicomponent Media

A. VELTRI,*[a,b] A. V. SUKHOV,[c] R. CAPUTO,[b] L. DE SIO,[d] M. INFUSINO[a,b] AND C. P. UMETON[b]

[a] Colegio de Ciencias e Ingeniería, Universidad San Francisco de Quito, Quito, Ecuador; [b] LICRYL (Liquid Crystals Laboratory, IPCF-CNR), and Center of Excellence CEMIF.CAL Department of Physics University of Calabria, 87036, Arcavacata di Rende (CS), Italy; [c] Institute for Problems in Mechanics, Russian Academy of Science, Moscow 119526, Russia; [d] Beam Engineering for Advanced Measurements Company, Winter Park, Florida 32789, USA
*Email: alessandro.veltri@gmail.com

## 5.1 Introduction

The last ten years have witnessed the development of diverse techniques of multicomponent curing and their applications to realize a wide range of optical and photonic devices.[1–10] The basic idea is very simple: when a mixture of two components, one of which is a photosensitive monomer, is exposed to a light pattern at the appropriate frequency, a polymer is formed in the higher-intensity regions and the passive component will be pushed *via* diffusion into the darker regions.[1–3]

The advantages of this technique are: (i) it is a one-step procedure easily integrable in a fabrication line (ii) it is relatively less expensive, compared to similar techniques, (iii) it can translate any light pattern into a structured material.

This very simple idea hides the much more complex physics on the interplay between photochemistry and diffusion for controlling the key parameters of these two processes (*e.g.* the curing rate and the diffusivity) allows very different morphologies to be realized using the same materials. An example of this is given in Chapter 11 *"POLICRYPS: A Multi-Purpose, Application-Oriented Platform"* of this book, where the difference between an HPDLC grating and a POLICRYPS are discussed. Namely, while an HPDLC grating consists in a sequence of polymeric slices alternated to films of liquid-crystal droplets dispersed in a polymer matrix, the POLICRYPS presents a sequence of regularly aligned LC films separated by isotropic polymer slices.

Both HPLC and POLICRIPS grating have been realized using the same mixtures of monomers and LC. The difference in the two procedures was in choosing different working parameters, such as the curing intensity and the temperature of the sample during the curing process.

This strong dependence of the final morphology on the physical parameters motivated the development of a model able to describe the interplay between photochemistry and diffusion, while taking into account the different mobility of long and short polymer chains. In this chapter we will present the details on how this model has been set up.

## 5.2 Photochemistry

We consider a sample of length $L$ in $z$ and infinite in $x$ and $y$, filled with three chemical species: liquid crystals, LC in concentration $(C)$, Monomers M in concentration $(M)$, and Initiators I in concentration $(I)$. For what concerns the chemical reactions, we chose to use the classical reaction scheme for radical polymerization:[11]

$$I + h\nu \rightarrow I^* \tag{5.1}$$

$$I^* \rightarrow 2\dot{R}_0 \tag{5.2}$$

describing the ignition stage of the system;

$$\dot{R}_0 + M \xrightarrow{k_3} \dot{R}_1 \tag{5.3}$$

$$\dot{R}_n + M \xrightarrow{k_4^n} \dot{R}_{n+1} \tag{5.4}$$

describing the prolongation stage;

$$\dot{R}_0 + \dot{R}_0 \xrightarrow{k_5} I \tag{5.5}$$

$$\dot{R}_n + \dot{R}_0 \xrightarrow{k_6^n} P_N \quad (N = n) \tag{5.6}$$

$$\dot{R}_i + \dot{R}_j \xrightarrow{k_7^{i+j}} P_M \quad (M = i+j) \tag{5.7}$$

describing the termination stage. The first two equations take into account the photoexcitation reaction of an initiator molecule I and the production of a radical molecule $\dot{R}_0$ that govern the whole process. Reactions (5.3) and (5.4) represent the $\dot{R}_i$ radical chain growth, by the addition of a monomer molecule $M$, Reaction (5.5) reproduces initiator molecules by two radical reactions, diminishing the concentration of radicals. Reactions (5.6) and (5.7) are terminations that give the final product $P_i$, i.e. the ith-order polymer chain. Here, we do not consider chain transfer, inhibition and crosslinking reactions. In order to simplify the treatment we make some assumptions:

**Assumption 1.** *The concentration of initiators I remains constant during the whole process.*

The validity of such an assumption depends on the average chain length and hence on the curing intensity; it is well justified, however, by the fact that, for practical reasons, $I$ is in excess in every commercially available monomer mixture.

**Assumption 2.** *The constants of both prolongation and termination stages are independent of the length of the radical chains involved.*[11] *Then we write:*

$$k_3^n = k_4 = k_p \ (p = prolongation), \ k_5 = k_6^N = k_7^M = k_t \ (t = termination).$$

**Assumption 3.** *Radicals and radical chains can be considered immobile.*

The reactivity of radicals and radical chains is in fact very strong, so that any variation in concentration for these species is almost due to chemical reactions. In other words, in the case of radicals and radical chains the diffusive term can be neglected with respect to the chemical one.

**Assumption 4.** *The total radical concentration can be considered constant during the process.*[11]

This assumption is justified by the typical variation time for radicals concentration that is very short with respect to all other typical times of the system.

Reactions (5.1) and (5.2) rules the production of radicals ($\dot{R}_0$) by light exposition of an initiator $I$. Therefore, the rate of increase of $R_0$ concentration is proportional to the intensity $W$ of the incident radiation and to the initiator concentration $I$ through an initiation constant $g$:

$$\frac{d\dot{R}_0}{dt} = gWI \ (\text{production term due to light})$$

On the other hand, reactions (5.3), (5.5) and (5.6) produce a radical depletion. Summing all terms, we have the balance equation for the steady state concentrations:

$$\frac{d\dot{R}_0}{dt} = gWI - k_p \dot{R}_0 M - k_t \sum_{n=0}^{\infty} \dot{R}_n$$

Using eqn (5.4)–(5.6) we can obtain for radical chains ($\dot{R}_n$) the following equation:

$$\frac{d\dot{R}_n}{dt} = k_p(\dot{R}_{n-1} - \dot{R}_n)M - k_t \sum_{n=0}^{\infty} \dot{R}_r \dot{R}_n \quad (n > 1)$$

In writing this expression we have used assumption 4. Summarizing all these equations, taking into account that $R_\infty = 0$ (the radical chains concentration decreases by increasing the chain length) and introducing the sum of radical concentrations

$$R = \sum_{n=0}^{\infty} \dot{R}_n$$

we obtain

$$\frac{d\dot{R}_0}{dt} = gWI - k_p \dot{R}_0 M - k_t \dot{R}_0 R \tag{5.8}$$

$$\frac{d\dot{R}_n}{dt} = k_p(\dot{R}_{n-1} - \dot{R}_n)M - k_t R \dot{R}_n \quad (n < 1) \tag{5.9}$$

We consider eqn (5.9) and take the sum on n from $1$ to $\infty$ for both members; then sum it with eqn (5.8). Thus, we obtain:

$$\frac{dR}{dt} = gWI - k_t R^2 \tag{5.10}$$

a result that directly comes from assumption 2. Hence, the production of $N$ radical chains $R_{n+1}$ costs the depletion of $N$ $R_n$, no matter what the monomer concentration $M$ is.

Taking into account assumption 4 we obtain

$$R = \frac{1}{k_t}\sqrt{k_t gW(x)I} \tag{5.11}$$

Using eqn (5.9) and assumption 4, and dividing by $\dot{R}_{n-1}$, we obtain

$$\frac{\dot{R}_n}{\dot{R}_{n+1}} = \frac{k_p M}{k_p M + k_t R}$$

the right-hand term does not depend on $\dot{R}_n$ or $\dot{R}_{n-1}$ so we can define

$$\gamma = \frac{k_p M}{k_p M + k_t R}. \tag{5.12}$$

In this way $\gamma$ is the well-known inverse kinetic chain length, *i.e.* the ratio of termination and prolongation rates.

*Inhomogeneous Photopolymerization in Multicomponent Media* 91

$$\frac{\dot{R}_n}{\dot{R}_{n+1}} = \gamma \tag{5.13}$$

or

$$\dot{R}_n = \gamma \dot{R}_{n-1} = \gamma^2 \dot{R}_{n-2} = \cdots = \gamma^n \dot{R}_0$$

Using the definition of $R$ and taking into account that $\gamma < 1$

$$R = \sum_{n=0}^{\infty} \dot{R}_n = \dot{R}_0 \sum_{n=0}^{\infty} \gamma^n = \dot{R}_0 \frac{1}{1-\gamma} \Rightarrow \dot{R}_0 = (1-\gamma)R$$

By substituting, we get

$$\dot{R}_n = R(1-\gamma)\gamma^n \tag{5.14}$$

The monomer concentration can be obtained in a similar way. Contributions to its rate come from eqn (5.3) and (5.4) and obviously they are all negative.

Taking into account these contributions and expression (5.11) for R, we get:

$$\left(\frac{dM}{dt}\right)_R = \frac{k_p}{k_t}\sqrt{k_t g W(x) I M} \tag{5.15}$$

where the footer $R$ indicates that this rate is due to the polymerization reactions only.

For polymer chains $P_N$ produced in reactions (5.6) and (5.7), we can write:

$$\left(\frac{dP_N}{dt}\right)_R = \frac{k_t}{2}\sum_{n=0}^{N} \dot{R}_n \dot{R}_{N-n} \tag{5.16}$$

We introduce now a critical length $N_0$. Not all the reaction products, in fact, can be considered immobile, since short polymer chains are certainly mobile. So, we introduce $N_0$ as the number of monomers for a chain starting from which we can think of the chain as immobile. In fact, diffusion in condensed phases is an activation-type process and the diffusion coefficient depends in a very abrupt, exponential manner upon the molecular mass. Thus, we may introduce such a critical length instead of treating the real dependence of mobility upon the chain length. We can define now

$$p = \sum_{N=N_0}^{\infty} NP_N$$

representing the concentration of all the monomers in immobile polymer chains. By deriving in time we get:

$$\frac{dP}{dt} = \sum_{N=N_0}^{\infty} N \frac{dP_N}{dt}$$

with no footer because there is no diffusion for immobile objects. Using eqn (5.14) and (5.16) we can write

$$\frac{dP}{dt} = \frac{k_t}{2} \sum_{N=N_0}^{\infty} \sum_{n=0}^{\infty} NR^2(1-\gamma)^2 \gamma^n \gamma^{N-n}$$

and

$$\frac{dP}{dt} = \frac{k_t R^2}{2}(1-\gamma)^2 \sum_{N=N_0}^{\infty} (N+1)N\gamma^N$$

The sum of the series finally gives:

$$\frac{dP}{dt} = \frac{k_t R^2}{2}\left[N_0^2(1-\gamma) + N_0(1-\gamma) + \frac{2\gamma}{1-\gamma}\right]\gamma^N \quad (5.17)$$

## 5.3 Mass Transport

First, it is necessary to define active and passive volumes. We call "passive" the volume $\Delta V$ occupied by the immobile polymer chains, while we call "active" the remaining volume $V - \Delta V$. The situation in the active volume is as follows:

- due to expulsion from the passive volume, there is, at the sides of the polymer slides, a concentration of $LC$ molecules higher than in the rest of the sample; these molecules will therefore start to diffuse in the active region;
- on the other hand, the formation of polymer clusters in the passive volume produces a diminishing in the concentration of monomer molecules; the molecules of this kind will therefore start to diffuse from the active volume, in order to fill the gap.

The species directly involved in diffusion are therefore monomers and liquid crystals. If $(M^0)$ and $(C^0)$ are their initial concentrations, we can define the total concentration as $T = M^0 + C^0$

On the other hand,

$$\frac{\Delta V}{V} = \frac{N_M^P \times V_{\text{mol}}}{N_{\text{tot}} \times V_{\text{mol}}} = \frac{N_M^P / V}{N_{\text{tot}} / V} = \frac{P}{T}$$

where $N^P{}_M$ is the number of monomers in polymer chains, $N_{\text{tot}}$ is the total of liquid crystal and monomers in the volume and $V_{\text{mol}}$ the volume occupied by one monomer. If we indicate with $S$ a surface perpendicular to the sample walls and parallel to the $yz$ plane, the flux of molecules diffusing in the $x$ direction through $S$ in the unit time from a little volume $\delta V$, is given by:

$$\Phi_D^S = -DS\nabla M$$

*Inhomogeneous Photopolymerization in Multicomponent Media*

where $D$ is the Fick diffusion constant. In our case, the passive volume $\Delta V$ affects the area $S$ that diffusion can take place through. Indeed, a section of $\Delta V$ is a fraction of $S$, thus the effective area available for diffusion will be

$$S' = S\left(1 - \frac{P}{T}\right)^\alpha$$

where $\alpha$ represents the growth ratio between volume and surface of the polymer walls, *i.e.* $\alpha = 1$ for linearly aligned polymer chains and $\alpha = 2/3$ in the case of random aligned polymer chains. The same consideration holds for concentrations and taking into account that the diffusion is along $x$, the expression of the effective flux in the unit time through the unit surface is:

$$\Phi_D^S = -DS\left(1 - \frac{P}{T}\right)^\alpha \frac{\partial}{\partial x} \frac{M}{\left(1 - \frac{P}{T}\right)}$$

The number of particles flowing in the $x$ direction in a time interval $dt$ can be written as:

$$dN_M = -V \frac{\partial \Phi_D(x)}{\partial x} dt$$

thus the rate of monomer concentration due to diffusion processes, is given by:

$$\left(\frac{dM}{dt}\right)_D = -\frac{\partial \Phi_D(x)}{\partial x} = D\frac{\partial}{\partial x}\left[\left(1 - \frac{P}{T}\right)^\alpha \frac{\partial}{\partial x}\frac{M}{\left(1 - \frac{P}{T}\right)}\right] \quad (5.18)$$

footer $D$ means that this is only the diffusive term. The treatment for $LC$ is, in fact the same; therefore, with eqn (5.15) and (5.17), we finally have all the coupled equation of our system:

$$\frac{dM}{dt} = D\frac{\partial}{\partial x}\left[\left(1 - \frac{P}{T}\right)^\alpha \frac{\partial}{\partial x}\frac{M}{\left(1 - \frac{P}{T}\right)}\right] - \frac{k_p}{k_t}\sqrt{k_t W(x) I} M \quad (5.19)$$

$$\frac{dC}{dt} = D\frac{\partial}{\partial x}\left[\left(1 - \frac{P}{T}\right)^\alpha \frac{\partial}{\partial x}\frac{C}{\left(1 - \frac{P}{T}\right)}\right] \quad (5.20)$$

$$\frac{dP}{dt} = \frac{k_t R^2}{2}\left[N_0^2(1-\gamma) + N_0(1-\gamma) + \frac{2\gamma}{1-\gamma}\right]\gamma^{N_0} \quad (5.21)$$

As for initial conditions, they are:

$$P(0, x) = 0; \quad L(0, x) = L_0; \quad M(0, x) = M_0$$

## 5.4 The System of Equations

It is useful now to rewrite the equations in a reduced form that enables an immediate insight into the features of the phenomena under consideration. We introduce the relative concentrations of components $\sigma = C/T$, $\mu = M/T$ and $\nu = P/T$, while, to have a dimensionless time, we use the typical time of depletion of monomers for chemical reactions that results from eqn (5.19): $\tau = k_p / k_p \sqrt{k_t g W(x) I}$. In order to produce a grating, we have to use a modulated radiation given by the interference of two beams of intensity $I_1$ and $I_2$:

$$W(x) = W_0 (1 + m \sin qx)$$

where $W_0 = I_1 + I_2$ is the total intensity, $m = (\sqrt{I_1 I_2})/(I_1 + I_2)$ is the visibility of the fringes and $q = 2\pi/\Lambda$, where $\Lambda$ is the fringe spacing. Introducing a dimensionless coordinate $\xi = qx$, we have:

$$\frac{d\mu}{d\tau} - B \frac{\partial}{\partial \xi}\left[(1-\nu)^\alpha \frac{\partial}{\partial \xi}\left(\frac{\mu}{1-\nu}\right)\right] + (1 + m \sin \xi)^{\frac{1}{2}} \mu = 0 \quad (5.22)$$

$$\frac{d\nu}{d\tau} - \frac{G}{2}(1 + m \sin \xi) \frac{\partial}{\partial \xi}\left[N_0^2(1-\gamma) + N_0(1+\gamma) + \frac{2\gamma}{1-\gamma}\right]\gamma^{N_0} = 0 \quad (5.23)$$

$$\frac{d\sigma}{d\tau} - B \frac{\partial}{\partial \xi}\left[(1-\nu)^\alpha \frac{\partial}{\partial \xi}\left(\frac{\sigma}{1-\nu}\right)\right] = 0 \quad (5.24)$$

where

$$B = \frac{4\pi^2 D k_t^{\frac{1}{2}}}{g W_0 I \Lambda^2}, \quad G = \frac{(g W_0 I k_t)^{\frac{1}{2}}}{k_M T}$$

and

$$\gamma = \frac{\mu}{\mu + G\sqrt{1 + m \sin \xi}}$$

The initial conditions have now the form:

$$\mu(0, \xi) = \mu_0 = M_0/T \,;\; \sigma(0, \xi) = \sigma_0 = C_0/T \,;\; \nu(0, \xi) = 0.$$

## 5.5 Analytical Solutions

In order to find analytical solutions of the system, we consider a particular curing regime. If the curing intensity $W_0$ is very high, the polymerization reaction is very fast in comparison with the diffusion process and the system is in the so-called fast-curing regime. In this case it is possible to neglect

Inhomogeneous Photopolymerization in Multicomponent Media

evolution terms for concentration due to diffusion, and our system can be considered an all-chemical one. $W$ starts from eqn (5.15)

$$\dot{R}_n = R(1-\gamma)\gamma^n$$

From eqn (5.12) we see that high $W$ implies high $R$ and from eqn (5.13) high $R$ implies low $\gamma$; so high $W$ implies low $\gamma$. We show that in this case we have short chains. In a volume containing a concentration of radicals of about 0.6, we can realize a plot showing the concentration of monomers embedded in long chains for $\gamma = 0.8$ (white circles) and $\gamma = 0.2$ (black triangles) (Figure 5.1). It is evident that for low $\gamma$ it is hard to produce long chains; in other words, in the fast-curing regime polymerization take place in the low-$W$ fringes.

A physical interpretation of this result can be the following: We have a starting situation with initiators $I$ and monomers $M$. When the process starts, in the bright fringes, radiation creates a great number of radicals $R_0$, so for a free radical it is more probable to find another radical and recombine in a new initiator than form a long polymer chain. On the other hand, in the dark fringes, a very low number of radicals are present and so, before a chain is closed, a lot of monomers can be captured, see Figure 5.2.

This corresponds to the case $B \ll 1$ and we can analyze it, by neglecting the diffusion term in the dynamic equation of the monomer concentration (5.23). Then we can write:

$$\frac{d\mu}{d\tau} + (1 + m \sin \xi)^{\frac{1}{2}}\mu = 0$$

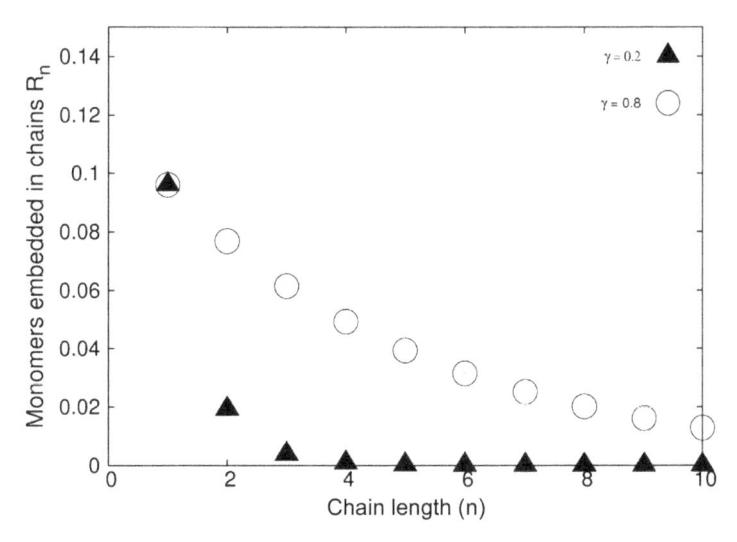

**Figure 5.1** Concentration of radical chains in function of the number of monomers for chains, for $\gamma = 0.2$ and $\gamma = 0.8$.

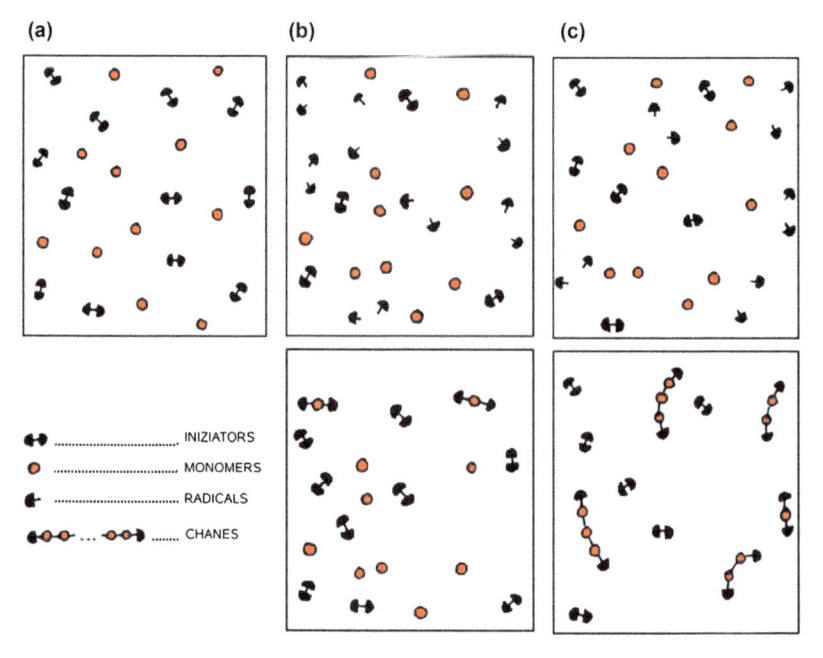

**Figure 5.2** (a) Initial mixture: initiator I and monomers M. (b) Bright fringes: a great number of radicals $R_0$ and consequently a great number of short chains are produced. (c) Dark fringes: a few number of radicals are produced and so, the production of long chains is favored.

which gives

$$\mu(\tau) = \mu_0^{-\alpha\tau} \tag{5.25}$$

where $\mu_0 = \mu(0) = M_0/Z$ and $a(\xi) = (1 + m \sin \xi)^{1/2}$. Taking into account that

$$(1-\gamma)\gamma a = -\frac{d\gamma}{dt}$$

we can easily express the variation of $\nu$ with respect to $\tau$ by rewriting eqn (5.24) as:

$$\frac{d\nu}{d\tau} = -\frac{Ga^2}{2}\left[N_0^2\gamma^{N_0-1} + N_0\frac{1+\gamma}{1-\gamma}\gamma^{N_0-1} + \frac{2\gamma^{N_0}}{(1-\gamma)^2}\right] \tag{5.26}$$

This equation can be easily integrated to obtain the expression for the steady-state distribution $\nu_\infty$

$$\nu_\infty = \frac{\mu_0}{2}\gamma^{N_0-1}[N_0 + 1(1-N_0)\gamma_0] \tag{5.27}$$

with conditions:

$$\gamma_0 = \gamma(\tau=0) = \frac{1}{1+\frac{Ga}{\mu_0}}, \quad \text{and} \quad \gamma(\tau \to \infty) = 0$$

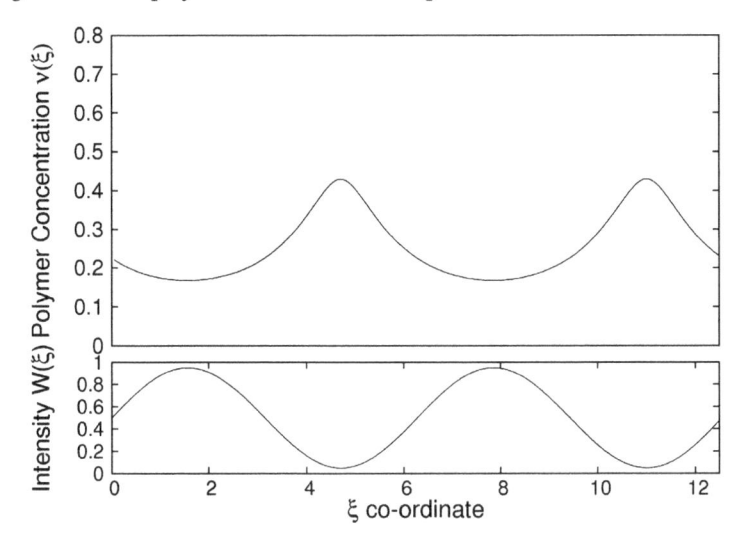

*Inhomogeneous Photopolymerization in Multicomponent Media* 97

**Figure 5.3** The final polymer concentration is represented as a function of the $\xi$ coordinate. The plot of the curing intensity $W(\xi)$ is reported as a reference.

This situation is well represented in Figure 5.3, where it is evident that polymerization takes place in the low-intensity fringes.

## 5.6 Numerical Solutions

If we are not in the fast-curing regime, the system (5.22)–(5.24) does not admit an analytical solution and we have to find a numerical approach. For this reason, we explored the solutions by means of a Runge–Kutta scheme for temporal derivatives and a central derivative scheme for spatial derivatives. Simulations have been carried out with $\mu_0 = 0.7$ as initial monomer concentration, which is close to typical experimental values.[2,6]

In order to check our results, the curves of the polymer concentration obtained for different $\tau$ values have been compared with the analytical solution derived in the fast-curing regime (Figure 5.4). Results show that, for $\tau = 15$, the numerical curve is well superimposed to the analytical one (which corresponds to $\tau \to \infty$).

The spatial distribution of the curing intensity is also presented in Figure 5.4 for comparison. In the frame of the same figure, we report a picture referring to the scanning electron microscope (SEM) analysis of a PDLC grating realized in these experimental conditions; the width of polymer slides is in good agreement with theoretical predictions. Figure 5.5 is concerned with the main features of our model: We have singled out the variable that determines the DE of gratings, that is the modulation $\Delta\sigma$ of the liquid-crystal concentration across the fringes, calculated as the first Fourier component of $\sigma(\xi)$.[3] The surface indicating $\Delta\sigma$ is plotted as a function of $B$ and $G$ parameters for $N_0 = 4$ and shows that two regions of high $\Delta\sigma$ values

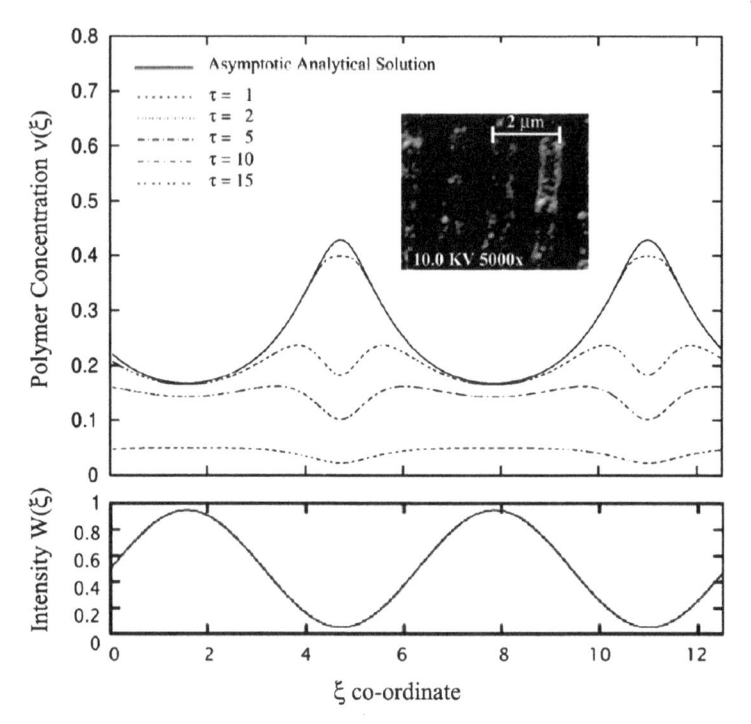

**Figure 5.4** Comparison between analytical and numerical trends of the polymer concentration $\nu$ vs. the spatial coordinate $\xi$ for different $\tau$ values.

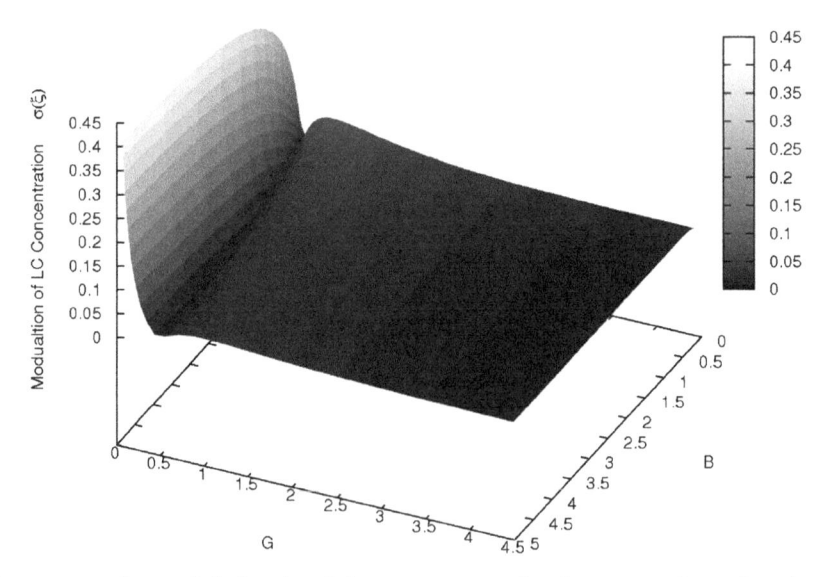

**Figure 5.5** The modulation $\Delta\sigma$ of the LC concentration is presented as a function of parameters $B$ and $G$ in a 3D surface. The blue zone identifies the H-PDLC grating zone while the red zone identifies the POLICRYPS gratings formation.

are present. This means that there are two possible ways of obtaining good gratings. For $B \ll 1$, the polymerization reaction is very fast in comparison with diffusion. Starting from a uniform distribution of initiators and monomers, when the radiation creates a high number of radicals in the bright fringes, for these molecules it is more probable to recombine and form a new initiator molecule, than to meet monomers and form a long polymer chain.

On the other hand, the low number of radicals created in the dark fringes can capture a great number of monomers before a chain in closed; this corresponds to the case already presented in Figure 5.4 (fast curing). Increasing $G$ corresponds to increase the light intensity $W_0$, and hence to make shorter chains (decrease of $\Delta\sigma$). When all chains contain less than $N_0$ monomers, no polymer chains can be considered immobile, and the formation of polymer slides is prevented. $G \ll 1$ with $B > 1$ means $k_p \ll k_t$. In these conditions it is more probable for a chain to be closed by a radical than to get a new monomer. There is no difference in chain length between dark and bright fringes and, because of the great number of available radicals, polymerization takes place mainly in the bright fringes. The situation is

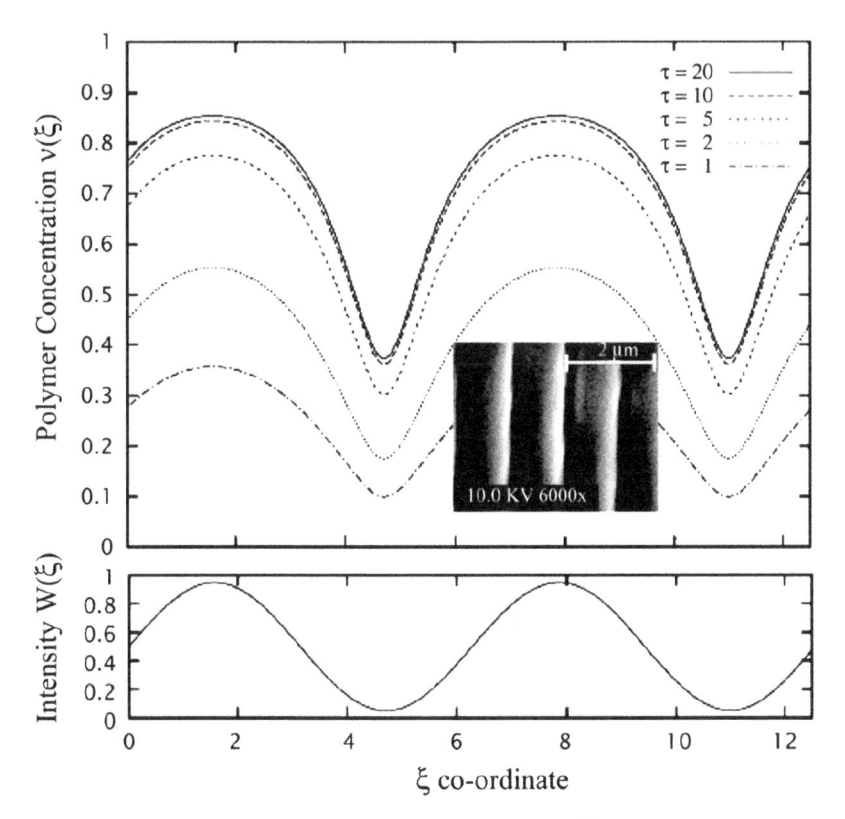

**Figure 5.6** Spatial trend of the polymer concentration $\nu(\xi)$ in a slow curing region of the $BG$ plane.

illustrated in Figure 5.6: Also in this case, the agreement with theoretical predictions is good; the curing intensity $W(\xi)$ is reported for a reference. The picture in the frame refers to the SEM analysis of a POLICRYPS grating, which has been realized exactly in these experimental conditions.

For $G \gg 1$ and $B \gg 1$ we have a high impinging intensity and a noticeable diffusion: The curing process becomes quite homogeneous and there is no modulation of the polymer concentration. In general, it can be seen that $BG = (4\pi k_t/k_p T)D/\Lambda^2$ is constant with respect to the curing intensity $W_0$; this means that an increase in $W_0$ implies a decrease of $B$ and an increase of $G$ moving along a particular hyperbola defined by parameters $\Lambda$ and $D$. An increase of $D$ or a decrease of $\Lambda$, however, implies an increase of $B$ while $G$ remains constant.

## 5.7 Conclusions

We summarize our results in the following considerations:

- The main features of the curing process of diffraction gratings in polymer-based liquid-crystalline composite materials can be explained by taking into account that, contemporarily to the formation of polymer chains, also diffusion of monomer and LC molecules occurs.
- It is possible to recognize two different regimes: The first one (fast curing) refers to a polymerization process that is faster than diffusion, so that polymer chains grow (in the dark fringes of the curing interference pattern) before monomer diffusion can take place. In the second regime (slow curing), the polymerization reaction is slow enough to enable monomer molecules to diffuse across the fringes (for a given fringe spacing) before they react. This regime is the best candidate to produce good holographic gratings, since it allows the highest values of the concentration modulation $\Delta\sigma$ to be obtained.
- Diffusion is the main mechanism that determines which kind of grating (POLICRYPS or PDLC) is going to be formed. In order to produce a PDLC grating, both low $B$ and average $G$ values are needed; this corresponds to the low average intensity and low temperature (low diffusion) that in general are used to make PDLC gratings.[1] It is also known that the actual technique for the production of POLICRYPS gratings needs a high temperature (high diffusion) during the curing process[3] (high $B$ and low $G$ values, in our model). These considerations lead us to interpret the two maxima in the $\Delta\sigma$ surface as the curing regimes in which PDLC or POLICRYPS gratings are realized, respectively.
- Finally, we point out that POLICRYPS gratings can be obtained also by using a very low intensity. In a study in 2003,[3] we suggested that, in this case, the curing time, which depends on intensity, would have been too long. However, a more recent study[12] showed that this solution is

actually practicable to achieve the slow-curing regime and thus the realization of POLICRYPS-like morphologies.

On the one hand, this model was realized with the idea to understand the physical-chemical processes in play during the formation of both H-PDLC and POLICRYPS diffraction gratings, identifying the parameters that control the final morphology. On the other hand, it is, in principle, usable to describe a wider range of multicomponent inhomogeneous curing processes: a 2D version of the proposed equations is easy to implement, which can allow modeling of the holographic realization of 2D structures in polymer and liquid crystal,[10] the same for a 3D version, which could be used to discuss advanced techniques of plasmonic curing;[13] One can also consider to add multiple passive species with different diffusivity or add terms to discuss nucleation processes, making this approach a very useful tool with a yet undisclosed potential.

# References

1. T. J. Bunning, L. V. Natarajan, V. P. Tondiglia and R. L. Sutherland, Holographic polymer-dispersed liquid crystals (H-PDLCs), *Annu. Rev. Mater. Sci.*, 2000, **30**, 83–115.
2. R. Caputo, A. V. Sukhov, C. P. Umeton and R. F. Ushakov, Formation of a grating of submicron nematic layers by photopolymerization of nematic-containing mixture, *J. Exp. Theor. Phys.*, 2000, **91**, 1190–1197.
3. A. Veltri, R. Caputo, C. Umeton and A. V. Sukhov, Model for the photoinduced formation of diffraction gratings in liquid-crystalline composite materials, *Appl. Phys. Lett.*, 2004, **84**, 3492–3494.
4. R. Caputo, A. V. Sukhov, N. V. Tabyrian, C. P. Umeton and R. F. Ushakov, Mass transfer processes induced by inhomogeneous photopolymerisation in a multicomponent medium, *Chem. Phys.*, 2001, **271**, 323–335.
5. R. Caputo, L. De Sio, A. V. Sukhov, A. Veltri and C. P. Umeton, Development of a new kind of switchable holographic grating made of liquid crystal films separated by slices of polymeric material (policryps), *Opt. Lett.*, 2004, **29**, 1261–1263.
6. R. Caputo; C. P. Umeton; A. Veltri; A. V. Sukhov and N. Tabiryan Holographic diffraction grating, process for its preparation and opto-electronic devices incorporating it, European Patent Request 1649318; *US Pat.* 2007/0019152A1, 2007.
7. R. Caputo, I. Trebisacce, L. De Sio and C. P. Umeton, Jones matrix analysis of dichroic phase retarders realized in soft matter composite materials, *Opt. Exp.*, 2010, **18**, 5776–5784.
8. R. Caputo, I. Trebisacce, L. De Sio and C. P. Umeton, Phase modulator behavior of a wedge-shaped POLICRYPS diffraction grating, *Mol. Cryst. Liq. Cryst.*, 2011, **549**(1), 29–36.

9. R. Caputo, L. De Sio, J. Dintinger, H. Sellame, T. Scharf and C. P. Umeton, Realization and characterization of POLICRYPS-like structures including metallic subentities, *Mol. Cryst. Liq. Cryst.*, 2012, **553**(1), 111–117.
10. M. Infusino, A. De Luca, V. Barna, R. Caputo and C. P. Umeton, Periodic and aperiodic liquid crystal-polymer composite structures realized via spatial light modulator direct holography, *Opt. Exp.*, 2012, **20**, 23138.
11. P. W. Atkins, *Physical Chemistry*, Oxford University Press, Oxford, 1987.
12. W. Huang, Y. Liu, Z. Diao, C. Yang, L. Yao, J. Ma and L. Xuan, Theory and characteristics of holographic polymer dispersed liquid crystal transmission grating with scaffolding morphology, *Appl. Opt.*, 2012, **51**, 4013.
13. C. Deeb, R. Bachelot, J. Plain, A. L. Baudrion, S. Jradi, A. Bouhelier, O. Soppera, P. K. Jain, L. Huang, C. Ecoffet, L. Balan and P. Royer, Quantitative Analysis of Localized Surface Plasmons Based on Molecular Probing, *ACS Nano*, 2010, **4**, 4579–4586.

CHAPTER 6

# Microfabrication Processes and Applications of Liquid Photosensitive Materials

CHIN-TAI CHEN

Department of Mechanical Engineering, National Kaohsiung University of Applied Sciences, 415, Chien-Kung Road, Kaohsiung 80778, Taiwan, Republic of China
Email: chintai@kuas.edu.tw

## 6.1 Introduction

Light as well-known electromagnetic radiation exhibits energy carried by photons with a certain wavelength and speed that is referred to as the wave–particle duality in quantum mechanics (QM). The power of light increases as its wavelength decreases so that invisible ultraviolet (UV) light with short wavelengths of 10–400 nm (energy per photon in the range between 3 eV and 124 eV) are of particular interest for academics and industries in many practical applications and researches. Moreover, the light with visible wavelengths of 400–700 nm can be emitted coherently in time to yield light amplification by stimulated emission of radiation (called a laser). Thus, pulsed lasers nowadays may generate a power of 1 mW–3 kW in a femtosecond (*i.e.* $10^{-15}$ s), yielding 6.2 eV–18.6 MeV for various applications such as micromachining and cutting.

On the other hand, a liquid as a collection of vibrating atoms constrained by intermolecular bonds can flow to exhibit no fixed shape under a definite volume, in which it is generally considered as one of the fundamental states of matter in fluidics. Vaporization of liquid water, for example, often occurs when it absorbs sufficient separation energy of ~4.5 eV to break hydrogen bonds of the $H_2O$ molecules near its free surface. Further, some solidified matters (*e.g.*, those that may contain hydrogen bonds sensitive to photons) from the original liquid state may absorb lighting energy to become dissolved in specifically selected solutions. Some different liquid or solid matter sensitive to exposure can strengthen, instead, their intermolecular bonds after absorption of light, in which the exposed portion become solidified or insoluble in specific solution. In common, therefore, the photons of light can play an interesting and significant role to substantially change the bonding strength between molecules within different liquid matter.

Incorporating the different energy of light with liquid materials, it can create a variety of formations of the original material states or properties. Hence, numerous present technologies and applications in demand have been found on such a fusion of liquid and light, which is elaborated in the following sections.

## 6.1.1 Fusion of Liquid and Light

As depicted in Figure 6.1, the energy level ($\nu$) of light and the viscous state ($\mu$) of liquid can be analogized and thus matched together. In other words, a liquid solution composed of photosensitive materials can change its viscosity after absorption of light with matching energy level. In fact, many functional chemical materials made of photosensitive monomers or polymers have been invented and used widely.[1–3] Photoresist, as one such material with photosensitive properties, has been studied, developed and applied in associated technologies and devices including microelectronics,[4,5] microsensors[6,7] and actuators,[8,9] microfluidics,[10,11] micro-optics,[12] and medical devices.[13]

In addition, many solid structures in free form can be created from their original liquid states directly by using suitable focused lasers that actually provide well-matched energy for triggering the polymerization of monomers and prepolymers *in situ*.[14,15] Thus, various free-form components with different sizes from small to large dimensions can be rapidly prototyped by light exposure to photosensitive materials. Consequently, three-dimensional

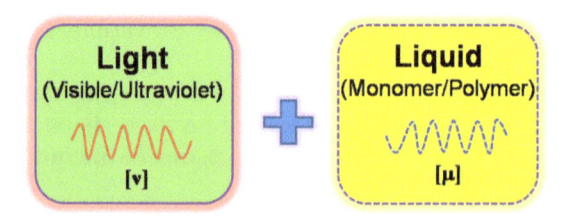

**Figure 6.1** Fusion of liquid and light generates various functional solid components to find many different applications in sciences and technologies including microelectronics, microsensors and actuators, microfluidics, micro-optics, and medical devices.

*Microfabrication Processes and Applications of Liquid Photosensitive Materials* 105

(3D) photofabrication[16] for complicated objects, such as photonic crystals,[17] tissue-engineering scaffolds,[18] and conical structures[19] have been explored and developed using different approaches in recent years. These industrial demands and technical development for such studies and applications are described below.

### 6.1.2 Demand and Development of Technologies

The mechanism and control for interaction between photosensitive materials and light energy are so complicated and significant in demand that multidisciplinary technologies and integrations have to be built up and developed in various applications. As illustrated in Figure 6.2, there are many basic elements of current photosensitive technologies required for real implementation of various applications (*i.e.* Apps. A, B, C, and D here). Those basic elements are comprised of the use for specific materials (*e.g.*, liquids sensitive to certain light power), delicate manufacturing processes (*e.g.*, microfabrication), precision control machines (for any operational motion), fabricated devices (microstructures and parts) and underlying systems (integrated).

Further, as demonstrated earlier in the literature, the applications primarily include electronics, fluidics, biologics, and optics. Hence, their fundamental principles of those elements for each of various applications must be clearly understood and provided, which are fully discussed in the following section.

## 6.2 Fundamental Principles

These liquid materials sensitive to certain light wavelengths can be first described and featured through functional principles in this section. They are generally characterized by strong different interactions of photon

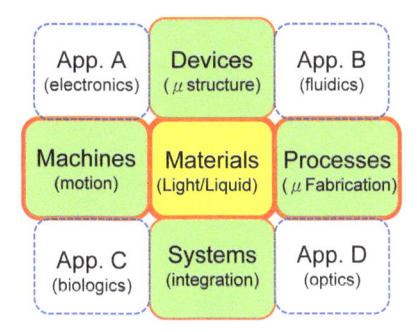

**Figure 6.2** Basic elements of photosensitive technologies required for realization of various applications (Apps. A, B, C, and D) comprise the use of specific functional materials (light/liquid), delicate manufacturing processes (microfabrication), precision control machines (motion), fabricated devices (microstructure) and systems (integrated).

energies with molecules in the nanometer scale.[20] Among the interactions, one polymerization process can be then induced within a number of monomers, consequently forming solid structures from a liquid. On the basis of the transformation mechanism, many small and large objects can be finally designed and fabricated in practice.

### 6.2.1 Characterization of Materials

When photosensitive materials are appropriately irradiated[21] by energetic light beam, they tend to be transformed greatly for their bonding strength of atoms. As demonstrated in Figure 6.3, the neighboring two atoms $m_1$ and $m_2$ are changing their force of bond (symbolized by $[\mu]$) as being in exposure to the amount of photons with suitable energy (symbolized by $[\nu]$).

In this way, the exposed atoms may be debonded and segmented, as shown in Figure 6.3(a). Therefore, those materials after exposure are soluble in certain chemical solution to be removed. In contrast, two atoms in the same neighborhood can be closely linked with a strong bond, while absorbing enough photonic energy, as shown in Figure 6.3(b). Consequently, those materials with enough exposure of light become solidified and immobile.[22] The latter is significantly associated with the chemical formation of polymers, which is further described below.

### 6.2.2 Photoinduced Polymerization

In fact, based on this bonding mechanism, a cluster of the identical molecules can be simultaneously linked together *in situ*. As illustrated in Figure 6.4(a), each of monomers, including $M_1$, $M_2$, $M_3$ and up to $M_n$ with small molecular weight, are linearly linked as a chain in one dimension and

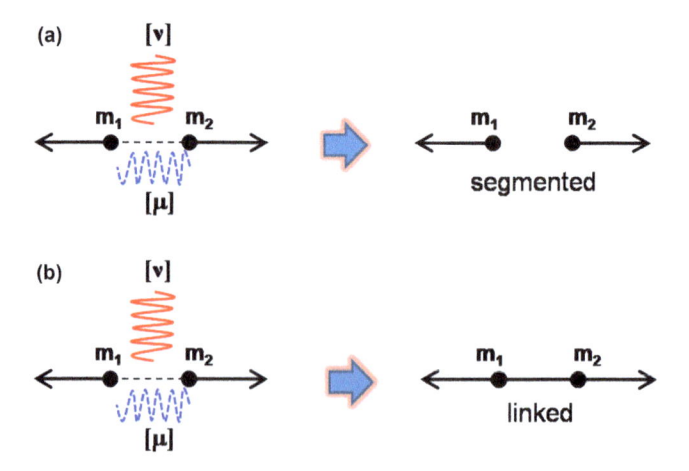

**Figure 6.3** Photosensitive material transformation in liquid induced by exposure of light: (a) bonding segmentation of some molecules (b) strong linkage of some molecules, if light wavelength and power appropriate.

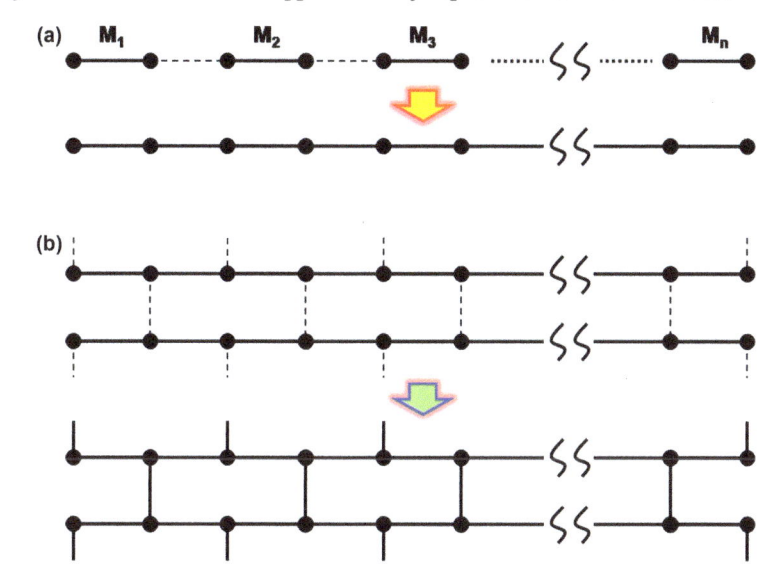

**Figure 6.4** Representation of polymerization process through monomers to polymers: (a) linear linkage of small molecules $M_1$, $M_2$, $M_3$ and $M_n$ forming a chain in one dimension (b) crosslinkage of small molecules generating a two- to three-dimensional macromolecule.

thus transferred into a polymer with large molecular weight. Correspondingly, this kind of polymerization process for certain monomers is able to be produced by crosslinkage in two or three dimensions, as shown in Figure 6.4(b).

Through the polymerization process induced by photonic energy,[3] many complicated three-dimensional (3D) structures have been designed and generated in the micro- or macroscale using the photosensitive materials over decades.[23,24] In particular, many of new development and research studies in recent years are led into miniaturization of objects, which is a very attractive and important topic, as discussed below.

### 6.2.3 Miniaturization of Objects

Since photons can feature short wavelengths in a range of micro- to nanometers (*i.e.* μm–nm scales), they are excellent candidates to be applied for the fabrication of three-dimensional (3D) microsized objects with high precision.[2,4] For instance, as demonstrated in Figure 6.5, using a laser with a wavelength of 780 nm and a pulse width of less than 140 fs, the fabrication of high aspect ratio (HAR) 3D microstructures such as a grid tower (grid width ∼771 nm and tower height ∼459 μm) and Tokyo Skytree (feature width ∼1.2 μm and height ∼293 nm) were made of photosensitive material by Obata *et al.*[25] This recent work of microfabrication demonstrated here is promising for future applications.

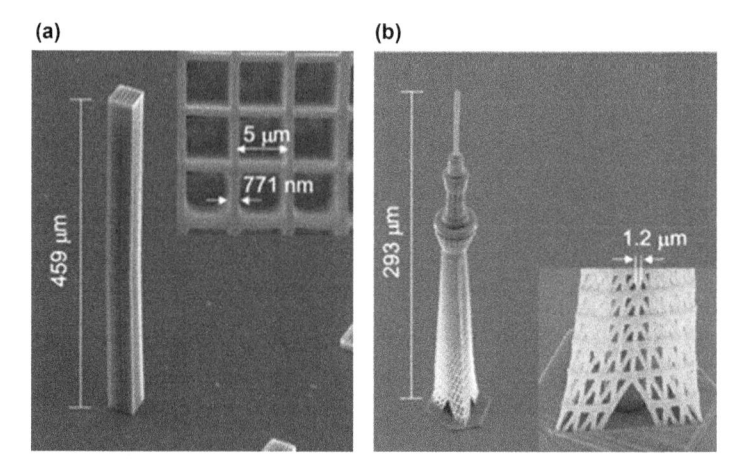

**Figure 6.5** Micro-objects are fabricated in use of photosensitive materials: (a) SEM image of 6×6 grid structure (b) SEM image of "Tokyo Skytree"[25] (reproduced with permission from Nature Publishing Group).

In addition, more various micro- to nanoscale technologies related with photosensitive materials have been currently explored and developed. They are, no doubt, playing an important role in present industrial applications, as collectively described in the following section.

## 6.3 Microfabrication Schemes

To satisfy a variety of practical needs in industrial applications, many techniques and schemes for fabrication at the microscale have been proposed and developed in the past decades.[26,27] To highlight those related with the use of photosensitive materials, four of the microfabrication schemes are discussed here, including photolithography (PL), soft lithography (SL), light stereolithography (LS) and inkjet printing (IJP) schemes. The former two schemes (PL and SL) basically require solid masks for fine patterning, while the latter two (LS and IJP) are substantially maskless, as detailed below.

### 6.3.1 Photolithography (PL)

Photolithography (PL) that transfers patterns into photosensitive materials by photomasks is currently the most common and important technique in the industries.[4–9] It generally requires one photoresist (PR) as photosensitive material to receive the image through masking under exposure of light. For example, as illustrated in Figure 6.6, a complete photolithography process typically comprises preparation of a substrate, spin coating of a PR, light exposure under a mask, curing of a PR, development of a PR, etching of a substrate and removal of a PR.

In this case of PL using a positive photoresist (PPR), the exposed area of the PR can be removed, and thus its corresponding area of the substrate is

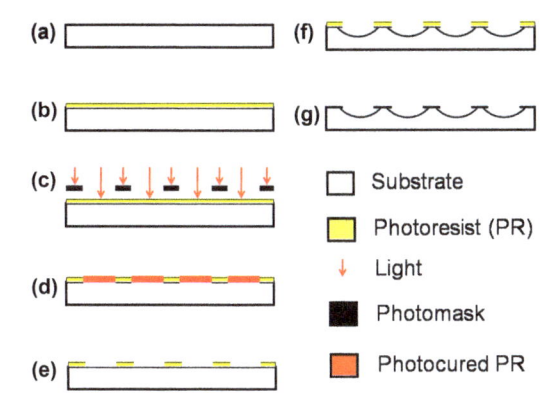

*Microfabrication Processes and Applications of Liquid Photosensitive Materials* 109

**Figure 6.6** Representation of typical photolithography procedure using the positive photoresist: (a) preparation of the substrate (b) spin coating of photoresist (c) light exposure under a photomask (d) photocured photoresist (e) development of the photoresist (f) etching of the substrate (g) removal of the photoresist.

etched by suitable etchants. It implies that the PPR molecules are debonded and soluble after exposure, as mentioned earlier. Conversely, a negative-photoresist (NPR) molecules are strengthened and immobile after exposure; Hence, if the NPR is used here, the exposed area is removed and etched (not shown here). Accordingly, the identical pattern is transferred precisely through the PR and the mask.

### 6.3.2 Soft Lithography (SL)

On the other hand, soft lithography (SL) recently developed by Whitesides and coworkers[28–30] presents a different scheme of microfabrication from the previous PL scheme, in which a soft and bendable substrate can be generated and patterned with complex microstructures. It mainly features the use of transparent polydimethysiloxane (PDMS) liquid to create a duplicate of image transferred from a master substrate. This scheme has been widely applied in nanoimprinting techniques.[31] Here, this master substrate that can be made by silicon (Si) and others (*e.g.*, glass) is substantially responsible for a mask of pattern previously used in PL. Namely, as illustrated in Figure 6.7, a basic scheme of soft lithography include three steps with preparation of a master substrate, pattern transfer by pouring of PDMS and release after PDMS curing.

It is clear in SL here that the PDMS curing for polymerization and solidification is carried out by thermal energy, instead of light exposure previously in PL. This feature means that the PDMS molecules are closely bonded and crosslinked together through thermal curing. However, it is possible to use photosensitive materials such as resists of OrmoClear[32] and NOA[33] replacing PDMS to form patterns as well.

### 6.3.3 Light Stereolithography (LS)

To be exempt from any solid (substrate) masks required in previous PL and SL, light stereolithography (LS) developed by Ovsianikov *et al.* directly employs light beams such as lasers that can be focused and scanned into small-sized spots in three-dimensional (3D) space.[2,9,15,18] Conceptually, as demonstrated in Figure 6.8, the light beam is emitted and focused at a small area of the

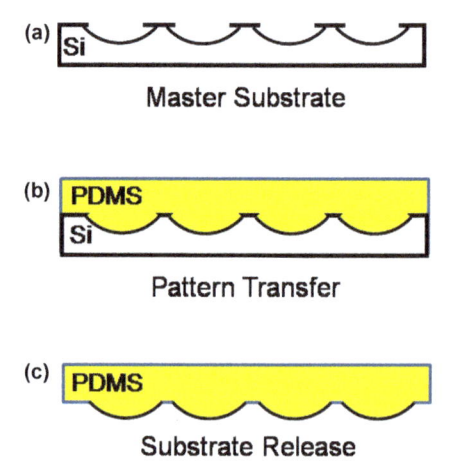

**Figure 6.7** Configuration of typical soft photolithography using polydimethylsiloxane (PDMS) liquid material onto a silicon (Si) substrate: (a) preparation of the master substrate (b) pattern transfer of pouring PDMS (c) substrate release after PDMS curing.

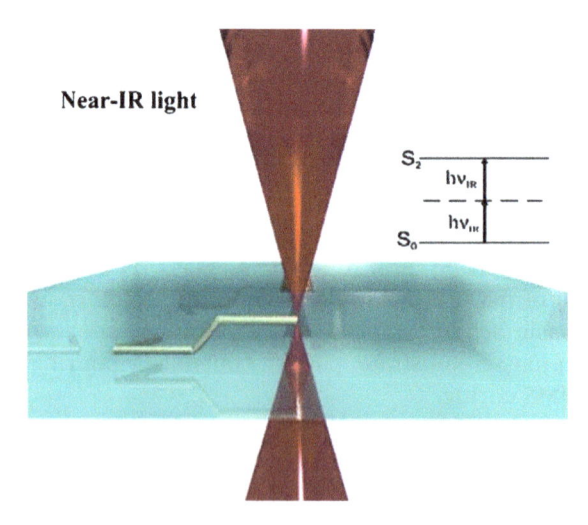

**Figure 6.8** Two-photon polymerization process initiated by light absorption in a high-intensity region that is limited to the focal volume of the light beam[18] (reproduced with permission from Elsevier B.V.).

photosensitive liquid monomers.[18] As this light intensity (*i.e.* exposure dose) at the focus is sufficient, a polymerization process of the liquid monomer is therefore initiated. In this way, any virtual images digitally generated from 3D computer-aided drawing (CAD) can be dynamically transferred into the materials exposed by virtue of optical projection and motion systems.

Based on light-beam writing in this LS scheme, there have been several similar approaches developed as named after two-photon polymerization[2,18,34–37,39] (2PP), single-photon polymerization[9] (1PP), rapid prototyping,[15,37] holographic lithography,[38] and direct laser writing.[39] In common, using different suitable materials, they can accomplish versatile micro- to macromachining of complicated 3D geometries to submicrometer precision.

### 6.3.4 Inkjet Print (IJP)

Alternatively, inkjet printing (IJP) that generates and deposits small droplets at the picoliter (pl) scale has been considered for a long time as one of direct writing means on substrate surfaces in free form. Interfacial properties between liquid droplets and solid surfaces are deeply investigated for controlling the formation of liquid deposit.[40,41] Microscale two-dimensional (2D) arrays of resist droplets can therefore be formed, dried and cured (thermal or light exposure) for imaging transfer on different substrates.[11,42] More importantly, composite structures made of multimaterials generated from different printheads have been demonstrated recently;[43] in this case, it shows a promising route of overcoming the single material limitation in previous LS schemes.

As the LS scheme does in microfabrication, this IJP scheme can also perform maskless patterning in 3D space. It can be generally carried out by using solid powders as well as liquid resist binder.[44] For instance, as demonstrated in Figure 6.9, one layer of powder is first spread onto a flat surface held on a stage; then another thin layer of liquid binder deposited digitally from a scanning printhead is added. This kind of powder-add-binder (PAB) step is thus repeated layer by layer by moving down the stage until the 3D

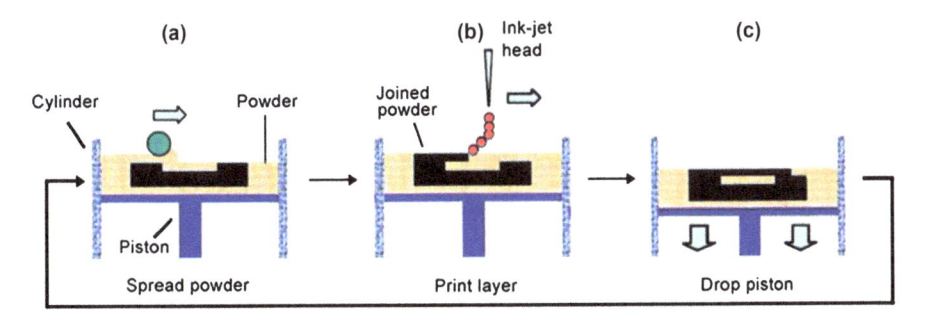

**Figure 6.9** Three-dimensional (3D) drop-on-demand (DOD) inkjet-printing steps for fabrication of free-form objects: (a) spread powder (b) print layer (c) drop piston[43] (reproduced with permission from Elsevier B.V.).

pattern is completed. Finally, the mixture of powder and binder is cured with thermal or light exposure, consequently forming a solid 3D part by PAB. Even complex hollow structures with high curvature may be formed as well.

## 6.4 Applications in Microdevices

Derived from the four major schemes of microfabrication, there have been numerous devices explored and presented for different applications. Especially focusing on recent developments at the microscale, they can be mainly categorized into the following types: microactuators, microsensors, microfluidic components, optical components, medical devices and others. These various types of devices are very significant and interesting to be further discussed as follows.

### 6.4.1 Microactuators

An actuator at the micrometer scale that can deliver motion of solids or fluids required as powered is certainly one of the most important parts in a mechanism.[5,8,9,45] It is the first main kind of microtransducers at present. As demonstrated in Figure 6.10, the micromotor[5] illustrated by Zhang *et al.* is

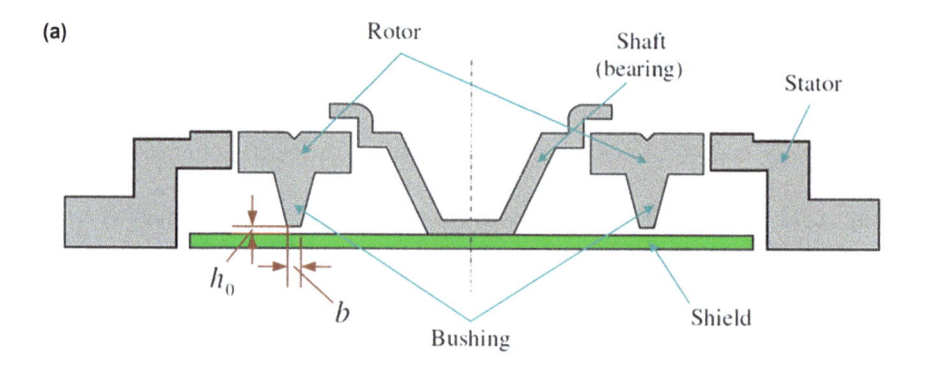

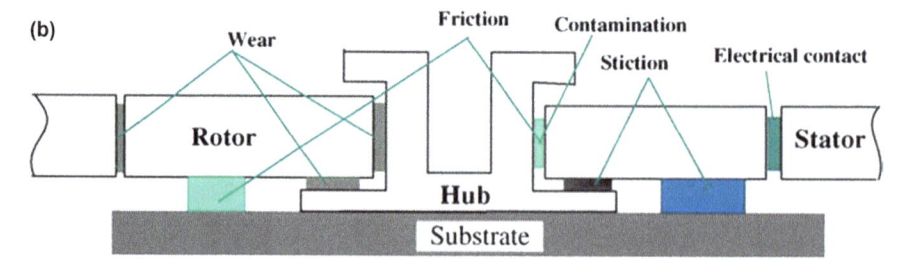

**Figure 6.10** An electrostatic micromotor as one of the microactuators based on a surface micromachining process (SMP): (a) schematic illustration of the rotor and the stator with a bearing shaft (b) practical tribological issues concerning the durable operation of the micromotor[5] (reproduced with permission from Elsevier B.V.).

typically driven by electrostatic force between the small rotors and stators. These microstructures are generally created using silicon wafers through a surface micromachining process (SMP) including a series of patterns for the motor, the rotor and the bearing shafts. Here, the patterning of those parts can be precisely carried out on the top surface of the substrate by using photolithography (PL). When compared to their weight at the microscale, this electrostatic force provides a highly efficient power source to drive the rotor under relatively low electric consumption. However, it may fail due to a lack of sufficient lubrication of the bearing over a period of time.

Besides, more types of microactuators for various working functions have been proposed and developed by means of thermal, piezoelectric, magnetic and shape-memory alloy (SMA) actuations. In addition, bulk micromachining processes (BMP) can be implemented as well. Though the PL is still dominant nowadays for producing those microactuators, the remaining schemes such as SL, LS and IJP described earlier have also to be explored and applied in the meantime as well.

### 6.4.2 Microsensors

A microsensor is the second type of transducer demonstrating equal importance with the previous microactuator. It is designed naturally with miniaturized structures[6,7] for receipt of different signals targeted in physical, chemical or biological ways. For example, as depicted in Figure 6.11, a microfabricated probe, in which two metal (*e.g.*, gold and palladium)

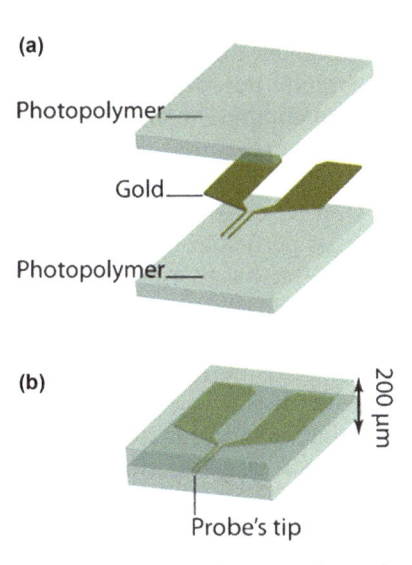

**Figure 6.11** Photographs of an arrayed electrowetting microprism: (a) a top view with an active area and fan-out electrodes (b) a simple demonstration of the prism array under two-axis illumination[46] (reproduced with permission from IOP Publishing Ltd).

electrodes are sandwiched in between two photopolymer substrates (*e.g.*, SU-8 and polyurethane), is specially proposed for electrochemical measurement and biosensing of specific target proteins.[46] Instead of the silicon-based substrate used previously in the micromotor, this microsensor is formed onto polymer-based substrates, although the same PL scheme is applied here. In this case, nanometer-thick and micrometer-wide electrodes fabricated here are able to permit electrical connection to the external world with a tiny current of several nA. Such a probe may be well functioned to potentially detect proteins generated from living cells in the future.

Again, more kinds of detection principles have been implemented for various microsensors that include capacitive, thermal, piezoresistive, piezoelectric and magnetic methods. It is also noted here that polymer-based substrates are particularly attractive for use of biological materials embedded within devices. In addition, the SL scheme can also be suitably adopted in such applications. As a result, fluid often plays a significant role for carrying the targeted materials, which are usually seen in microscale channels of fluidic components, as described below.

### 6.4.3 Microfluidic Components

A microfluidic component is basically designed for delivery of fluid within tiny channels at the microscale.[10,11] Many materials such as chemical and biological ones are able to be effectively transported within the microfluidic devices, while the electrical charges are delivered in microelectronic chips. Hence, it plays a significant role seen in many applications of microfluidics in the 21st century as microelectronics has done in various industries in the past fifty years. For instance, as demonstrated in Figure 6.12, a microbioreactor composed of channel networks is designed and fabricated here for cultivation of the cells.[47] Also, these microfluidic channels and chambers are simply made of two microstructured transparent PDMS layers based on the SL scheme, as described earlier. In this case, the fetal human hepatocytes (FHHs) and Hep G2 cells are injected and delivered within the bioreactor for observation of perfusion culture through an optical microscope. As expected, similar PDMS-based microfluidic devices fabricated by a SL scheme can be found for biological engineering in the future.[48]

Furthermore, a variety of microfluidic components such as mixers, valves and pumps can be built to generate microfluidic microsystems, in which actuation and sensing mechanism are enabled within a multilayer functional chip. In addition to the biological materials, it can also be applied for transporting light through the fluid to visualize them *in vitro*, which is further discussed in the following.

### 6.4.4 Optical Components

Conventional optical devices at the microscale such as filters, mirrors and lenses have been investigated and developed in optical microsystems,[12,49] in

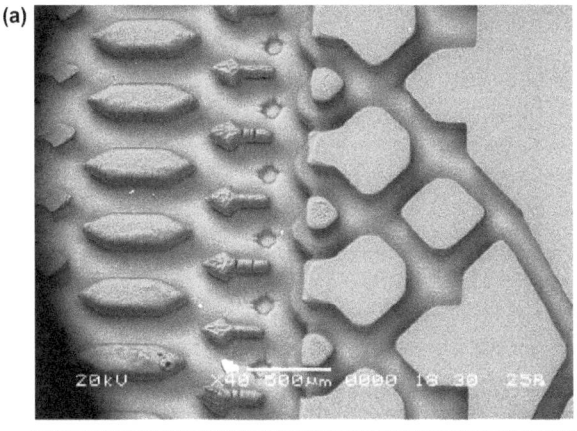

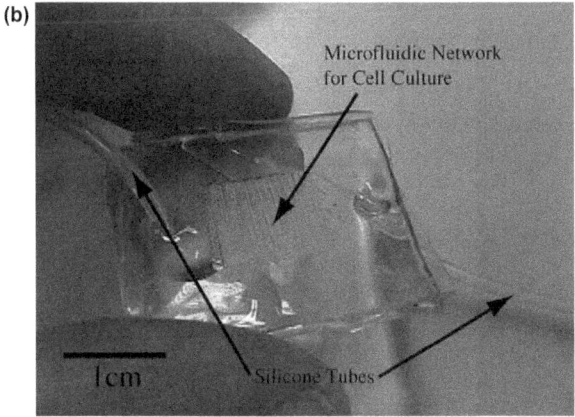

*Microfabrication Processes and Applications of Liquid Photosensitive Materials*  115

**Figure 6.12** Perfusion culture in microfluidic environments: (a) a SEM of the microstructured PDMS layer (b) a photograph of the bioreactor[47] (reproduced with permission from Elsevier B.V.).

which many of them are microfabricated by means of PL, SL and IJP. However, optofluidics that enables fusing fluid with light here presents a novel emerging field of science and engineering for different applications in recent years. As demonstrated in Figure 6.13, the LED light with three different colors is shone and transported within the structured (*e.g.*, rectangular and circular) guide made of OrmoClear polymer.[50] This photosensitive polymeric liquid core is initially printed with the outer shell of a viscoelastic aqueous Pluronic F127 copolymer solutions, effectively forming the core–shell filamentary structure. After photocuring with ultraviolet light, the F127 copolymer is removed, and the OrmoClear polymer is solidified and released. It shows optical losses of 0.2–0.5 dB cm$^{-1}$ for the printed waveguides with a radius of curvature $R = 4$–20 mm.

Accordingly, the optofluidics may open a route to better visualization of the materials within fluidic channels, in particular for those biological molecules carried by liquid in biochips. In other words, the lighting of materials offers

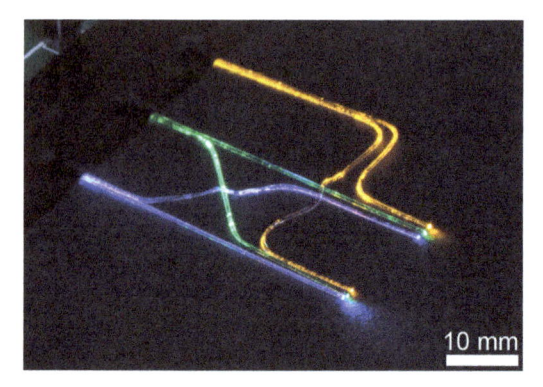

**Figure 6.13** Visible color LED light guided from one side to the other through the UV-cured OrmoClear liquid formation removed from the fugitive shells: a photograph of six-waveguide network coupled to three different colors of LED light[50] (reproduced with permission from John Wiley and Sons).

good visibility for real-time observation, in which biological information is of much interest for these medical devices, as discussed below.

### 6.4.5 Medical Devices

There are many medical devices including the microsensors and actuators, which are intimately associated with microfluidics, as mentioned earlier. Nevertheless, the fabricated microneedles are most elementary ones for medical life applications among them.[13] Transdermal drug delivery can be carried out through the hollow or solid microneedles that are inserted into the micrometer layers of skin.[51] In addition, here the fluorescence imaging method can generally provide good visualization of the targeted materials for measurement of micropattern.[52] Hence, as illustrated in Figure 6.14, the Ormocer cylindrical pillars made of polymers contain neuroblast-like cells observed by fluorescence imaging.[53] The green cytoskeleton and blue nuclei are clearly colored in the fluorescent micrographs. The microneedle-like structures[53,54] are directly fabricated through induced polymerization of the biomaterials by a LS scheme, as defined previously. Other means including PL, SL and IJP schemes could be applied for different substrates of materials as well.

Likewise, more medical devices incorporating three major scientific fields of biologics, optics and fluidics (BOF) may evolve into complex systems, which are composed of the interplay between various lights and biomaterials delivered by water-based liquids. In this case, it can present therefore a promising application of biological engineering in the near future.

### 6.4.6 Miscellaneous

In addition to those described previously, there are many diverse structures or components proposed for similar applications including blasted holes,[55]

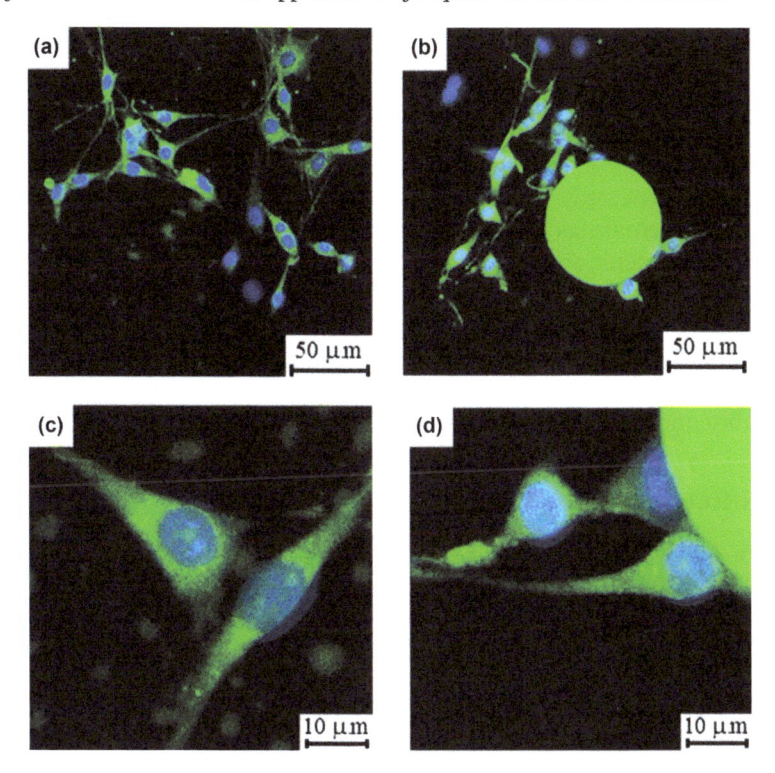

**Figure 6.14** Fluorescent micrographs at several different magnifications of B35 neuroblast-like cells on an Ormocer pillar 48 h after seeding, where the cytoskeleton and nuclei are denoted by green and blue colors, respectively: (a), (b) at 50-micrometer scale and (c), (d) at 10-micrometer scale[53] (reproduced with permission from Elsevier B.V.).

microstamper,[56] suspended structures,[57] ball valves,[58] nanocomposite gearwheels,[59] and organic thin-film-transistors (TFTs).[60] Moreover, most of them are fabricated in principle by combinations of the schemes as mentioned, although different materials such as powders,[55,57,58] polystyrene,[56] and SU8-resist[59] are especially used in those applications. In addition, complex-structured devices, such as self-assembled monolayers (SAMs)[60] and total ossicular replacement prostheses (TORPs),[61] are developed using similar approaches. These schemes can be applied with a variety of specific materials and complicated functions within the complicated devices.[62–64]

## 6.5 Conclusion and Outlook

In brief, photosensitive liquid materials that interplay with photonic energy have been explored and developed successfully over fifty years. At present, these materials play a significant role in a variety of microfabrication schemes including photolithography (PL), soft lithography (SL), light stereolithography (SL) and inkjet printing (IJP). Using those schemes, many components and/or devices such as microactuators, microsensors, microfluidic components, optical components and medical devices are designed and produced in industrial applications. Moreover, incorporating different schemes and materials, it is able to generate more complicated structures for multifunctional devices in the future.

As can be expected, the research and development of photosensitive materials will remain as one fundamental of many core technologies at the nano- to micrometer scale in the following decade. Among current engineering and science, a great diversity of materials and devices incorporating the interplay of mechanics, electronics, fluidics, optics and biologics together may evolve in different applications. As a whole, there is no doubt that this current trend of combinational technologies is directed towards more novel fabrication schemes and smaller devices than ever.

# References

1. T. A. Ameel, R. O. Warrington, R. S. Wegeng and M. K. Drost, *Energy Convers. Manage.*, 1997, **38**, 969.
2. A. Ovsianikov, J. Viertl, B. Chichkov, M. Oubaha, B. MacCraith, I. Sakellari, A. Giakoumaki, D. Gray, M. Vamvakaki, M. Farsari and C. Fotaki, *ACS Nano*, 2008, **2**, 2257.
3. J.-K. Chen and C.-J. Chang, *Materials*, 2014, **7**, 805.
4. K. E. Petersen, *Proc. IEEE*, 1982, **70**, 420.
5. W. Zhang, G. Meng and H. Li, *Microelectron. Reliab.*, 2005, **45**, 1230.
6. A. B. Frazier, R. O. Warrington and C. Friedrich, *IEEE Trans. Ind. Electron.*, 1995, **42**, 423.
7. M. R. Gongora-Rubio, P. Espinoza-Vallejos, L. Sola-Laguna and J. J. Santiago-Aviles, *Sens. Actuators, A*, 2001, **89**, 222.
8. A. Bertsch, S. Zissi, J. Y. Jezequel, S. Corbel and J. C. Andre, *Microsyst. Technol.*, 1997, **3**, 42.
9. S. Maruo and K. Ikuta, *Sens. Actuators, A*, 2002, **100**, 70.
10. N. T. Nguyen, *Flow Meas. Instrum.*, 1997, **8**, 7.
11. C.-T. Chen, K.-H. Wu, C.-F. Lu and F. Shieh, *J. Micromech. Microeng.*, 2010, **20**, 055004.
12. H. Schmidst, D. Yin, J. P. Barber and A. R. Hawkins, *IEEE J. Sel. Top. Quantum*, 2005, **11**, 519.
13. S. Henry, D. V. McAllister, M. G. Allen and M. R. Prausnitz, *J. Pharm. Sci.*, 1998, **87**, 922.
14. C. K. Chua, S. M. Chou and T. S. Wong, *Int. J. Adv. Manuf. Technol.*, 1998, **14**, 146.
15. A. Bertsch, P. Bernhard, C. Vogt and P. Renaurd, *Rapid Prototyping J.*, 2000, **6**, 259.
16. A. Ovisanikov, A. Ostendorf and B. N. Chichkov, *Appl. Surf. Sci.*, 2007, **253**, 6599.
17. M. Malinauskas, V. Purlys, M. Rutkauskas, A. Gaidukeviciute and R. Gadonas, *Lith. J. Phys.*, 2010, **50**, 201.

18. A. Ovisanikov, M. Malinauskas, S. Schlie, B. Chichkov, S. Gittard, R. Narayan, M. Lobler, K. Sternberg, K.-P. Schmitz and A. Haverich, *Acta Biomater.*, 2011, **7**, 967.
19. D. S. Correa, M. R. Cardoso, V. Tribuzi, L. Misogyuti and C. R. Mendonca, *IEEE J. Sel. Top. Quantum*, 2012, **18**, 176.
20. R. M. Nyffenegger and R. M. Penner, *Chem. Rev.*, 1997, **97**, 1195.
21. S. K. Chakarvarti and J. Vetter, *Radiat. Meas.*, 1998, **29**, 149.
22. C. R. Chatwin, M. Farsari, S. Huang, M. I. Heywood, R. C. D. Young, P. M. Birch, F. Claret-Tournier and J. D. Richardson, *Int. J. Adv. Manuf. Technol.*, 1999, **15**, 281.
23. S. Monneret, C. Provin, H. Le Gall and S. Corbel, *Microsyst. Technol.*, 2002, **8**, 364.
24. S. Yang, M. Megens, J. Aizenberg, P. Wiltzius, P. M. Chaikin and W. B. Russel, *Chem. Mater.*, 2002, **7**, 2831.
25. K. Obata, A. El-Tamer, L. Koch, U. Hinze and B. N. Chichkov, *Light: Sci. App.*, 2013, **2**, e116, DOI: 10.1038/lsa.2013.72.
26. H. Yang and S.-W. Kang, *Int. J. Mach. Tool Manuf.*, 2000, **40**, 1065.
27. E. Belloy, A. Sayah and M. A. M. Gijs, *Sens. Actuators, A*, 2001, **92**, 358.
28. S. Brittan, K. Paul, X.-M. Zhao and G. Whitesides, *Phys. World*, 1998, **May**, 31.
29. Y. Xia and G. M. Whitesides, *Angew. Chem., Int. Ed.*, 1998, **37**, 550.
30. J. Hu, R. G. Beck, R. M. Westervelt and G. M. Whitesides, *Adv. Mater.*, 1998, **10**, 574.
31. L. J. Guo, *J. Phys. D: Appl. Phys.*, 2004, **37**, R123.
32. D. J. Lorang, D. Tanaka, C. M. Spadaccini, K. A. Rose, N. J. Cherepy and J. A. Lewis, *Adv. Mater.*, 2011, **23**, 5055.
33. C.-T. Chen and H.-P. Huang, *J. Micromech. Microeng.*, 2013, **23**, 095009.
34. C. A. Coenjarts and C. K. Ober, *Chem. Mater.*, 2004, **16**, 5556.
35. M. Farsari, G. Filippidis, K. Sambani, T. S. Drakakis and C. Fotakis, *J. Photochem. Photobiol., A*, 2006, **181**, 132.
36. A. Ovisanikov and B. N. Chichkov, *Int. J. Appl. Ceram. Technol.*, 2007, **4**, 22.
37. A. Ovsianikov, B. Chichkov, O. Adunka, H. Pillsbury, A. Doraiswamy and R. J. Narayan, *Appl. Surf. Sci.*, 2007, **253**, 6603.
38. E. Stankevicius, M. Gedvilas, B. Voisiat, M. Malinauskas and G. Raciukaitis, *Lith. J. Phys.*, 2013, **53**, 227.
39. S. Turunen, E. Kapyla, M. Lahteenmaki, L. Yla-Outinen, S. Narkilahti and M. Kellomaki, *Opt. Laser Eng.*, 2014, **55**, 197.
40. R. Lipowsky, P. Lenz and P. S. Swain, *Colloids Surf., A*, 2000, **161**, 3.
41. Y. Xia, D. Qin and Y. Yin, *Curr. Opin. Colloid Interface Sci.*, 2001, **6**, 54.
42. C.-T. Chen and C.-T. Chuang, *Micro Nano Lett.*, 2012, **7**, 733.
43. E. Sutanto, Y. Tan, M. S. Onses, B. T. Cunningham and A. Alleyne, *Manuf. Lett.*, 2014, **2**, 4.
44. Y. Lu and S. C. Chen, *Adv. Drug Delivery Rev.*, 2004, **56**, 1621.
45. L. Hou, J. Zhang, N. Smith, J. Yang and J. Heikenfeld, *J. Micromech. Microeng.*, 2010, **20**, 015044.

46. B. P. Corgier and D. Juncker, *Sens. Actuators, B*, 2011, **157**, 691.
47. E. Leclerc, Y. Sakai and T. Fujii, *Biochem. Eng. J.*, 2004, **20**, 143.
48. E. Leclerc, K. S. Furukawa, F. Miyata, Y. Sakai, T. Ushida and T. Fujii, *Biomaterials*, 2004, **25**, 4683.
49. C.-T. Chen and K.-Z. Tu, *Sens. Actuators, A*, 2012, **188**, 367.
50. D. J. Lorang, D. Tanaka, C. M. Spadaccin, K. A. Rose, N. J. Cherepy and J. A. Lewis, *Adv. Mater.*, 2011, **23**, 5055.
51. R. F. Pereira, C. C. Barrias, P. L. Granja and P. J. Bartolo, *Nanomedicine*, 2013, **8**, 603.
52. H. Zhu, X. Zhou, F. Su, Y. Tian, S. Ashili, M. R. Holl and D. R. Meldrum, *Sens. Actuators, B*, 2012, **173**, 817.
53. A. Doraiswamy, C. Jin, R. J. Narayan, P. Mageswaran, P. Mente, R. Modi, R. Auyeung, D. B. Chrisey, A. Ovsianikov and B. Chichkov, *Acta Biomater.*, 2006, **2**, 267.
54. E. Belloy, A. Sayah and M. A. M. Gijs, *Sens. Actuators, A*, 2001, **92**, 358.
55. Z. Zhang, Z. Wang, R. Xing and Y. Han, *Polymer*, 2003, **44**, 3737.
56. A.-G. Pawlowski, E. Belloy, A. Sayah and M. A. M. Gijs, *Microelectron. Eng.*, 2003, **67–68**, 557.
57. C. Yamahata, F. Lacharme, Y. Burri and M. A. M. Gijs, *Sens. Actuators, B*, 2005, **110**, 1.
58. S. Jiguet, M. Judelewicz, S. Mischler, A. Bertch and P. Renaud, *Microelectron. Eng.*, 2006, **83**, 1273.
59. A. Benor, A. Hoppe, V. Wagner and D. Knipp, *Thin Solid Films*, 2007, **515**, 7679.
60. A. Ovisanikov and B. N. Chichkov, *Int. J. Appl. Ceram. Technol.*, 2007, **4**, 22.
61. H. C. Chiamori, J. W. Brown, E. V. Adhiprakasha, E. T. Hantsoo, J. B. Straalsund, N. A. Melosh and B. L. Pruitt, *Microelectron. J.*, 2008, **39**, 22.
62. G.-W. Huang and C.-T. Chen, *Micro Nano Lett.*, 2012, 7, 1080.
63. H. Durou, D. Pech, D. Colin, P. Simon, P.-L. Taberna and M. Brunet, *Microsyst. Technol.*, 2012, **18**, 467.
64. J. Horak, C. Dincer, H. Bakirci and G. Urban, *Sens. Actuators, B*, 2014, **191**, 813.

CHAPTER 7
# *UV-Cured Functional Coatings*

M. SANGERMANO,*[a] I. ROPPOLO[b] AND M. MESSORI[c]

[a] Politecnico di Torino, Dipartimento di Scienza Applicata e Tecnologia, C.so Duca degli Abruzzi 24, 10129, Torino, Italy; [b] Istituto Italiano di Tecnologia, Center for Space Human Robotics, Corso Trento 21, 10129, Torino, Italy; [c] Università di Modena e Reggio Emilia, Dipartimento di Ingegneria "Enzo Ferrari", Via Vignolese 905/A, 41125, Modena, Italy
*Email: marco.sangermano@polito.it

## 7.1 Introduction

The UV-induced polymerization of multifunctional monomers has found a large number of industrial applications,[1] mainly in the production of films, inks and coatings on a variety of substrates including paper, metal and wood. Moreover, it has demonstrated scope for more high-tech applications such as coating of optical fibers and fabrication of printed circuit boards.

Part of the reason for the growing importance of UV-curing techniques, both in industrial and academic research, is a peculiar characteristic[2,3] that induces fast transformation of a liquid monomer into a solid polymer film having distinctive physical–chemical and mechanical properties. It can be considered environmental friendly owing to the solvent-free methodology, and is usually carried out at room temperature, thus conferring added energy-saving advantages.

During a UV-curing process, radical or cationic species are generated by the interaction of UV light with a suitable photoinitiator, and subsequently a radical chain-growth polymerization or a cationic ring-opening polymerization process take place. In this chapter we will outline a literature survey where UV-curing techniques have been used for the production of functional coatings.

---

RSC Smart Materials No. 13
Photocured Materials
Edited by Atul Tiwari and Alexander Polykarpov
© The Royal Society of Chemistry 2015
Published by the Royal Society of Chemistry, www.rsc.org

Today, multifunctionality in a coating formulation is essential since coatings not only provide protection for the surface, but they should also provide additional functions such as high scratch resistance, electrical conductivity, high adhesion properties and so on.

## 7.2 Scratch-Resistant UV-Cured Coatings and Multifunctional Coatings

The achievement of scratch-resistant coatings could find important applications[4] in fields such as automotive gloss coatings,[5] optical lenses,[6] and flooring varnishes.[7] There are two main ways to improve the scratch resistance of organic coatings, either by optimizing the polymer component by using more rigid structures, or by adding inorganic fillers with the achievement of organic–inorganic hybrid materials.

The inorganic filler can be simply embedded into the polymeric matrix having only weak bond interactions (hydrogen bonds, van der Waals interactions, *etc*.) responsible for the cohesion to the whole structure; these materials are classified as belonging to Class I organic–inorganic hybrids. In Class II hybrid materials, the two phases are linked together through strong chemical bonds (covalent or ionic-covalent bonds).[8] Several synthetic strategies can be adopted for the preparation of organic–inorganic hybrid materials according to the so-called top-down or bottom-up approaches, respectively. The top-down approach corresponds to the assembling or the dispersion of well-defined nanobuilding blocks that consist of perfectly calibrated preformed objects that keep their integrity in the final material, while the bottom-up approach generally corresponds to very convenient soft chemistry based routes including conventional sol–gel chemistry.[9]

Inorganic oxides with particle size under 100 nm represent a promising group of fillers for the top-down approach and they showed an important enhancement on mechanical behavior of the crosslinked polymers already at very low loading.[10] Nanoparticles (NPs) such as $SiO_2$, $ZrO_2$ and $TiO_2$ have been embedded in photocurable resins.[11,12] Among them, $SiO_2$ has been the first type of NPs produced and studied in several polymer systems. The photocuring behavior of UV-curable acrylic-based coating systems filled with silica NPs have been studied by photo-DSC and FTIR spectroscopy.[13] It was shown that the cure rate increased gradually by increasing the silica content in the photocurable formulation, whereas it decreased above the concentration of 10 wt%. These results suggest that the presence of silica NPs below 10 wt% can accelerate the curing process due to the synergistic effect of silica NPs as an effective flow or diffusion-aid agent during the photopolymerization process, which improve the mobility of propagating chains, as well as suggest that $SiO_2$ NPs can lengthen the path of the UV-light by partial scattering or reflection. However, a decrease of photocuring reactivity occurred above 10 wt% of silica content, which is attributable to aggregation of silica NPs due to their high surface energy.

Silica NPs were employed, in the range between 5 and 15 wt%, in order to obtain organic–inorganic hybrid coatings *via* cationic UV curing of an epoxy based system.[14] The influence of the presence of $SiO_2$ on the rate of polymerization was investigated by real-time FT-IR. The silica nanofiller induced both a bulk and a surface modification of UV-cured coatings with an increase in $T_g$ values, modulus values and surface hardness by increasing the amount of silica into the photocurable monomer. Similarly, epoxy-based coatings containing $TiO_2$ NPs were prepared.[15]

One of the main limits of the top-down approach is that commercially available powders often contain aggregates that cannot be destroyed during processing and therefore deteriorate some important properties like transparency. Furthermore, the high volume fraction content will adversely modify the rheology of the formulation with an important deterioration in processability. In this respect, the bottom-up approach, by the *in situ* generation of the inorganic domains through a sol–gel process could represent an appealing strategy to avoid NP aggregation and the increase of the formulation viscosity. Silica, titania and zirconia inorganic domains were formed within UV-curable epoxy matrices using this method.[15–17] These coatings showed improved mechanical behavior and achieved high surface hardness.

In order to compare the hybrid materials obtained either *via* a top-down or a bottom-up approach, photocured acrylic coatings (bisphenol-A ethoxylate dimethacrylate–methacryloyloxypropyl-trimethoxysilane mixture, BM) containing silica inorganic domains were prepared by dispersing preformed silica NPs or by *in situ* generation *via* a sol–gel process, and their scratch behavior compared.[18] Excellent scratch-resistant coatings characterized by high critical load and high recovery were obtained by UV and sol–gel dual-curing process. On the contrary, coatings with very poor scratch resistance characterized by large plastic deformation, severe cracking and weak recovery were obtained by dispersing preformed NPs of silica into the acrylic resin.

The penetration depth and the residual depth patterns of the investigated UV-cured films are reported in Figure 7.1. The introduction of preformed silica NPs (samples BM05-$SiO_2$ and BM10-$SiO_2$, with a silica content of 5 and 10 wt%, respectively) produced a deterioration of scratch-resistance properties with a marked increase of both penetration and residual depth. This could be due both to the formation of high stress concentration around the particles as well as to a weak compatibility between the filler and the matrix. These two phenomena could lead to debonding and void formation close to the particles with a deterioration of the scratch-resistance behavior instead of an improvement. On the contrary, in the case of samples reinforced with *in situ* generated silica (samples BM15 and BM30, with a silica content of 15 and 30 wt%, respectively), a consistent improvement of penetration resistance by increasing the content of the silica precursor (tetraethoxysilane, TEOS) was noted in terms of penetration depth and residual depth. The better results for dual-cured hybrid coatings with respect to the coatings obtained with dispersed preformed silica NPs are well evidenced also by the

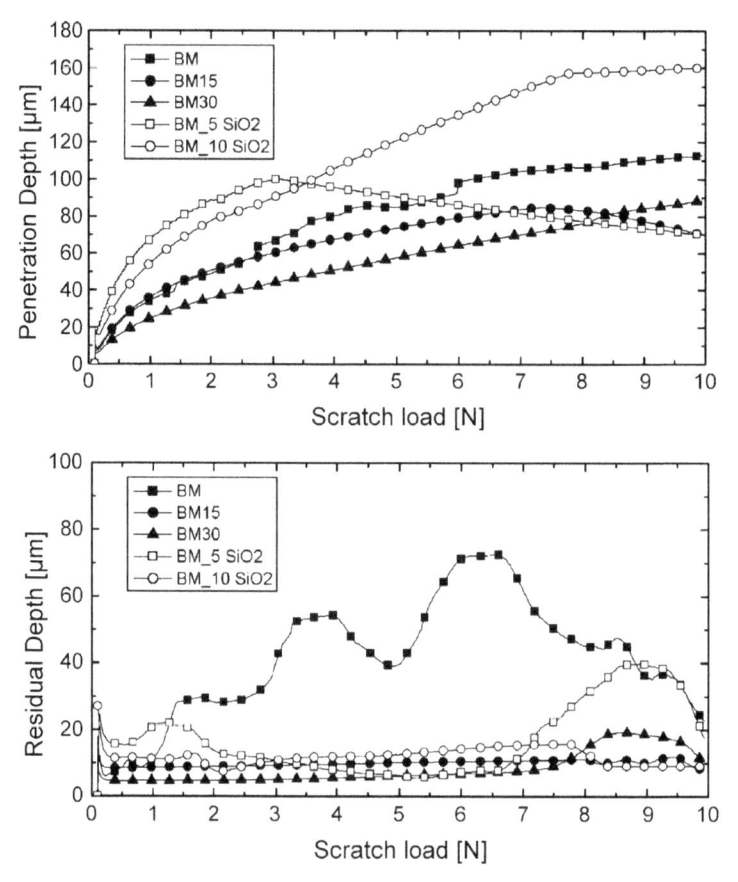

**Figure 7.1** Penetration depth and residual depth in a progressive load scratch test (reprinted from ref. 18, with permission).

analysis of the optical images collected after the scratch test reported in Figure 7.2.

Multifunctional coatings were also investigated by enhancing at the same time the scratch resistance and the conductivity. Transparent UV-cured antistatic hybrid coatings, with improved scratch resistance, have been prepared by the sol–gel method, based on commercially available acrylic resin.[19] The addition of TEOS and silane-functionalized compatibilizer greatly improved the abrasive resistance of the pure acrylic resin, while the addition of an intrinsically conductive polymer allowed formulation of the coating with conductivities in the antistatic region. Similarly, organic–inorganic antistatic hybrid acrylic coatings were prepared by adding trialkoxy-silyl ammonium salt to the photocurable formulations in order to introduce an antistatic additive that could be covalently linked to the hybrid network through a co-condensation reaction involving the alkoxy groups.[20] When the samples were cured under a nitrogen atmosphere, a complete conversion of acrylic double bonds was achieved after 90 s of irradiation

UV-Cured Functional Coatings 125

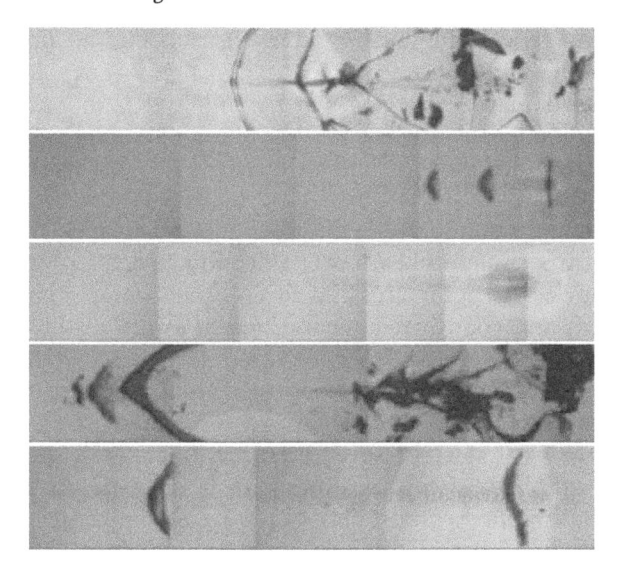

**Figure 7.2** Optical micrographs of the damage after scratch test for different coatings (from the top: BM, BM15, BM30, BM05-SiO$_2$ and BM10-SiO$_2$ (reprinted from ref. 18, with permission).

both for the acrylic resin and the formulations containing the ammonium salt. A consistent improvement of penetration resistance by increasing alkoxy-ammonium salt content was noted. The inorganic component and antistatic additives increase the $\varepsilon'$ and $\sigma_{AC}$ values of the hybrid coatings and decrease the resistivity ones, showing its efficiency for increasing the antistatic properties of coatings, improved with respect to the pure acrylic resin.

Another example of multifunctionality achieved by filler dispersion is reported in the literature where a simple, cheap and easily scaled-up process for obtaining a multifunctional coating with both IR-reflective and hydrophobic properties is described.[21] The system is based on a UV-curable epoxy coating containing IR-reflective fillers and a silicone additive to modify the surface properties with a good hydrophobicity enhancement. These coatings can be applied on solar devices in order to limit their overheating under operating conditions. The optical properties of the obtained coatings have been evaluated by UV-vis-NIR spectroscopy showing that the films act as good "shields" in the NIR region, giving rise to a reduction of the absorbed heat and of the working temperature of the cells. A pronounced increase of hydrophobicity for the films containing the silicone additive was also observed.

## 7.3 Conductive UV-Cured Coatings

Polymer films are a promising candidate for the realization of flexible electronics, due to their easy processability, low cost and availability.[22] The main limit in their application is related to the very high surface resistivity.

There are many reports in the literature focused on the preparation of conductive polymeric coatings.

In general, two main strategies can be followed to obtain films with resistivity in the range $10^4$–$10^3$ Ω cm: (i) addition of conductive fillers up to percolation threshold into polymeric matrix,[23–25] (ii) use of intrinsic conductive polymers.[26]

One way to obtain conductive coatings by photochemical means is the dispersion of carbon nanotubes (CNT) as conductive fillers in UV-curable epoxy formulations.[27] Highly crosslinked epoxy networks characterized by high gel content and good mechanical properties were readily obtained by UV irradiation. The surface resistivity of cured films was investigated and it was shown that the material was transformed from an electrical insulating into a dissipative one already at 0.025 wt% CNT content.

More recently, the effect of several carbon fillers, exfoliated graphite, functionalized graphene sheets, multiwalled carbon nanotubes (MWCNTs), and oxidized multiwalled carbon nanotubes, were compared on the curing process and physical properties of a cationically photocurable epoxy resin.[28] The DC conductivity for the nanocomposites showed an important increase in the presence of the carbon fillers. All the systems showed an electrical percolation threshold, but with MWCNTs it was attained at a lower concentration (<0.1 wt%, see Figure 7.3).

An alternative approach concerns the concomitant redox and polymerization processes by using certain metal salts such as silver and gold salts at

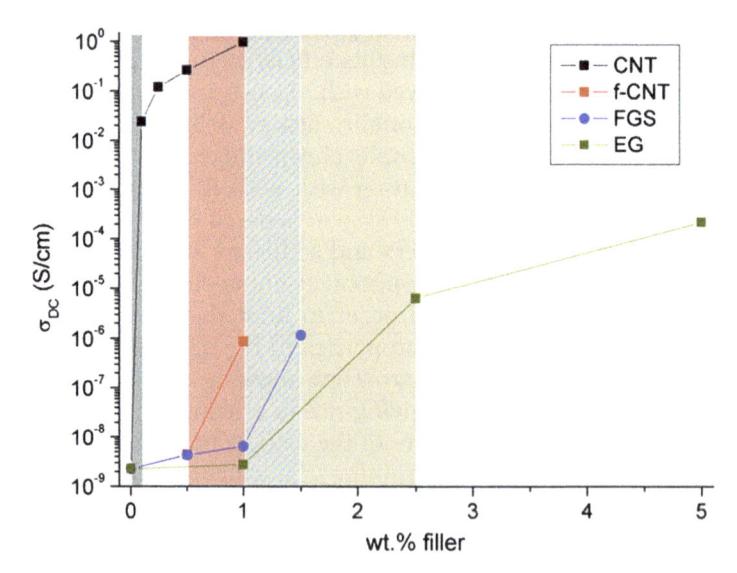

**Figure 7.3** Values of the DC electrical conductivity of the epoxy nanocomposites as a function of the nanofiller content (exfoliated graphite, EG; functionalized graphene sheets, FGS; multiwalled carbon nanotubes, CNT; oxidized multiwalled carbon nanotubes, f-CNT) (reprinted from ref. 28, with permission).

their higher oxidation states and free-radical photoinitiators. The approach was successfully applied[29–33] to both free-radical and cationic systems based on acrylates and epoxides, respectively. In the process, while metal NPs were formed by electron transfer from the photochemically generated free radical to the salts, polymerizations were initiated by free radicals or carbocations thus formed depending on the type of monomers used in the formulation. The resistivity *versus* metal precursor content was explained on the basis of a specific percolation model, valid for subthreshold 3-dimensional systems, where the conductive filler in the form of spherical particles is dispersed into a continuous dissipative matrix.[34]

In another study, *in situ* photoinduced oxidation of pyrrole to polypyrrole was successfully employed to form an acrylic UV-crosslinked network characterized by high surface conductivity.[35]

Similarly, conductive epoxy-polythiophene network films were obtained by simultaneous photoinduced step-growth and cationic ring-opening polymerization processes, as reported by Sangermano *et al.*[36] The conductive polythiophene component was formed by the electron-transfer reaction between phenyliodinium radical cations and thiophene followed by proton release and successive coupling reactions. In a parallel process, the protons started cationic polymerization of the bis-epoxy monomer to yield cross-linked epoxy network conductive coatings.

## 7.4 Photoluminescent UV-Cured Coatings

Luminescent materials have long attracted attention due to their wide variety of applications such as light-emitting diodes, lasers, fluorescence sensors, display devices and biological probes.[37–40] UV-curing technology could be an interesting technique for the preparation of photoluminescent films.

Transparent light-emitting hybrid materials were produced by UV curing of acrylic resins containing TEOS as silica precursors and a photoluminescent copper iodide cluster $[Cu_4I_4L_4]$.[41,42] Optical measurements were performed on UV-cured films, showing high transparency of the materials up to a content of 30 wt% of TEOS and, most of all, a bright luminescence with a maximum of emission centered around 565 nm (yellow-orange). The emission spectra for the investigated films are reported in Figure 7.4.

The possibility was demonstrated to obtain new advanced materials in which functional properties such as photoluminescence and scratch resistance were successfully conjugated in a hybrid film.

Since the emission depends on the rigidity of the hosting matrix, the molecular copper iodide cluster was dispersed in a series of photopolymerizable acrylic matrices (mixtures of bisphenol-A ethoxylate diacrylate BEDA and bisphenol-A ethoxylate dimethacrylate BEMA) exhibiting different glass-transition temperatures.[43] The photoluminescence characterization of the composite materials showed a red shift of the emission band by decreasing the $T_g$ value of the polymeric matrix. A change of the emitted color ranging from yellow to orange is thus observed (see Figure 7.5). This effect is

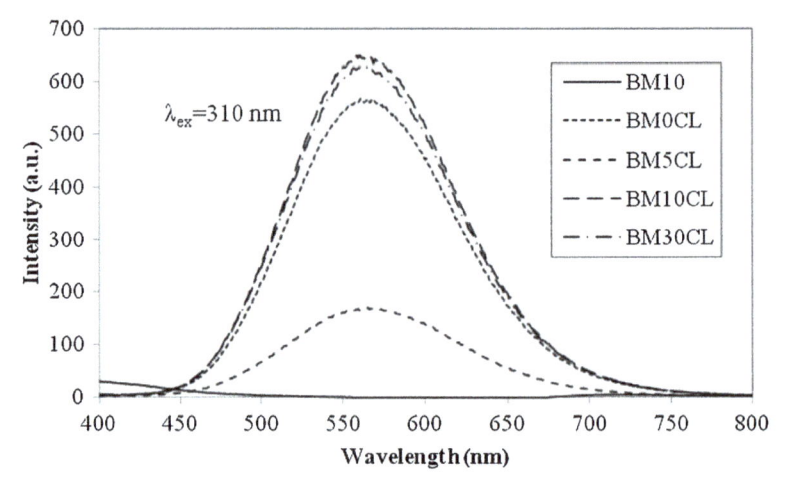

**Figure 7.4** Emission spectra of hybrid dual-cured films with and without 2 wt% of copper iodide cluster (BMxCL and BMx, respectively) and with different TEOS content (x indicates the weight content of TEOS in the initial formulation) (reprinted from ref. 41, with permission).

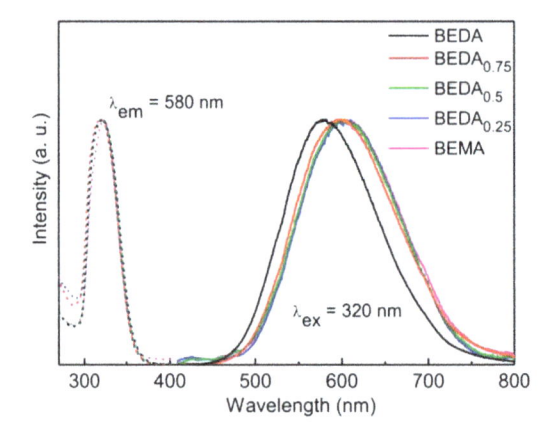

**Figure 7.5** Luminescence spectra of UV-cured films with 2 wt% of copper iodide cluster (BEDA, BEDAn and BEMA) with normalized emission (solid lines) and corresponding excitation spectra (dotted lines) (n indicates the weight fraction of BEDA in BEDA–BEMA mixtures) (reprinted from ref. 43, with permission).

based on the luminescence rigidochromism of the copper iodide cluster that emits at lower energy by increasing the medium flexibility.

Light/emitting composite materials were produced *via* UV-curing by embedding $Gd_2O_3:Eu^{3+}$ nanorods in different epoxy matrices obtained through cationic UV-curing.[44] The effects of the polymer matrix and concentration of the photoactive filler ($Gd_2O_3:Eu^{3+}$ nanorods) on the $Eu^{3+}$ luminescence properties were studied. The composite materials present original photoluminescence properties and demonstrate the potential of UV-curing as a technique to develop smart photoactive coatings.

## 7.5 UV-Cured Coatings in Photocatalysis

Heterogeneous photocatalysis is largely employed to achieve a complete oxidation of organic and inorganic species.[45–47] It takes advantage of some semiconductors, which can be used as photocatalysts both suspended in the water effluent to be treated or immobilized on various types of supports. $TiO_2$ is by far the most preferred material thanks to its high photocatalytic activity, chemical/photocorrosion stability and nontoxicity.[48] If titania will maintain its photocatalytic activity even when dispersed into a polymeric coating, this could be an interesting field of applications in coating industry.

For this, the UV-curing technique was chosen because of its peculiarity and also because of its increasing importance in coatings applications. By cationic UV-curing of epoxy resin new hybrid catalyst coating has been synthesized encapsulating $TiO_2$ NPs into the polymeric matrix.[49] It was shown that the presence of titania NPs did not significantly influence the curing process. The photocatalytic efficiency of these new catalysts was evaluated by studying the degradation of some organic compounds under UV-light. Methylene blue was employed as a model molecule to investigate the photoactivity toward organic molecules directly adsorbed on the coating surface. Under irradiation, complete dye degradation was achieved within 90 min (see Figure 7.6). Phenol and 3,5-dichlorophenol were used as target molecules for studying photoactivity toward organic compounds dissolved in aqueous solution. Also in these cases, complete degradation of the organic compounds was achieved.

Graphene oxide (GOx) can also be classified as a semiconductor, and thanks to this behavior it can be used in order to activate photo-oxidative processes in a similar way as happens with the well-known $TiO_2$. In literature, mixture of GOx and $TiO_2$ were already proven to be photoactive.[50] Therefore, new photocatalytic materials were prepared by dispersing GOx into a cycloaliphatic epoxy resin and the formulations were crosslinked by UV-irradiation.[51] The activity of these new photocatalytic materials was evaluated in aqueous phase; photoactivity was determined by following the photodegradation of phenol during irradiation in the presence of epoxy crosslinked films containing GOx. The target molecule was degraded completely in about 72 h of irradiation when the crosslinked film contains 3 phr of GOx. The disappearance of the target molecule could occur through two concomitant effects: the photodegradation and the adsorption process: both phenomena will be useful for eliminating pollutants from water.

Also, epoxy films containing single-wall carbon nanotubes in the range between 0.1 and 0.3 wt% were prepared by means of UV-induced polymerization and the achieved materials were used as photocatalysts on adsorbed, aqueous, and gas phases. The efficiency in aqueous-phase photodegradation was determined by following the photodegradation of phenol and

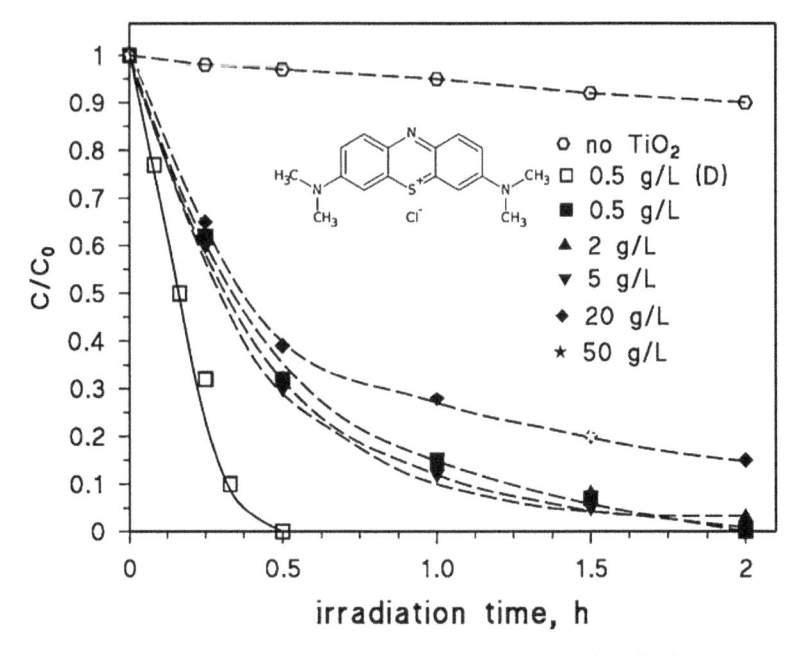

**Figure 7.6** Methylene blue degradation as a function of irradiation time in the presence of dispersed TiO$_2$ (D) or coatings containing increasing TiO$_2$ concentrations (reprinted from ref. 49, with permission).

3,5-dichlorophenol. Both molecules were degraded completely in about 72 h of irradiation. The photo-oxidative activity in gas phase was evaluated using nitrogen monoxide as a probe molecule: a lower efficiency in the conversion of gas-phase pollutants with respect to an absorbed one or pollutants in liquid phase was evidenced, but it was shown that the new proposed catalyst material is suitable for both oxidative (degradation of phenol or methylene blue) and reductive reactions. Once again, this is a clear example of multifunctionality, where the filler will perform both a mechanical reinforcement on the crosslinked film while operating as a photocatalyst.

## 7.6 Conclusions

This chapter aimed to review the use of UV-curing technology for the preparation of multifunctional coatings. Scratch-resistant UV-cured coatings were achieved either *via* a top-down or a bottom-up approach. The key challenge is the homogeneous dispersion and distribution of the inorganic domains in order to get highly scratch-resistant coatings without compromising the durability of the crosslinked material. Multifunctionality can be achieved if together with the inorganic domains antistatic additives are added to the photocurable formulation, obtaining crosslinked films characterized by high scratch resistance and antistatic properties. Conductive UV-cured films can be obtained by adding conductive fillers up to a percolative content: both CNTs and functionalized graphene sheets were added to a UV-curable formulation, all the cured systems showed an electrical percolation threshold. Good conductivity was also achieved by the *in situ* metal NPs formation through a photoreduction process. Photoluminescent UV-cured films were prepared by dispersing photoluminescent copper iodide clusters into a UV-curable acrylic resin. As an alternative, light-emitting composite materials were produced *via* UV curing by embedding $Gd_2O_3$:$Eu^{3+}$ nanorods in different epoxy matrices obtained through cationic UV curing. Finally, titania, GOx and CNTs were used as photocatalysts showing a good photocatalytic efficiency when dispersed into UV-cured films.

In brief, the UV-curing process has been demonstrated to be a good technique for the preparation of multifunctional coatings that could find advanced and innovative applications in fields such as automotive gloss coatings, flooring varnish, scratch-resistant coatings, flexible electronics, conductive coatings, light-emitting diodes, lasers, fluorescence sensors, display devices and heterogeneous photocatalysts.

# References

1. P. Dufour, in *Radiation Curing in Polymer Science and Technology – Vol I: Fundamentals and Methods*, ed. J. P. Fouassier and J. F. Rabek, Elsevier Science Publishers, London & New York, 1993, p. 1.
2. L. Schlegel and W. Schabel, in *Radiation Curing in Polymer Science and Technology – Vol I: Fundamentals and Methods*, ed. J. P. Fouassier and J. F. Rabek, Elsevier Science Publishers, London & New York, 1993, p. 119.
3. C. Decker, *Prog. Polym. Sci.*, 1996, **21**, 593–650.
4. L. Lin, G. S. Blackman and R. R. Matheson, *Prog. Org. Coat.*, 2000, **40**, 85–91.
5. U. Schulz, V. Wachtendorf, T. Klimmasch and P. Alers, *Prog. Org. Coat.*, 2001, **42**, 38–48.
6. G. Schottner, K. Rose and U. Posset, *J. Sol-Gel Sci. Technol.*, 2003, **27**, 71–79.
7. C. Flosbach and W. Schubert, *Prog. Org. Coat.*, 2001, **43**, 123–130.
8. F. Bauer, H. Ernst, U. Decker, M. Findeisen, H. J. Glasel, H. Langguth, E. Hartmann, R. Mehnert and C. Peuker, *Macromol. Chem. Phys.*, 2000, **201**, 2654–2659.
9. C. J. Brinker and G. W. Shrerer, *Sol-Gel Science: The Physics and Chemistry of Sol-Gel Processing*, Academic Press, San Diego, 1990.
10. C. Sanchez, G. J. D. A. Soler-Illia, F. Ribot, T. Lalot, C. R. Mayer and V. Cabuil, *Chem. Mater.*, 2001, **13**, 3061–3083.
11. H. J. Glasel, F. Bauer, H. Ernst, M. Findeisen, E. Hartmann, H. Langguth, R. Mehnert and R. Schubert, *Macromol. Chem. Phys.*, 2000, **201**, 2765–2770.
12. S. Sepeur, N. Kunze, B. Werner and H. Schmidt, *Thin Solid Films*, 1999, **351**, 216–219.

13. J. D. Cho, Y. B. Kim, H. T. Ju and J. W. Hong, *Macromol. Res.*, 2005, **13**, 362–365.
14. M. Sangermano, G. Malucelli, E. Amerio, A. Priola, E. Billi and G. Rizza, *Prog. Org. Coat.*, 2005, **54**, 134–138.
15. M. Sangermano, G. Malucelli, E. Amerio, R. Bongiovanni, A. Priola, A. Di Gianni, B. Voit and G. Rizza, *Macromol. Mater. Eng.*, 2006, **291**, 517–523.
16. E. Amerio, M. Sangermano, G. Malucelli, A. Priola and B. Voit, *Polymer*, 2005, **46**, 11241–11246.
17. M. Sangermano, B. Voit, F. Sordo, K. J. Eichhorn and G. Rizza, *Polymer*, 2008, **49**, 2018–2022.
18. E. Amerio, P. Fabbri, G. Malucelli, M. Messori, M. Sangermano and R. Taurino, *Prog. Org. Coat.*, 2008, **62**, 129–133.
19. M. E. L. Wouters, D. P. Wolfs, M. C. van der Linde, J. H. P. Hovens and A. H. A. Tinnemans, *Prog. Org. Coat.*, 2004, **51**, 312–320.
20. M. Sangermano, D. Foix, G. Kortaberria and M. Messori, *Prog. Org. Coat.*, 2013, **76**, 1191–1196.
21. I. Roppolo, N. Shahzad, A. Sacco, E. Tresso and M. Sangermano, *Prog. Org. Coat.*, 2014, **77**, 458–462.
22. Q. D. Ling, D. J. Liaw, C. X. Zhu, D. S. H. Chan, E. T. Kang and K. G. Neoh, *Prog. Polym. Sci.*, 2008, **33**, 917–978.
23. A. Chiolerio, L. Vescovo and M. Sangermano, *Macromol. Chem. Phys.*, 2010, **211**, 2008–2016.
24. M. Karttunen, P. Ruuskanen, V. Pitkanen and W. M. Albers, *J. Electron. Mater.*, 2008, **37**, 951–954.
25. A. Chiolerio and M. Sangermano, *Mater. Sci. Eng., B*, 2012, **177**, 373–380.
26. K. Reuter, S. Kirchmeyer and A. Elschner, *Handbook of Thiophene-Based Materials*, John Wiley & Sons, Ltd, 2009, pp. 549–576.
27. M. Sangermano, S. Pegel, P. Potschke and B. Voit, *Macromol. Rapid Commun.*, 2008, **29**, 396–400.
28. M. Martin-Gallego, M. Hernandez, V. Lorenzo, R. Verdejo, M. A. Lopez-Manchado and M. Sangermano, *Polymer*, 2012, **53**, 1831–1838.
29. M. Sangermano, Y. Yagci and G. Rizza, *Macromolecules*, 2007, **40**, 8827–8829.
30. Y. Yagci, M. Sangermano and G. Rizza, *Polymer*, 2008, **49**, 5195–5198.
31. Y. Yagci, M. Sangermano and G. Rizza, *Chem. Commun.*, 2008, 2771–2773.
32. M. Sangermano, I. Roppolo, V. H. A. Camara, C. Dizman, S. Ates, L. Torun and Y. Yagci, *Macromol. Mater. Eng.*, 2011, **296**, 820–825.
33. Y. Yagci, O. Sahin, T. Ozturk, S. Marchi, S. Grassini and M. Sangermano, *React. Funct. Polym.*, 2011, **71**, 857–862.
34. M. Sangermano and A. Chiolerio, in *Nanoparticles Featuring Electromagnetic Properties: From Science to Engineering*, ed. P. A. A. Chiolerio, Research Signpost, Kerala, India, 2012, pp. 85–103.
35. V. S. Ijeri, J. R. Nair, C. Gerbaldi, R. S. Gonnelli, S. Bodoardo and R. M. Bongiovanni, *Soft Matter*, 2010, **6**, 4666–4668.
36. M. Sangermano, F. Sordo, A. Chiolerio and Y. Yagci, *Polymer*, 2013, **54**, 2077–2080.

37. T. Justel, H. Nikol and C. Ronda, *Angew. Chem., Int. Ed.*, 1998, **37**, 3085–3103.
38. J. C. G. Bunzli and C. Piguet, *Chem. Soc. Rev.*, 2005, **34**, 1048–1077.
39. B. Hotzer, I. L. Medintz and N. Hildebrandt, *Small*, 2012, **8**, 2297–2326.
40. G. Blasse and B. C. Grabmaier, *Luminescent Materials*, Springer, Berlin, 1994.
41. I. Roppolo, M. Messori, S. Perruchas, T. Gacoin, J. P. Boilot and M. Sangermano, *Macromol. Mater. Eng.*, 2012, **297**, 680–688.
42. I. Roppolo, E. Celasco, A. Fargues, A. Garcia, A. Revaux, G. Dantelle, F. Maroun, T. Gacoin, J. P. Boilot, M. Sangermano and S. Perruchas, *J. Mater. Chem.*, 2011, **21**, 19106–19113.
43. I. Roppolo, E. Celasco, M. Sangermano, A. Garcia, T. Gacoin, J. P. Boilot and S. Perruchas, *J. Mater. Chem. C*, 2013, **1**, 5725–5732.
44. I. Roppolo, M. L. Debasu, R. A. S. Ferreira, J. Rocha, L. D. Carlos and M. Sangermano, *Macromol. Mater. Eng.*, 2013, **298**, 181–189.
45. C. H. Kwon, H. M. Shin, J. H. Kim, W. S. Choi and K. H. Yoon, *Mater. Chem. Phys.*, 2004, **86**, 78–82.
46. F. Sayilkan, M. Asilturk, S. Erdemoglu, M. Akarsu, H. Sayilkan, M. Erdemoglu and E. Arpac, *Mater. Lett.*, 2006, **60**, 230–235.
47. P. S. Awati, S. V. Awate, P. P. Shah and V. Ramaswamy, *Catal. Commun.*, 2003, **4**, 393–400.
48. S. K. Lee, S. McIntyre and A. Mills, *J. Photochem. Photobiol., A*, 2004, **162**, 203–206.
49. P. Calza, L. Rigo and M. Sangermano, *Appl. Catal., B*, 2011, **106**, 657–663.
50. G. Williams, B. Seger and P. V. Kamat, *ACS Nano*, 2008, **2**, 1487–1491.
51. M. Sangermano, P. Calza and M. A. Lopez-Manchado, *J. Mater. Sci.*, 2013, **48**, 5204–5208.

CHAPTER 8

# Photoreactive Polymers For Microarray Chips

DI ZHOU,[a,b] PONNURENGAM MALLIAPPAN SIVAKUMAR,[c] TAE IL SON[c,d] AND YOSHIHIRO ITO*[a,c]

[a] Emergent Bioengineering Materials Research Team, RIKEN Center for Emergent Materials Science, 2-1 Hirosawa, Wako, Saitama 351-0198, Japan; [b] Jiangsu Key Laboratory of Advanced Functional Materials, School of Chemistry and Material Engineering, Changshu Institute of Technology, Jiangsu 215500, China; [c] Nano Medical Engineering Laboratory, RIKEN Advanced Science Institute, 2-1 Hirosawa, Wako, Saitama 351-0198, Japan; [d] Department of Biotechnology, Chung-Ang University, Anseong 456-756, Republic of Korea
*Email: y-ito@riken.jp

## 8.1 Introduction

Microarray technologies are a crucial tool for large-scale and high-throughput biological science and technology applications. These technologies are a typical result of the combination of biotechnological and microfabrication methods, and they allow the fast, easy, and parallel detection of thousands of addressable elements in a single experiment under identical conditions. Various biological interactions can be analyzed in terms of their interactomics, which include their genomics, proteomics, glycomics, cellomics, and metabolomics.

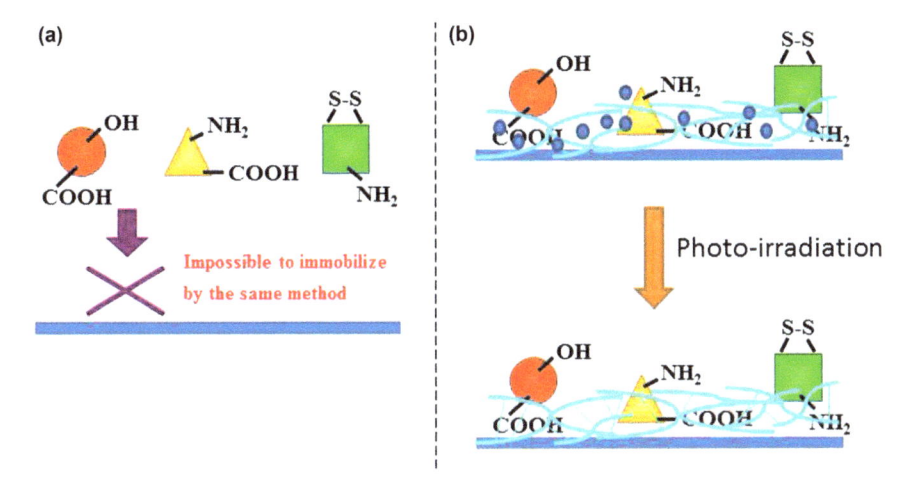

**Figure 8.1** (a) Schematic illustration of the difficulties associated with the covalent immobilization of different biological components on one chip using the same method. (b) Schematic illustration of the present novel photo-immobilization technique for the preparation of microarrays.
Reprinted with permission from ref. 1. Copyright (2014) John Wiley and Sons.

The microarray immobilization of biomolecules such as DNAs, peptides, and proteins on substrates has been achieved using several methods, including noncovalent adsorption (physical adsorption), and covalent attachment on chemically modified surfaces.[1] Although many types of microarray immobilization have been investigated, practical technologies have been developed only for DNA microarrays. One of the reasons for this is the difficulties associated with the immobilization of proteins. Because proteins have different functional groups, and because these groups differ in quantity and location from each other, it is very difficult to covalently immobilize different proteins, polysaccharides, antigens, and cells on one chip using the same method (Figure 8.1(a)). Furthermore, some biomolecules may not have enough functional groups, or their recognition sites may be used in the covalent immobilization process.

To overcome the difficulties associated with the covalent immobilization of different types of biological components using the same method, we developed a new photoimmobilization technique (Figure 8.1(b)) for the preparation of microarrays.[2] This photoimmobilization method made it possible to easily covalently immobilize various types of targets. Additionally, by using nonbiofouling polymers as matrices, it was possible to reduce the nonspecific interactions with the biological components. Because the nonbiofouling polymers did not nonspecifically interact with the biological components, they could be used to form an immobilization matrix to reduce the nonspecific interactions (N) and enhance the specific interactions (S), which resulted in a high S/N ratio (Figure 8.2).

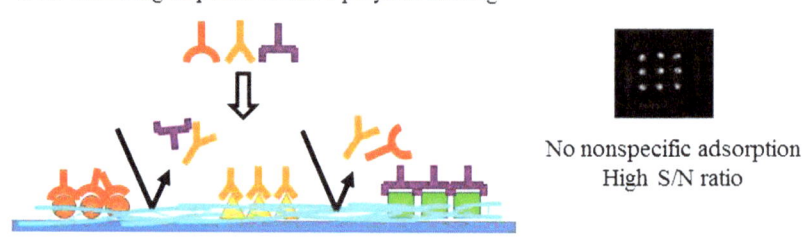

**Figure 8.2** (a) Reducing nonspecific interactions (N) and enhancing specific interactions (S) using the photoimmobilization technique presented here, resulting in a high S/N ratio. (b) Representative chemiluminescent images showed nonspecific adsorption and no nonspecific adsorption on non-biofouling polymer photoimmobilized surface.
Reprinted with permission from ref. 43. Copyright (2014) John Wiley and Sons.

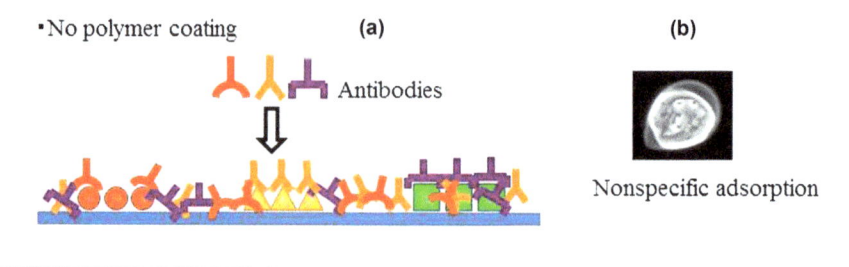

## 8.2 Photoreactive Polymers for Microarrays

The use of photolinkers is a popular method to attach biomolecules to substrate surfaces. Typically, photoreactive functional groups such as benzophenone,[3–5] diazirine,[6–24] and phenyl azide[25–32] moieties are used as photolinkers (Figure 8.3); these moieties produce photogenerated ketyl, carbene, and nitrene radicals, respectively, to form covalent links with target molecules and materials.[33] Because radical reactions occur on every organic material, including the biological molecules, the polymer matrix, and the chip surface, the photoimmobilization method does not require the use of any special functional groups such as amino, carboxyl, hydroxyl, or thiol groups, unlike other conventional immobilization methods. Different biological components can therefore be immobilized using the same method.

In addition to the efficient immobilization achieved using radical reactions, the reduction of the background noise of the chip surface caused by the nonspecific binding of proteins is also important for the assay sensitivity, *i.e.* the signal/noise ratio. To prepare such surfaces, amphiphilic polymers such as poly(ethylene glycol) (PEG) or poly(vinyl alcohol), and zwitterionic groups such as carbobetaine, phosphobetaine, and sulfobetaine

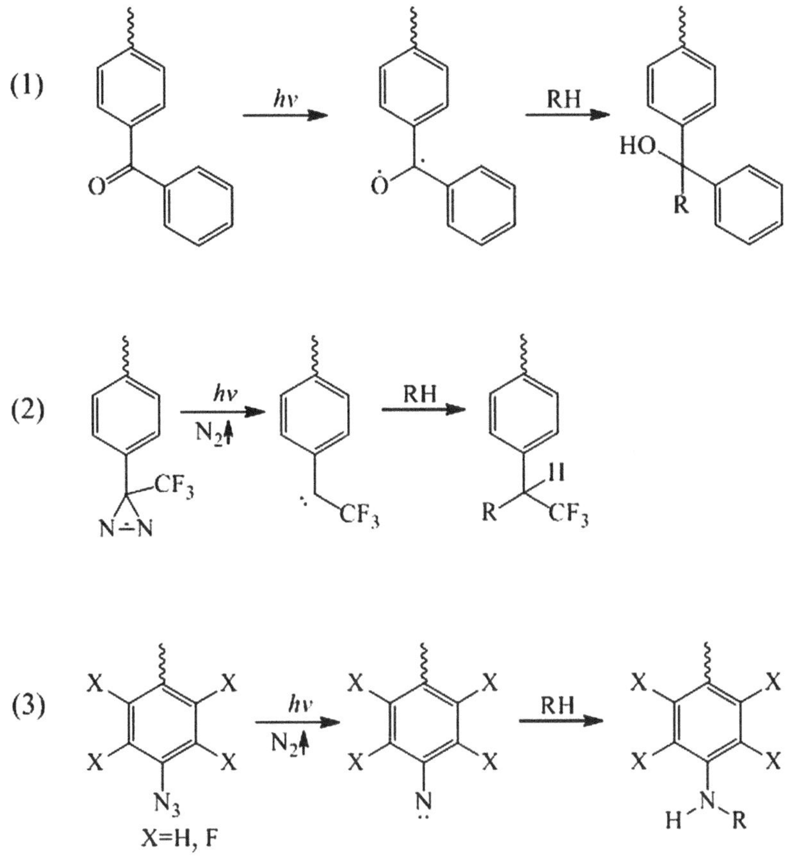

**Figure 8.3** Examples of photoreactive functional groups used in photoimmobilization.

have been used. We developed novel photoreactive and nonbiofouling polymers using postmodification or copolymerization, as shown in Figure 8.4.

## 8.2.1 Photoreactive Amphiphilic Polymers

PEG is amphiphilic, and polymers derivatized from PEG have nonionic hydrated grafted tails that make them hydrophilic. This prevents biofouling, and is helpful in reducing nonspecific protein interactions and ensuring the biocompatibility of the materials. We prepared the photoreactive PEG-containing polymer shown in Figure 8.4(A), and used this polymer to create nonadherent and nonbiofouling surfaces for microarray applications.[34,35] To achieve this, PEG-methacrylate was copolymerized with acryloyl-4-azidobenzoic acid in the presence of azobisisobutyronitrile, which acted as an initiator. The prepared polymer was coated and photoimmobilized on

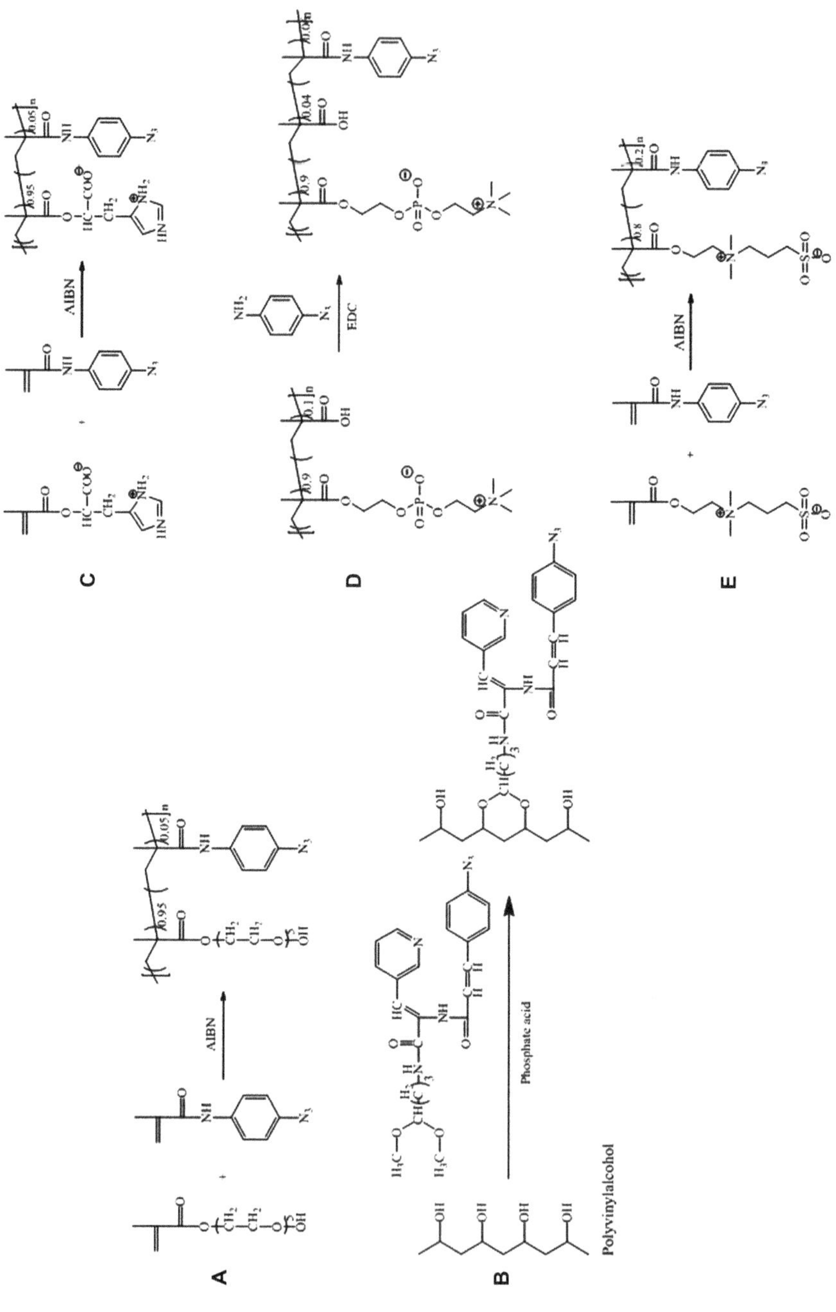

**Figure 8.4** Schemes for the preparation of the photoreactive and nonbiofouling polymers, and their structures.

plastic, glass, and titanium surfaces. Protein adsorption on the immobilized regions was reduced, and COS-7 cells did not adhere to the photoreactive PEG-immobilized regions.

Another amphiphilic polymer, polyvinyl alcohol modified with azidophenyl groups, was immobilized for cell culture using micropatterning, and was employed to prepare microarrays (Figure 8.4(B)).[36]

## 8.2.2 Photoreactive Zwitterionic Polymers

A polar histidine group containing photoreactive polymer was prepared by reacting two methacrylates, methacryloyl-L-histidine and 4-azidophenyl methacrylamide, as shown in Figure 8.4(C).[37] The polymer was photoimmobilized on polyester disks. Static contact-angle measurements showed that the polymeric surface was modified to be comparatively hydrophilic. Protein adsorption was reduced on the polymer-immobilized regions, and the spreading and adhesion of mammalian cells in these regions was reduced compared with that in the nonimmobilized regions.

2-Methacryloyloxyethyl phosphorylcholine (MPC) is a biomimetic polymer that acts to block thrombogenesis by reducing platelet adhesion and activation. Because phosphorylcholine is a lipid head found naturally in cell membranes, MPC is considered to reduce interactions with biological components such as proteins and cells. The photoreactive polymer was synthesized by coupling 4-azidoaniline with a copolymer containing MPC and methacrylic acid, as shown in Figure 8.4(D).[38,39] The synthesized polymer was coated on surfaces such as polyethylene and polypropylene. Micropatterning was carried out using photomasking, and there were significant differences between the protein adsorption and cell adhesion observed in the nonimmobilized and immobilized regions.

A sulfobetaine group containing photoreactive polymer was synthesized via the copolymerization of 4-azidophenyl methacrylamide and 2-(N-3-sulfopropyl-N,N-dimethylammonium) ethyl methacrylate, using AIBN as an initiator, as shown in Figure 8.4(E).[40] The synthesized polymer was coated and photoimmobilized on polymeric surfaces such as polystyrene and polyester. The surfaces were hydrophilic after they were coated with the polymer (this was determined using static contact-angle measurements); this indicated that the polar sulfobetaine groups were present on the surface. The presence of the zwitterionic sulfobetaine groups resulted in a significant reduction in the adhesion of proteins and mammalian cells on the photoimmobilized surfaces, compared with the nonimmobilized surfaces.

## 8.3 Applications of Photoimmobilized Microarrays

Some orientation typically occurs when conventional immobilization methods are used, because of the uneven distribution of functional groups on the biological molecules, as shown in Figure 8.5. When specifically modified proteins are used, the orientation of the immobilized proteins tends to occur. In contrast, it was possible to immobilize biological materials

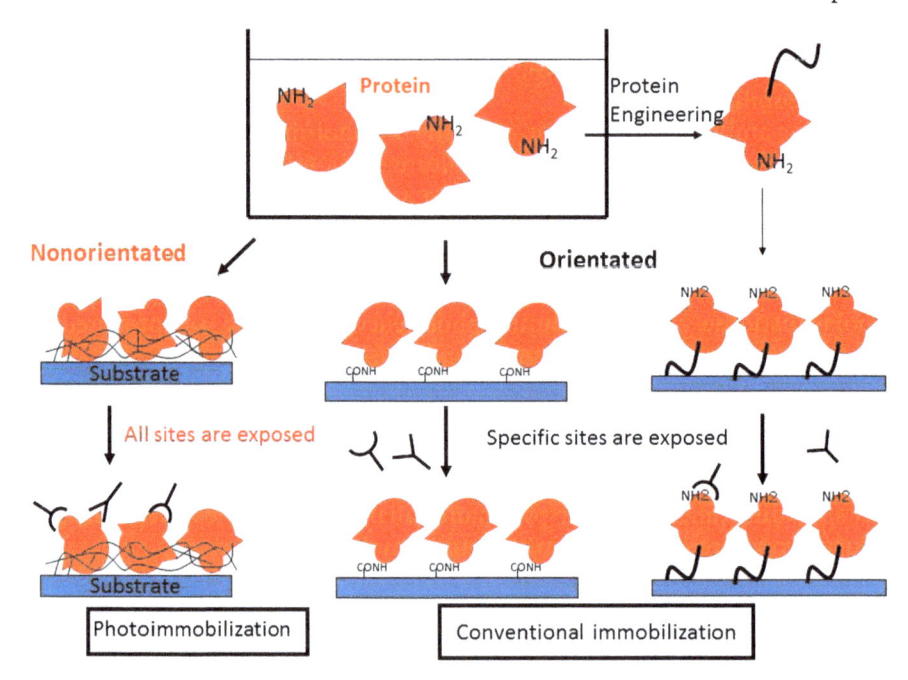

**Figure 8.5** Comparison of immobilization methods: Photoimmobilization leads to the random orientation of immobilized molecules, whereas other covalent immobilization methods lead to some orientation of the immobilized molecules, because of the uneven distribution of the functional groups on the molecules being immobilized. In genetically engineered proteins, which have peptide sequences adhered on the end chain, the recognition sites are limited to the remaining parts of the molecule.
Reprinted with permission from ref. 1. Copyright (2014) John Wiley and Sons.

without the occurrence of any molecular orientation using the photoimmobilization method (Figure 8.5). The photoimmobilization method was therefore suitable for the analysis of multiple interactions of the immobilized molecules with analytes, because the recognition sites on the immobilized molecules were not as limited.

## 8.3.1 Proteins

We microarray-immobilized proteins to examine the effects of the immobilized proteins on cellular functions, especially cell adhesion. Poly(acrylic acid) containing an azidophenyl group in its side chains was synthesized *via* the coupling reaction of poly(acrylic acid) with azidoaniline.[41] An aqueous solution containing the photoreactive polymer was coated on the surface of a polystyrene chip. After drying, proteins of various concentrations were spotted onto the polystyrene chip. Finally, the chip was irradiated with ultraviolet light.

Different amounts of bovine serum albumin (BSA) or fibronectin were microarrayed, and the adhesion of different cell types—including mouse leukemia monocytes RAW264, African green monkey COS-7 kidney cells, and rat pheochromocytoma PC12 cells was investigated. The number of cells that adhered to the photoreactive polymer-immobilized region was the same as the number of cells that adhered to the nontreated polystyrene. In contrast, the adhesion of COS-7 and PC12 cells to the BSA-immobilized regions was strongly suppressed. Although the immobilized fibronectin did not enhance the cell adhesion, it did enhance cell spreading for a short time. These cellular behaviors were similar to those observed on protein-adsorbed surfaces. This indicated that the photoimmobilized proteins had the same activity as the adsorbed proteins that are typically employed in the biological sciences.

A microarray chip was also fabricated using a poly(vinyl alcohol) polymer with azidophenyl groups incorporated in its side chains (AWP).[36] An AWP solution was spin coated on a glass slide, and then pattern-immobilized in the presence of a photomask. When the adsorption onto the glass slide was measured, the pattern formed was stable in organic solvents including methanol, acetone, and tetrahydrofuran. BSA, collagen, and fibronectin were photoimmobilized on AWP-precoated glass slides, and the cell adhesion was investigated.

Figure 8.6 shows results for the adhesion of mouse fibroblast STO cells. The STO cells adhered not only to the collagen- and fibronectin-spotted

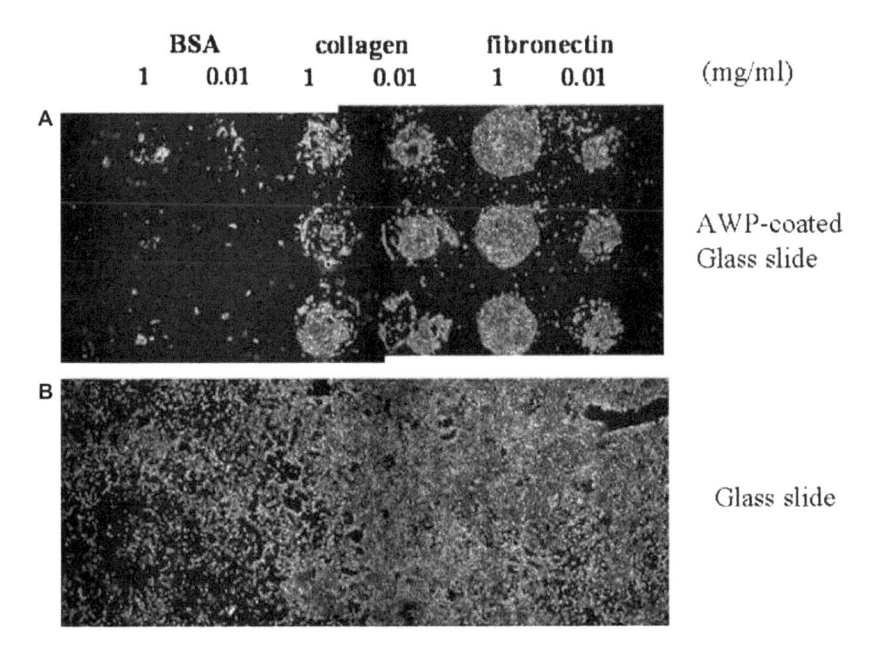

**Figure 8.6** Micrograph of mouse fibroblast STO cells adhered to bovine serum albumin (BSA)-, collagen-, and fibronectin-spotted AWP-coated glass slides (A), or glass slides only (B), over 2 h. The AWP-coated surface reduced the nonspecific adhesion of cells to the slide.
Reprinted with permission from ref. 1. Copyright (2014) John Wiley and Sons.

AWP-coated glass slides, but also to the glass surface. Human hepatocyte HepG2 cells, COS-7 cells, and RAW264 cells did not adhere to the AWP-coated surface. Thus, the AWP-coated surface provided good contrast micrographs for the cell-adhesion assay, and was useful for profiling the cell properties.

### 8.3.2 Allergens

In the photoimmobilization method, the immobilized materials were orientated more randomly than in conventional methods, as illustrated in Figure 8.5. The most suitable applications for this technique are therefore in basic research investigating various interactions between biological molecules, and in clinical diagnosis techniques for the detection of polyclonal antibodies in blood. Because a polyclonal antibody has many recognition sites, the immobilized antigen also provides many recognition sites. In oriented immobilization, some of these recognition sites must be covered. The photoimmobilization method therefore suggests itself as suitable for allergy diagnosis.

Allergy diagnosis was performed using a microarray biochip.[2] Additionally, we developed an automated microarray diagnostic system for specific IgE using photoimmobilized allergens, as shown in Figure 8.7(A).[35] An aqueous solution of a photoreactive PEG-based polymer was spin coated on a plate, and an aqueous solution of each allergen was microspotted on the coated plate and allowed to dry in air. Finally, the plate was irradiated with an ultraviolet lamp to achieve covalent immobilization. An automated machine using these plates was developed for an antigen-specific IgE assay. Initially, the patient serum was added to the microarray plate, and after the reaction of the microspotted allergen with the IgE, the adsorbed IgE was detected using a peroxidase-conjugated anti-IgE-antibody. The chemiluminescence intensity of the substrate decomposed by the peroxidase was detected automatically using a sensitive charge-coupled device camera. The results were comparable to those achieved using conventional specific IgE. Using this system, six different allergen-specific IgE were assayed using 10 IA of serum within a period of 20 min (Figure 8.7(B)).

### 8.3.3 Autoantigens

Autoimmune diseases such as rheumatoid arthritis, multiple sclerosis, and autoimmune diabetes are characterized by the production of autoantibodies that serve as useful diagnostic markers, surrogate markers, and prognostic factors. We devised an *in vitro* system to detect these clinically pivotal autoantibodies using a photoimmobilized autoantigen microarray.[42] Aqueous solutions of each autoantigen were mixed with a PEG methacrylate polymer and a photoreactive crosslinker, and the mixtures were microspotted on a plate and dried in air. Finally, the plate was irradiated with an ultraviolet lamp to achieve immobilization.

Photoreactive Polymers For Microarray Chips 143

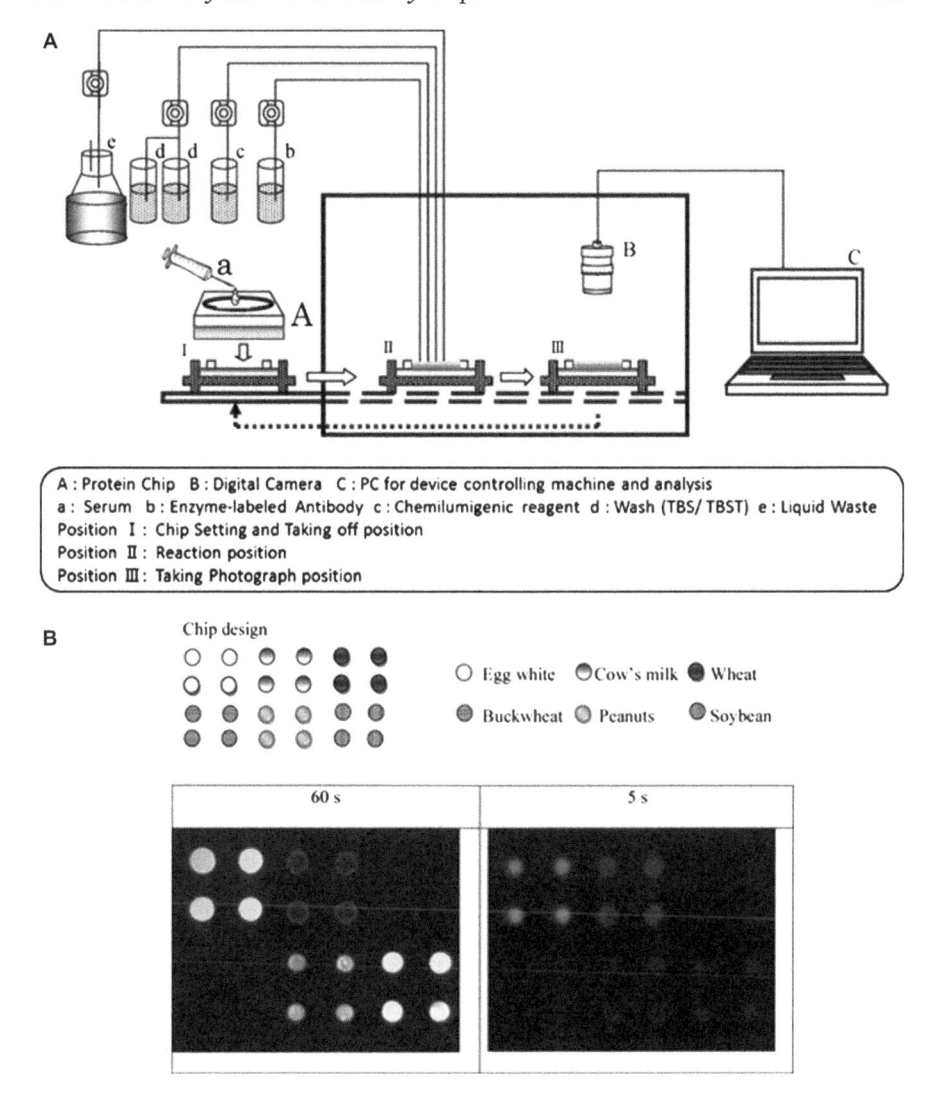

A : Protein Chip   B : Digital Camera   C : PC for device controlling machine and analysis
a : Serum   b : Enzyme-labeled Antibody   c : Chemilumigenic reagent   d : Wash (TBS/ TBST)   e : Liquid Waste
Position I : Chip Setting and Taking off position
Position II : Reaction position
Position III : Taking Photograph position

Chip design

○ Egg white   ◎ Cow's milk   ● Wheat
◉ Buckwheat   ◯ Peanuts   ◍ Soybean

**Figure 8.7** (A) Schematic diagram of the automated assay system. (B) Effect of prewashing on the preparation of the microarray immobilized plates. Chemiluminescence images captured by the CCD camera were mounted in the automated assay machine.
Reprinted with permission from ref. 35. Copyright (2014) Royal Society of Chemistry.

In the assay, patient serum was added to the microarray plate. The antigen-specific immunoglobulin G (IgG) adsorbed on the microspotted autoantigen was detected using a peroxidase-conjugated anti-IgG antibody. The chemiluminescence intensities of the substrate decomposed by the peroxidase were detected using a sensitive CCD camera. All autoantigens were immobilized stably using this method, and the stably

144                                                                    Chapter 8

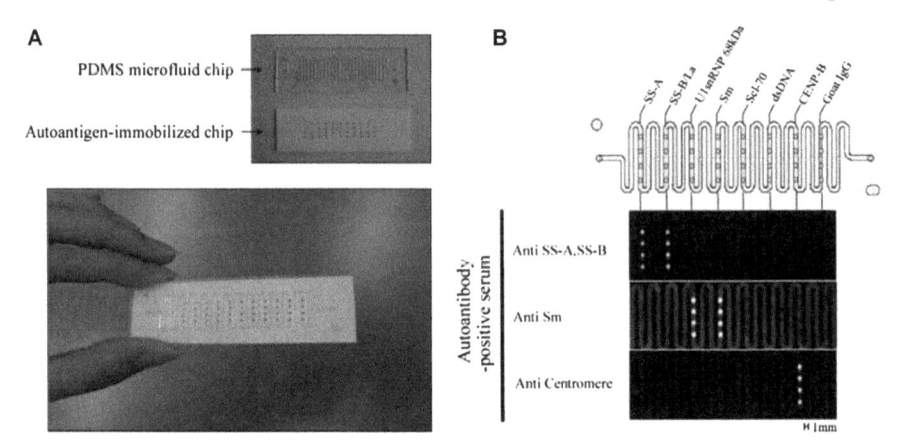

**Figure 8.8**  (A) Photographs of a polystyrene chip with dye microspots (for visualization), and a PDMS chip with a microfluid channel. The two plates were superimposed during the assay. (B) Chemiluminescence images captured by the CCD mounted in the automated assay machine.
Reprinted with permission from ref. 42. Copyright (2014) John Wiley and Sons.

immobilized autoantigens were used to screen for antigen-specific IgG. In addition, the plate was covered with a polydimethylsiloxane sheet containing microchannels (Figure 8.8), and automated measurements were carried out.

### 8.3.4  Viruses

The photoimmobilized microarray was applied for the detection of virus-specific IgGs.[43] Viruses were immobilized on the microarrays using a radical crosslinking reaction. A new photoreactive polymer containing perfluorophenyl azide and PEG methacrylate was prepared, and was then coated on plates. Inactivated measles, rubella, mumps, varicella-zoster, and recombinant Epstein–Barr virus were added to the coated plates; the resulting samples were irradiated with ultraviolet light to facilitate immobilization, as shown in Figure 8.9. Virus-specific IgGs in healthy human sera were assayed using these prepared microarrays, and the results obtained were compared with those obtained using conventional enzyme immunoassays (EIAs).

We observed a high correlation (0.79–0.96) between the results obtained using the automated microarray technique and those obtained using the EIAs. The microarray-based assay was more rapid, required smaller reagent and sample quantities, and was easier to conduct compared with conventional EIA techniques. This microarray format shows great potential for rapid and multiple serological diagnoses of viral diseases, and could be developed further for clinical applications.

*Photoreactive Polymers For Microarray Chips* 145

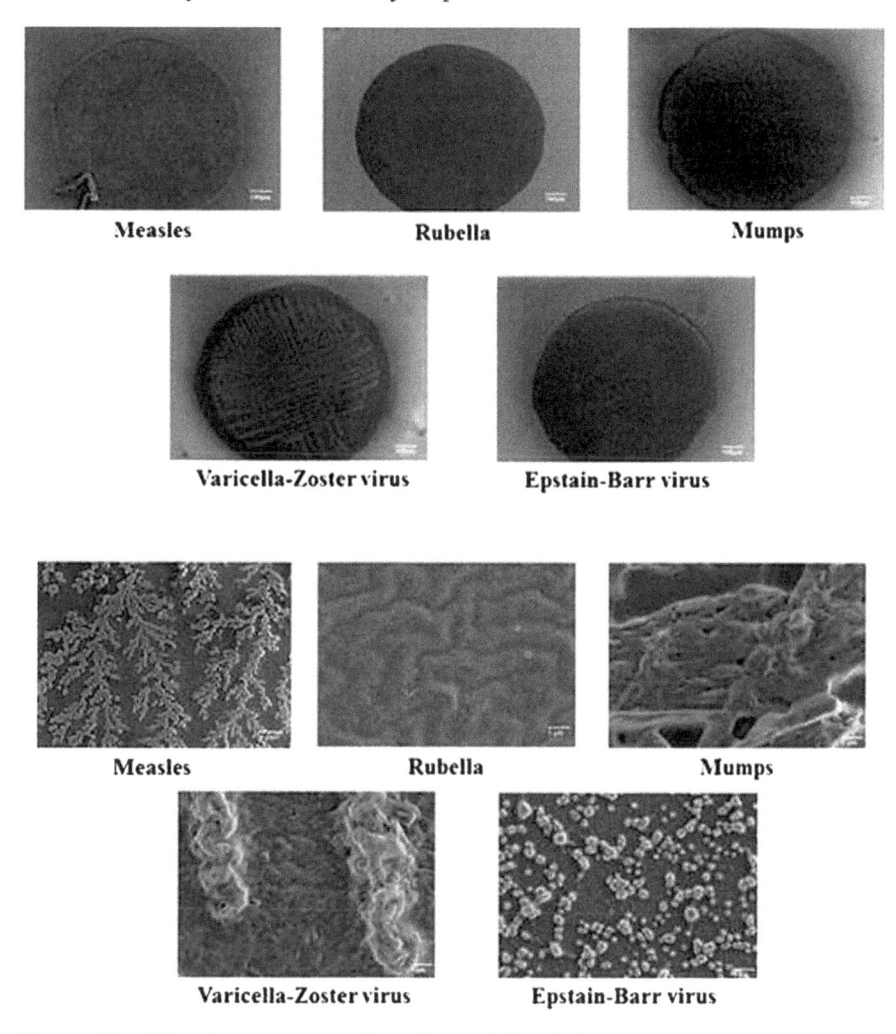

**Figure 8.9** SEM photomicrographs of immobilized viruses.

## 8.3.5 Cells

A cell microarray was prepared for a clinical analysis using photo-immobilization.[38] The analyte was cell-specific antibodies in blood. Typically, A-type blood does not contain anti-A-type antibodies; however, irregular antibodies are sometimes produced because of a previous transfusion, or pregnancy. Blood containing irregular antibodies cannot be used for transfusions, because the irregular antibodies aggregate with the cells of the patient receiving the transfusion. Blood is therefore typically checked for the presence of irregular antibodies before a transfusion. However, the check is performed using a blood cell aggregation assay; this is a time-consuming process, and skill is required to interpret the results. If a cell-array method was realized for the detection of irregular antibodies, the assay system could be automated, reducing the time required to obtain results; therefore, we microarrayed a panel of cells that expressed the A, B, O, and AB antigens. The cells were mixed with photoreactive polymers and then microspotted on a polystyrene dish that was irradiated with ultraviolet light. Antibody solution or serum was added to the microarray plate, and the resulting sample was incubated for a certain time. After washing, the array assay was used with human serum.

Figure 8.10 shows the results for the interactions of the cell microarray with antibodies. The O-type human serum reacted with A-, B-, and AB-type

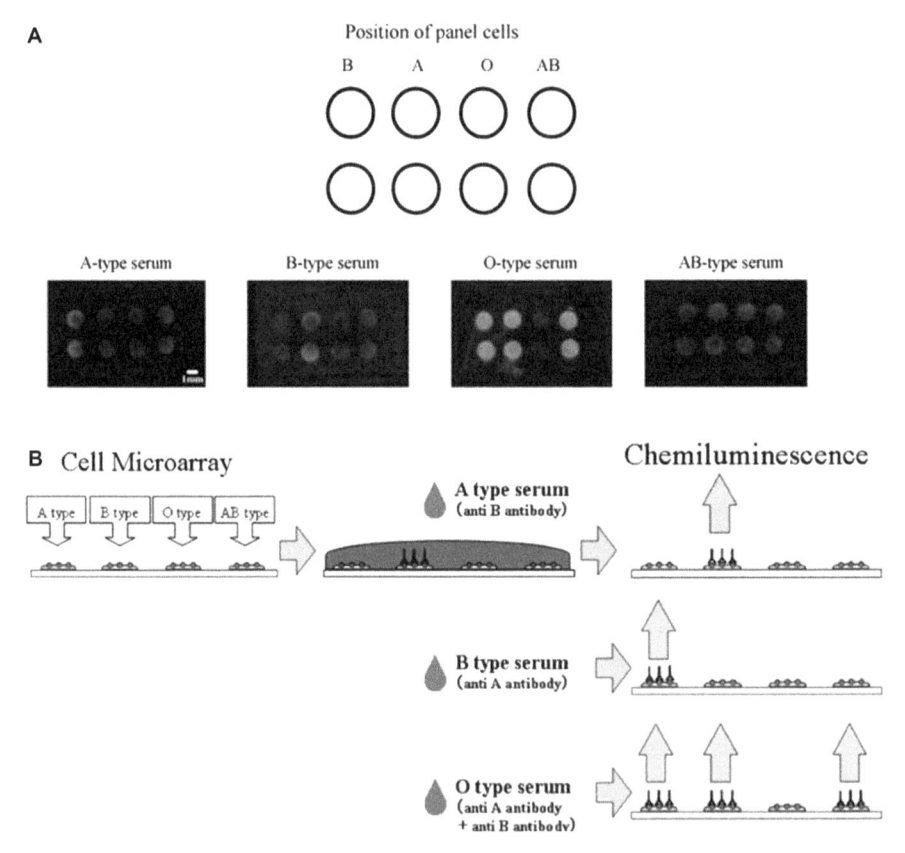

Figure 8.10 (A) Image of antibodies adsorbed onto blood cells microarrayed on a chip with photoreactive poly(phosphatidylcholine methacrylate). (B) Schematic illustration explaining the images shown in (A). The A-type serum contained anti-B antibodies that adsorbed on B-type blood cells, whereas the B-type serum contained anti-A antibodies that adsorbed on A-type cells. The O-type serum contained anti-A and anti-B antibodies that adsorbed on A, B, and AB cells, and the AB-type serum contained no antibodies for any of the cells.

Reprinted with permission from ref. 1. Copyright (2014) John Wiley and Sons.

cells, and the A- and B-type sera reacted with B- and A-type cells, respectively, although the reaction intensity was relatively low. This was because the A- and B-type sera contained anti-B and anti-A antibodies, respectively. The AB-type serum did not react with either type of cell, because it contained no antibodies for these antigens. Because the O-type serum contained anti-A and anti-B antibodies, this serum reacted with the whole panel of cells except the O cells. These results corresponded with the results of an aggregation test performed for conventional blood types.

## 8.4 Conclusion

The miniaturization of multiple biological assays has had a great impact on the development of biomedical technologies. Here, we employed photoreactive polymers for the microarray immobilization of various biological components, including proteins, allergens, autoantigens, viruses, and cells. Microarray technologies will increase in importance with the continuing progress of bioinformatics and clinical informatics, and this progress will produce new applications for these photoreactive polymers.

## References

1. Y. Ito, *Biotechnol. Prog.*, 2006, **22**, 924–932.
2. Y. Ito, T. Yamauchi and K. Omura, *Kobunshi Ronbunshu*, 2004, **61**, 501–510.
3. G. Sundarababu, H. Gao and H. Sigrist, *Photochem. Photobiol.*, 1995, **61**, 540–544.
4. C. L. Hypolite, T. L. McLernon, D. N. Adams, K. E. Chapman, C. B. Herbert, C. C. Huang, M. D. Distefano and W. S. Hu, *Bioconjugate Chem.*, 1997, **8**, 658–663.
5. O. Prucker, C. A. Naumann, J. Rühe, W. Knoll and C. W. Frank, *J. Am. Chem. Soc.*, 1999, **121**, 8766–8770.
6. H. Gao, E. Kislig, N. Oranth and H. Sigrist, *Biotechnol. Appl. Biochem.*, 1994, **20**, 251–263.
7. H. Gao, M. Sanger, R. Luginbuhl and H. Sigrist, *Biosens. Bioelectron.*, 1995, **10**, 317–328.
8. H. Sigrist, A. Collioud, J. F. Clemence, H. Gao, R. Luginbuhl, M. Sanger and G. Sundarababu, *Opt. Eng.*, 1995, **34**, 2339–2348.
9. H. Gao, R. Luginbuhl and H. Sigrist, *Sens. Actuators, B*, 1997, **38**, 38–41.
10. N. Barie, M. Rapp, H. Sigrist and H. J. Ache, *Biosens. Bioelectron.*, 1998, **13**, 855–860.
11. N. Barie, H. Sigrist and M. Rapp, *Analusis*, 1999, **27**, 622–629.
12. Y. Chevolot, J. Martins, N. Milosevic, D. Leonard, S. Zeng, M. Malissard, E. G. Berger, P. Maier, H. J. Mathieu, D. H. G. Crout and H. Sigrist, *Bioorg. Med. Chem.*, 2001, **9**, 2943–2953.
13. I. Caelen, H. Gao and H. Sigrist, *Langmuir*, 2002, **18**, 2463–2467.

14. S. Angeloni, J. L. Ridet, N. Kusy, H. Gao, F. Crevoisier, S. Guinchard, S. Kochhar, H. Sigrist and N. Sprenger, *Glycobiology*, 2005, **15**, 31–41.
15. N. Kanoh, S. Kumashiro, S. Simizu, Y. Kondoh, S. Hatakeyama, H. Tashiro and H. Osada, *Angew. Chem., Int. Ed.*, 2003, **42**, 5584–5587.
16. N. Kanoh, K. Honda, S. Simizu, M. Muroi and H. Osada, *Angew. Chem., Int. Ed.*, 2005, **44**, 3559–3562.
17. N. Kanoh, A. Asami, M. Kawatani, K. Honda, S. Kumashiro, H. Takayama, S. Simizu, T. Amemiya, Y. Kondoh, S. Hatakeyama, K. Tsuganezawa, R. Utata, A. Tanaka, S. Yokoyama, H. Tashiro and H. Osada, *Chem.-Asian J.*, 2006, **1**, 789–797.
18. N. Kanoh, M. Kyo, K. Inamori, A. Ando, A. Asami, A. Nakao and H. Osada, *Anal. Chem.*, 2006, **78**, 2226–2230.
19. N. Kanoh and H. Osada, *J. Synth. Org. Chem., Jpn.*, 2006, **64**, 639–650.
20. N. Kanoh, T. Nakamura, K. Honda, H. Yamakoshi, Y. Iwabuchi and H. Osada, *Tetrahedron*, 2008, **64**, 5692–5698.
21. T. Tomohiro, M. Hashimoto and Y. Hatanaka, *Chem. Rec.*, 2005, **5**, 385–395.
22. M. R. Bond, H. C. Zhang, P. D. Vu and J. J. Kohler, *Nat. Protoc.*, 2009, **4**, 1044–1063.
23. W. G. Huang, W. Sun, Z. Q. Song, Y. B. Yu, X. Chen and Q. S. Zhang, *Org. Biomol. Chem.*, 2012, **10**, 5197–5201.
24. H. Mehenni, V. Pourcelle, J. F. Gohy and J. Marchand-Brynaert, *Aust. J. Chem.*, 2012, **65**, 193–201.
25. T. Matsuda and T. Sugawara, *Langmuir*, 1995, **11**, 2272–2276.
26. M. Mizutani, S. C. Arnold and T. Matsuda, *Biomacromolecules*, 2002, **3**, 668–675.
27. J. C. Miller, H. P. Zhou, J. Kwekel, R. Cavallo, J. Burke, E. B. Butler, B. S. Teh and B. B. Haab, *Proteomics*, 2003, **3**, 56–63.
28. Y. Q. Guan, L. M. He, S. M. Cai and T. H. Zhou, *J. Mater. Sci. Technol.*, 2006, **22**, 200–204.
29. D. N. Kim, W. Lee and W. G. Koh, *J. Chem. Technol. Biotechnol.*, 2009, **84**, 279–284.
30. S. Kumar, D. K. Kannoujia, A. Naqvi and P. Nahar, *Biochem. Eng. J.*, 2009, **47**, 132–135.
31. P. Sharma, S. F. Basir and P. Nahar, *J. Colloid Interface Sci.*, 2010, **342**, 202–204.
32. J. Kim, D. Hong, S. Jeong, B. Kong, S. M. Kang, Y. G. Kim and I. S. Choi, *Chem.-Asian J.*, 2011, **6**, 363–366.
33. D. M. Dankbar and G. Gauglitz, *Anal. Bioanal. Chem.*, 2006, **386**, 1967–1974.
34. Y. Ito, H. Hasuda, M. Sakuragi and S. Tsuzuki, *Acta Biomater.*, 2007, **3**, 1024–1032.
35. Y. Ito, N. Moritsugu, T. Matsue, K. Mitsukoshi, H. Ayame, N. Okochi, H. Hattori, H. Tashiro, S. Sato and M. Ebisawa, *J. Biotechnol.*, 2012, **161**, 414–421.

36. Y. Ito, M. Nogawa, M. Takeda and T. Shibuya, *Biomaterials*, 2005, **26**, 211–216.
37. M. Sakuragi, S. Tsuzuki, H. Hasuda, A. Wada, K. Matoba, I. Kubo and Y. Ito, *J. Appl. Polym. Sci.*, 2009, **112**, 315–319.
38. Y. Ito, T. Yamauchi, M. Uchikawa and Y. Ishikawa, *Biomaterials*, 2006, **27**, 2502–2506.
39. T. Konno, H. Hasuda, K. Ishihara and Y. Ito, *Biomaterials*, 2005, **26**, 1381–1388.
40. M. Sakuragi, S. Tsuzuki, S. Obuse, A. Wada, K. Matoba, I. Kubo and Y. Ito, *Mater. Sci. Eng., C*, 2010, **30**, 316–322.
41. Y. Ito and M. Nogawa, *Biomaterials*, 2003, **24**, 3021–3026.
42. T. Matsudaira, S. Tsuzuki, A. Wada, A. Suwa, H. Kohsaka, M. Tomida and Y. Ito, *Biotechnol. Prog.*, 2008, **24**, 1384–1392.
43. P. M. Sivakumar, N. Moritsugu, S. Obuse, T. Isoshima, H. Tashiro and Y. Ito, *PLoS One*, 2013, **8**, e81726.

CHAPTER 9

# Boron/Phosphorus-Containing Flame-Retardant Photocurable Coatings

EMRAH ÇAKMAKÇI AND MEMET VEZIR KAHRAMAN*

Marmara University, Department of Chemistry, 34722, Istanbul/Turkey
*Email: mvezir@marmara.edu.tr

## 9.1 Flame Retardancy

According to the National Fire Protection Association (NFPA), 1 389 500 fires that caused 3005 civilian deaths, 17 500 civilian injuries, and $11.7 billion in property damage, were reported in the USA in 2011.[1] This sad situation is just an example of what fire can cause in the technologically advanced modern society. There is much more heart-wrenching fire hazards occurred in other parts of the world and also there are several tragic fire hazards in history.[2] Mankind could have never thought that stealing fire from gods would bring such disaster!

Today, although it can be said that there is a decreasing trend in fire hazards due to better education, better fire safety buildings and standards, wider use of smoke alarms and the use of flame retardants,[3] there is still a long way to go to make the world a safer place from fire.

Since all carbon-based materials are combustible and so are commercial plastics, in order to reduce the risk of fires they should be either replaced by inherently flame-retardant (FR) polymers or their flame retardancy should be increased by the addition of flame retardants or they should be modified to exhibit a certain level of flame resistance. By the way, it must be noted that when we use the words "flame retardant" for any material, it does not mean that this material will never burn. Flame retardants that impede the ignition of materials have been in use for thousands of years. It is known that in 360 BC, timbers were painted with vinegar to be resistant to fire. In 1735 Obadiah Wylde used a mixture of alum, borax and ferrous sulfate to impart flame resistance to paper and textiles and obtained the first patent on flame retardants.[4]

Today, there are several flame-retardant additives which are widely used in plastics. Up to now several reviews have been published explaining the combustion and fire resistance of polymers and also the effect of these additives in different polymers.[5–10] These flame-retardant chemicals can be roughly classified into two groups: halogenated flame retardants and nonhalogenated flame retardants. Although halogenated flame retardants are highly efficient, most of these flame retardants were banned or highly restricted due to the release of toxic gases during their combustion. Therefore, the use of other inorganic elements containing flame retardants has become of great importance and there are several papers and patents on the use of these materials in addition to several commercial products. Among these flame retardants, phosphorus-based additives and monomers are widely preferred and used.[11]

UV-curable coatings have a wide range of applications in the market due to their outstanding features like fast production rates, low energy consumption, good adhesion, environmental friendliness and good chemical and solvent resistance.[12] With 100% solid contents, photocurable coatings are free of volatile organic compounds (VOC) and this green technology reduces carbon emissions. Thus, they fit perfectly to today's demands. Today, the photocurable coating market is expanding at a rapid pace to new applications. As a result, new requirements emerge for photocurable materials. As one of the most critical requirements, flame retardancy for photocurable coatings is highly desired.

Before we start to summarize the recent developments in flame-retardant coatings there is one more important issue that we want to talk about: How do we determine the flame retardancy of a certain material? There are three most widely used techniques to measure the level of flame retardancy: LOI, UL94 and cone calorimetry. We strongly recommend that the reader consults the books or other reviews that are cited in this text to become informed about these techniques.

This chapter is devoted to the developments in P/B-containing flame-retardant photocurable coatings in the past 10–15 years. We intentionally did not include the patent literature. We must note that halogen-free flame-retardant photocurable coatings was previously reviewed by Randoux et al.,[13] and a review of the recent developments in flame-retardant coatings has also been published.[14] In this work we only focus on the synthesis of P/B-containing photocurable monomers, their flame retardancy and also the effect of some P/B-containing additives used in flame-retardant photocurable coatings. We also try to mention the coating performance of these flame-retardant coating materials wherever it was possible.

## 9.2 Phosphorus-Containing Flame-Retardant Photocurable Coatings

Phosphorus-containing materials are not only highly efficient flame retardants; they also have good effects on the mechanical, optical and adhesive properties of several polymeric materials. Phosphorus-based flame retardants generally work in the condensed phase by promoting high amounts of carbonaceous char. In this section we will try to summarize some of the phosphorus-containing monomers and additives used as flame retardants in photocurable coatings in the last decade. Scheme 9.1 displays the chemical structures of some of the monomers used in phosphorus-containing photocurable coatings.

Vinyl phosphonic acid (VPA) is a commercially available monomer that has good flame-retardant and adhesive properties in addition to its proton conductivity. Almost 13 years ago, Imamoglu and Yagci, prepared VPA

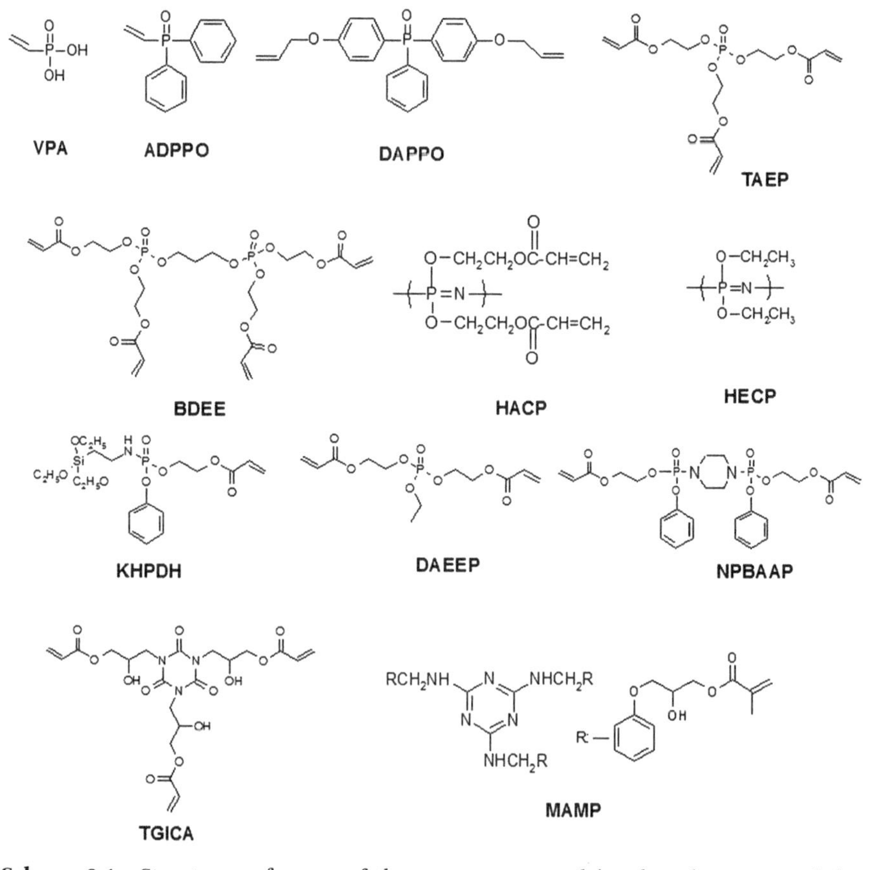

**Scheme 9.1** Structures of some of the monomers used in phosphorus-containing photocurable coatings.

containing urethane-acrylate-based UV-curable coatings.[15] The photocurable coating consisted of 60 wt% urethane acrylate oligomer, 10 wt% trimethylolpropane triacrylate (TMPTA), 15 wt% 1-,6-hexandioldiacrylate (HDDA), 10 wt% N-vinyl pyrrolidone (NVP) and 5 wt% photoinitiator. VPA was introduced to this system by replacing NVP and the concentration of VPA was varied. The flame retardancy of this coating material was studied by means of TGA and LOI. TGA under an air atmosphere showed that the char residue was increased with the increasing concentration of VPA. Acidic moieties in VPA reacted at higher temperatures and formed intermolecular anhydride linkages that caused an intumescent carbonaceous char (Scheme 9.2). They stated that this char layer acted as a thermal barrier that impeded the heat transfer to the combustible polymeric layer below and was responsible for the pronounced flame retardancy. Moreover, while the LOI of the base formulation was 17, it increased 2.5 times of the LOI value of the base formulation when 1 wt% VPA was added. This further increased to 76 when the VPA amount was doubled (2 wt%). At higher VPA concentrations; 5 and 9 wt%, the LOI value had even reached to 100.

In another study, Basturk et al., used VPA to functionalize epoxidized soybean oil (ESBO) for the preparation of flame-retardant UV-curable organic–inorganic hybrid coatings.[16] They first synthesized an aliphatic urethane acrylate oligomer. ESBO was functionalized with either VPA or

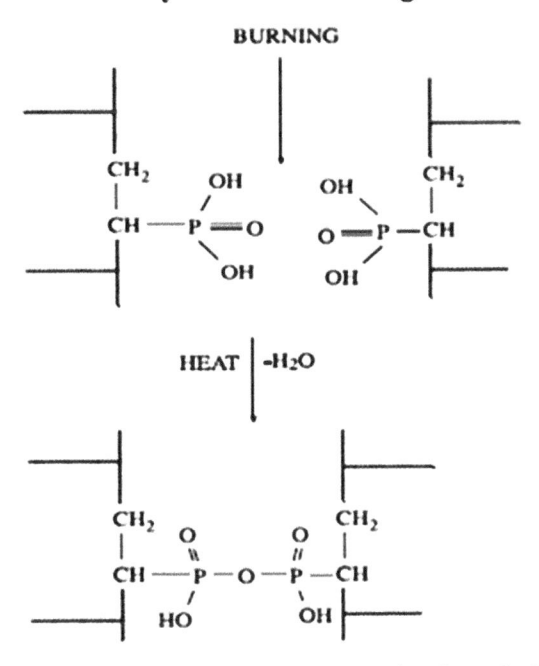

**Scheme 9.2** Char formation mechanism *via* intermolecular anhydride linkages.

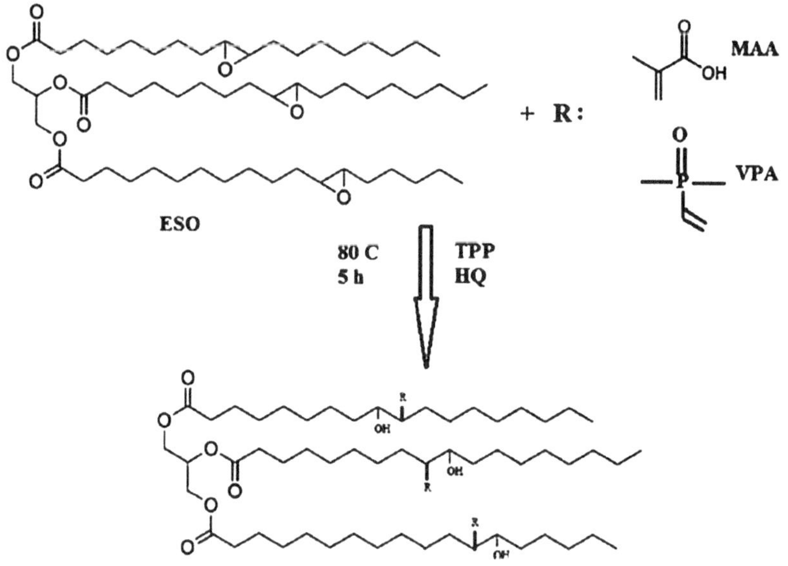

**Scheme 9.3** Functionalization of EBSO with MA or VPA.

methacrylic acid (MA). This functionalization is depicted in Scheme 9.3. Formulations were prepared by mixing the urethane acrylate oligomer with the functionalized ESBO and hybrid coatings were developed by adding a sol–gel precursor. A synergistic effect of phosphorus and silicon on the enhancement of char formation and the flame retardation was observed. Thermogravimetric analysis revealed that the char yield of the coating material increased with the addition of VPA functionalized ESBO. The LOI value increased from 19 to 23.5 when VPA functionalized ESBO was used with the sol–gel precursor.

Interestingly, there are not many studies on the flame retardancy of photocurable materials utilizing this useful monomer. This could be attributed to some of the undesired properties that VPA brings to photocurable coatings. It is known that allyl or vinyl groups do not tend to polymerize. Thus, the gel contents of the vinyl/allyl groups containing coating are generally low. For instance, the same paper by Imamoglu and Yagci, describes the decrease of gloss property of the coatings as the VPA concentration increases due to the migration of unreacted VPA monomers.[2] The decrease of gel content was also observed when VPA-functionalized ESBO was used.[16] Nevertheless VPA is an effective flame-retardant monomer for photocurable coatings.

Besides commercial phosphorus-containing monomers like VPA, several FR photocurable coating materials were prepared by using novel phosphorous monomers. Thanks to the wide potentialities of phosphorus chemistry, various monomers can be synthesized in high purity and with different phosphorus percentages. Sheng-Wu Zhu and Wen-Fang Shi

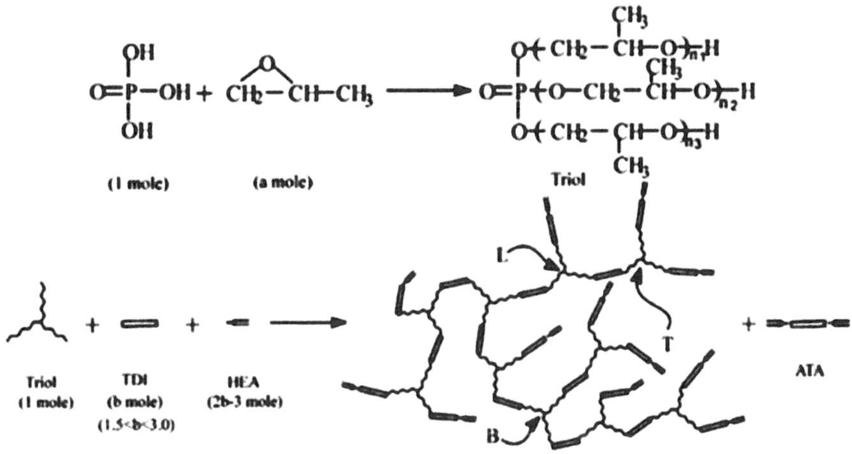

Scheme 9.4  Synthetic route to hyperbranched PU acrylates.

investigated the synergistic effect of P and N on the flame retardancy of hyperbranched PU acrylates.[17] First, they synthesized a polyol by reacting phosphoric acid and propylene oxide. Then, hyperbranched PU acrylates (Scheme 9.4) were prepared by reacting the phosphorus-containing polyol with toluene diisocyanate (TDI) and 2-hydroxylethyl acrylate (HEA).[17,18] Also, an aromatic urethane diacrylate was synthesized from the reaction of TDI and 2 moles of HEA. The hyperbranched polyol was mixed with a commercial aliphatic PU acrylate and the aromatic urethane diacrylate. They prepared several formulations by varying the amount of these acrylic oligomers in order to obtain coating materials with different P and N ratios.

The coating material containing 100% hyperbranched polyol displayed a LOI value of 27. This coating material contained theoretically 2.8% P and 5.8% N. The LOI value decreased with increasing amount of the commercial aliphatic PU acrylate and the LOI value of the photocured material that did not contain the phosphorous hyperbranched polyol was found to be 19.5. They proved that the presence of nitrogen has a significant effect on the fire retardation of the phosphorus-containing photocurable coatings. The effect of nitrogen and phosphorus percentages on the LOI of the coatings can be seen in Figure 9.1. Moreover, they investigated the degradation mechanism by TGA-FTIR and found that P–O–P bonds were formed, while P–O–C bonds were cleaved during combustion. They showed that the char that protects the underlying materials from fire forms through this mechanism.

The same authors also prepared novel methacrylated phosphates in order to increase the flame retardancy of epoxy acrylates.[19,20] Diethyl phosphate was reacted with glycidyl methacrylate (GA) and the resulting monofunctional methacrylate was denoted as MAP. MAP was then reacted with TDI to obtain a urethane diacrylate; MADP. The synthetic procedures of these two monomers are outlined in Scheme 9.5. A commercial epoxy acrylate, EB600 was mixed with MAP and MADP in different ratios and cured under UV light.

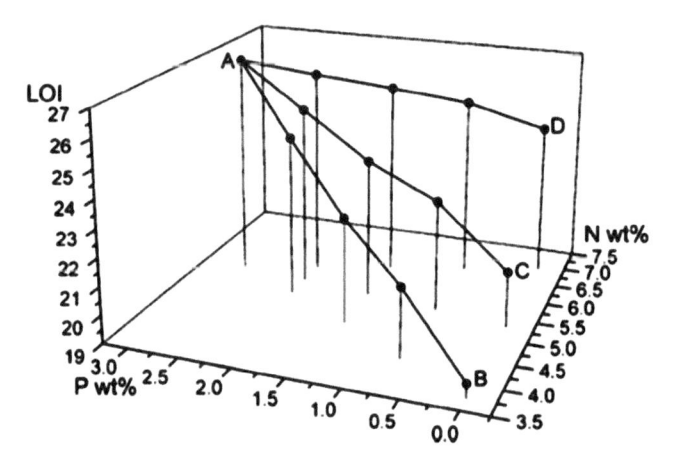

**Figure 9.1** The effect of N and P percentages on the LOI of hyperbranched PU acrylate based coatings.

**Scheme 9.5** Synthetic routes for MAP and MADP.

Benzoyl-1-hydroxyl-cyclohexanol (Irgacure 184) was used as photoinitiator. The thermal properties of the cured films were characterized by TGA, LOI and cone-calorimeter measurements. According to cone-calorimeter measurements, the heat-release rate (HRR) and the total heat-release rate (THR) that express the intensity of a fire, decreased when MAP and MADP were added to the epoxy acrylate. Moreover, the time of ignition ($T_{ig}$) was shortened with the increasing amounts of MAP and MADP. The early degradation of the phosphorous species in the UV-cured epoxy acrylate system decreased the ignition time but on the other hand they caused the formation of high amounts of char in the early stages of burning that covered the surface of the rest of the material and protected it from fire. Cone-calorimeter measurements also showed that the total smoke production (TSP) was lowered due to the presence of the phosphorous methacrylates. As expected, the LOI values and the char yields increased linearly with the increase in the P percentage in the epoxy acrylate.

In order to increase the flame retardancy of epoxy acrylates, Kahraman et al., prepared UV-curable formulations with two different phosphine oxide monomers:

Allyldiphenyl phosphine oxide (ADPPO) and 4,4′-bis(allyloxyphenyl) phenyl phosphine oxide (DAPPO).[21] As noted earlier, the low polymerization tendency of the allyl monomers again caused low double-bond conversion problems. ADPPO was found to dimerize instead of participating in the photoinitiated free-radical polymerization. On the other hand, DAPPO exhibited a better performance and the char residue of the epoxy acrylate was doubled and increased from 7.7% to 15.8% when 30 wt% DAPPO was used. Moreover, the presence of DAPPO also increased the onset degradation temperature and the maximum weight loss temperatures, respectively.

Later, we showed that the low polymerization tendency of allyl compounds can be enhanced by a thiol–ene polymerization system.[22] We mixed different amounts of ADPPO with trimethylolpropane tris(3-mercaptopropionate) and a commercial aliphatic urethane triacrylate (Scheme 9.6). Due to the unique characteristic of thiol–ene polymerization, we obtained photocured materials with high conversions and high gel contents. As a result, the char

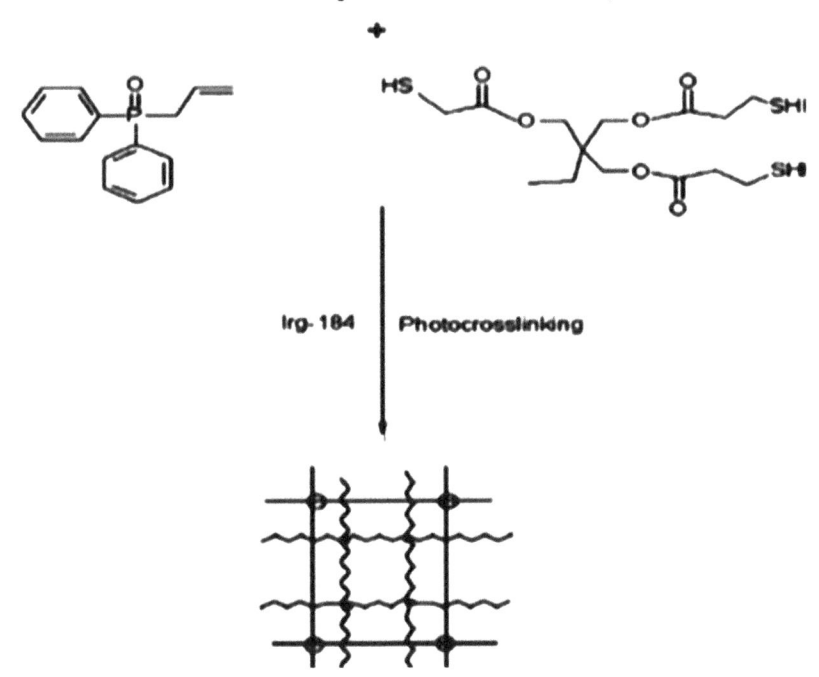

Scheme 9.6  Preparation of flame-retardant thiol–ene cured coatings.

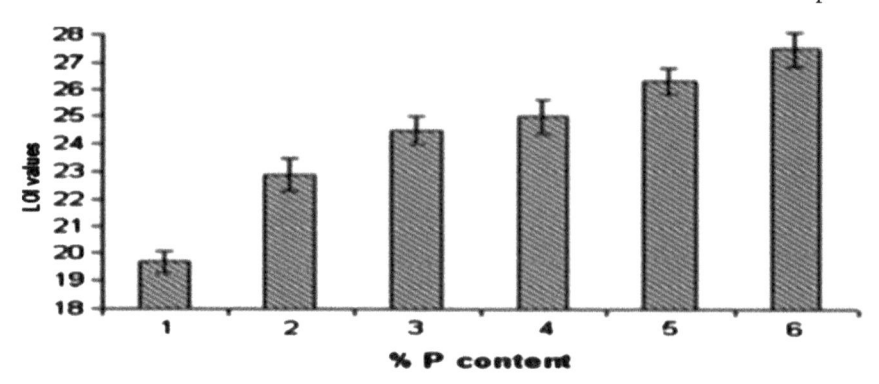

**Figure 9.2** LOI values *versus* P percentage in flame-retardant thiol–ene cured coating.

yields increased with the addition of ADPPO, in contrast to our previous study.[21] Also, the decreased onset degradation and maximum temperatures were observed for this study that again can be attributed to the increased flame retardancy. The early degraded phosphorous species caused high amounts of char formation and consequently increased the fire retardancy of the thiol–ene photopolymerized network. The LOI of the coatings increased as the P percentage was increased (Figure 9.2).The LOI value of the photocured coating with the 4.4% phosphorus content was 27.5, while the LOI of the ADPPO free network was 19.5.

In another study, Hongbo Liang and Wenfang Shi synthesized highly fire-retardant two-phosphate-based acrylates by reacting phosphorus oxychloride and hydroxylethyl acrylate: tri(acryloyloxyethyl) phosphate (TAEP) and di(acryloyloxyethyl)ethyl phosphate (DAEEP).[23] The degradation profile and the pyrolysis mechanism of these materials were investigated by TGA-FTIR and DP-MS measurements, respectively. The LOI value of the photocured DAEEP was 29 and the trifunctional acrylated phosphate; TAEP displayed 7% higher oxygen index than DAEEP. Results showed the formation of poly(phosphoric acid) at around 300 °C and thus the phosphorus content was increased in the char at this stage of burning. Later, they used the same monomers and investigated their effects on the thermal properties of epoxy and polyurethane acrylates.[24] These low-viscosity acrylates (TAEP and DAEEP) decreased the viscosity of the neat urethane and epoxy oligomers and increased the photopolymerization rate and double-bond conversions. The LOI value of the epoxy acrylate/urethane acrylate-TAEP/DAEEP systems increased as the amount of the FR increased. However, none of the photocured urethane acrylate-TAEP/DAEEP systems were able to pass the UL94 test. On the other hand, 50 wt% TAEP- or DAEEP-containing epoxy acrylates were found to exhibit LOI values above 28 and these systems were classified as UL94 V-0.

The group of Wenfang Shi published several papers on photocurable FR systems. They used TAEP in another study.[25] It was blended with a

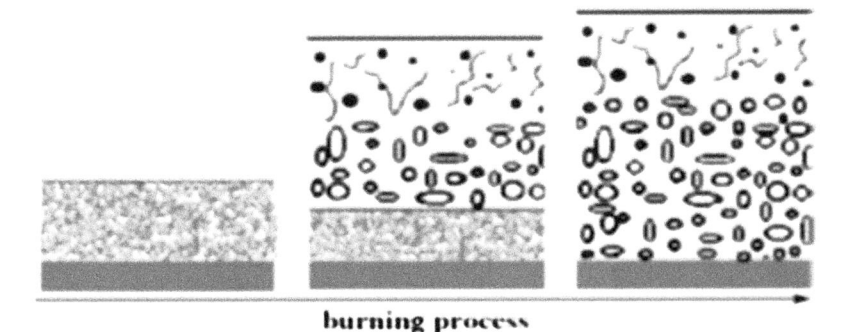

**Figure 9.3** Intumescent char formation mechanism. The solid lines and black bottom indicate the formed solid char; the blank ellipse and blank circles indicate the emitted gases; the black circles indicate the gases wrapped by the char and the gray part is the substrate.

methacrylated phenolic melamine (MAMP) and UV-curable polymeric materials were obtained. TAEP is a condensed-phase active phosphorus FR. MAMP generates volatile compounds on combustion, expands and works mainly on the gas phase. The idea behind using MAMP and TAEP together was to prepare an intumescent flame-retardant system. The term "intumescent" is used for flame retardants that expand and form carbonaceous foam upon heat exposure. A schematic representation of intumescent char formation is given in Figure 9.3. The flame retardancy of the system was studied by means of TGA, LOI and expansion calculations. When MAMP decomposes, it generates nonflammable gases that cause the char to expand. Thus, the insulating property of the char increases. According to the results, the TAEP–MAPM mixture at a certain ratio generated high amounts of char, a high degree of expansion and a LOI value of 41. Above this certain TAEP : MAPM ratio, the further increase in MAPM content resulted in a decrease in the fire-retardant properties due to the destructive effect of the volatile compounds on char thickness. Therefore, the authors concluded that the synergistic effect occurs between P and N at a certain ratio.

9,10-Dihydro-9-oxa-10-phosphaphenanthrene-10-oxide (DOPO) is a widely used flame retardant for many polymers and a precursor for the synthesis of functionalized monomers and additives. Chen *et al.*, found a smart way of synthesizing a novel flame-retardant coupling agent by using DOPO.[26] They first reacted DOPO with 3-glycidoxypropyltrimethoxysilane (GPTMS) and then the resulting –OH functional, P- and Si-containing monomer was reacted with TDI–HEA adduct in order to obtain both acrylate and methoxysilane groups bearing a coupling agent. The synthetic route for these monomers is depicted in Scheme 9.7. The monomer was mixed with 3 wt% Dorocure 1173 (photoinitiator) and 2 wt% dibutyl tin dilaurate (DBTDL) and it was dual cured first with UV light and then moisture. DBTLD was used as a catalyst for the curing of silane moieties. The LOI of the cured monomer was found to be as high as 48, owing to the synergistic effect of P,

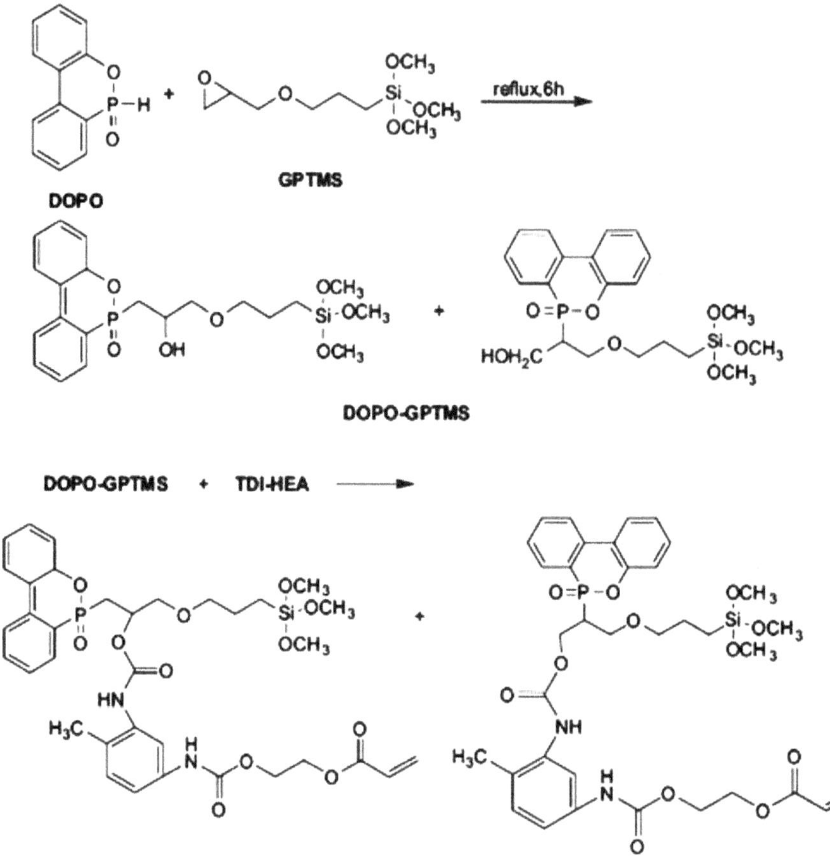

**Scheme 9.7** Preparation of DOPO containing flame-retardant coupling agent.

Si and also N (due to the presence of urethane linkages). They investigated the flame-retardant mechanism of this novel monomer by investigating the SEM images of the formed char and realized that a coherent and dense char was formed without gas bubbles, which led them to conclude that the condensed-phase flame-retardant mechanism was the dominant factor of flame retardancy. Similar to previous studies they confirmed the formation of poly(phosphoric acid) upon combustion by TGA-FTIR studies. This phosphorus- and silane-containing monomer was later used by the same group of authors and UV-curable intumescent flame-retardant resins were prepared by blending it with a six-armed star-shaped urethane acrylate.[27] According to Si-NMR results the hybrid system displayed high silanol conversion and a silica network was formed. The flame retardancy and the thermal properties of this organic–inorganic hybrid material were investigated by TGA-FTIR, LOI and cone-calorimeter measurements. TGA-FTIR studies supported the previous results of the degradation profile of the P- and Si-containing monomer. The LOI of the hybrid system first increased with the increase in N percentage then started to decrease due to the release of high amounts of volatile compounds that caused the char to worsen. The high flame retardancy of this system stems from the combination of different flame-retardant mechanisms of these three atoms. As the material burns, silanol groups start to form crosslinks and produce glass-like char while phosphorus groups also produce high amounts of phosphorus–carbon-rich char layer that keeps the rest of the material from burning. On the other hand, in contrast to Si, and P,N-containing species work in the gas phase and produce nonflammable gases that reduce the concentration of oxygen on the char layer and these gases cause char to expand that in turn decreases the heat transfer to the char-covered surface.[27] Cone-calorimeter results HRR value of the blend with the highest LOI value, decreased and the time of burning increased.

In another study, the same authors blended this six-armed star-shaped urethane acrylate with TAEP.[28] In both studies the highest LOI value was found to be 41. In contrast to the previous study they further investigated the flame retardancy of this new photocurable coating material by UL94 test. They found that neat TAEP and 3 and 6% N containing TAEP star-shaped urethane acrylate blends pass the UL94 V0 test. On the other hand, when the N content was 9%, the flame retardancy of this blend was decreased due to the corresponding decrease in P% and the factors that we discussed previously (such as; destruction of char by excessive volatile components).

In 2007 Chen *et al.*, published a paper where they once again demonstrated a clever strategy to prepare UV-curable flame-retardant coatings.[29] They synthesized a novel phosphate-based diacrylate, DMPE (Scheme 9.8). They investigated the double-bond conversion of DMPE by real time FTIR measurements and found that the 84% (Figure 9.4(a)) of the double bonds were consumed during the free-radical photopolymerization within 100 s. Moreover, this diacrylate exhibited a LOI value of 39 and under air atmosphere it yielded 53% and 36% of char at 600 °C and 800 °C, respectively (Figure 9.4(b)).

TAEP, which is one of the most widely used monomers for the preparation of photocurable FR coatings in recent years, has a slightly complicated synthetic process. Therefore, a similar phosphorous acrylate, TGMAP was synthesized from the reaction of GA and phosphoric acid.[30] The synthesis of TGMAP can be seen in Scheme 9.9. Its thermal properties were investigated by TGA and TGA-FTIR. This monomer yielded about 35% char at 700 °C under a nitrogen atmosphere. According to TGA results, TGMAP displayed a three-stage degradation profile in which first phosphate groups were decomposed, then the pyrolysis of aliphatic chains and acrylate moieties took place. Finally, the unstable structures were degraded in the char. The authors compared the char yield of the TGMAP with the char yield of TAEP and DAEEP and concluded that both the decomposition temperature and residual weight of TGMAP were higher. They compared their results with the results of ref. 23 (in this manuscript). However, in ref. 23 the TGA data was collected under an air atmosphere for TAEP and DAEEP while the TGA of

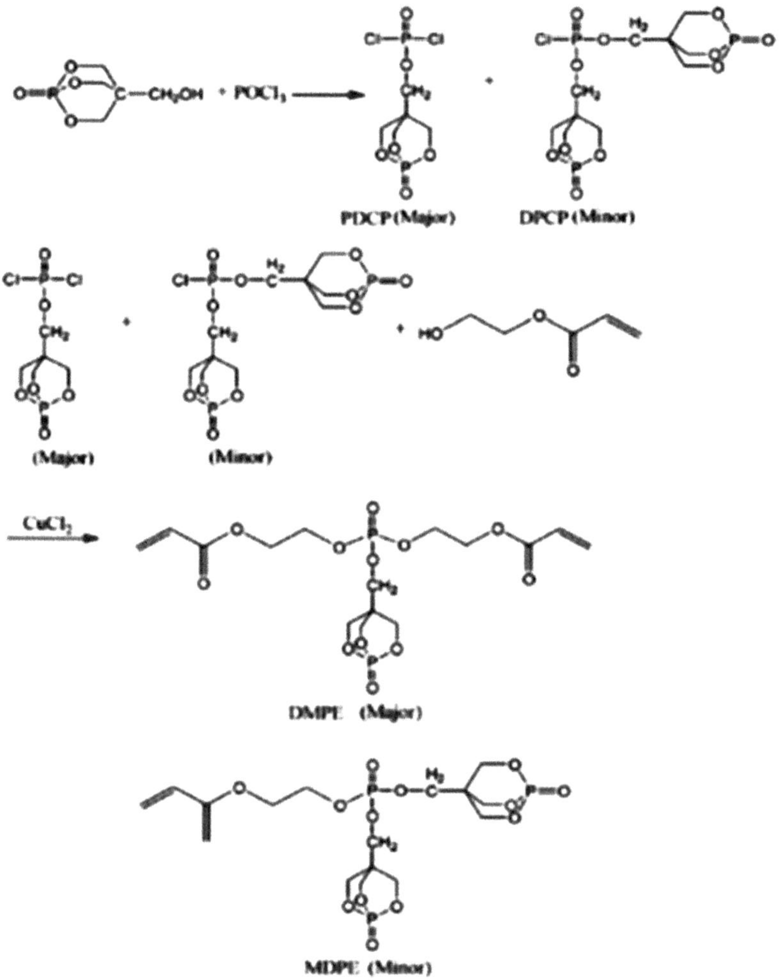

**Scheme 9.8** Preparation of DMPE.

TGMAP was performed under a nitrogen atmosphere. Thus, it is not possible to make a direct comparison between the results of these two papers. On the other hand, it is well known that polymeric materials degrade faster and fiercely under air atmosphere than they do under a blanket of nitrogen gas, due to thermo-oxidative degradation. Thus, it can be expected that the thermal stability of TGMAP would be lower than the thermal stability of TAEP and DAEEP in the presence of oxygen.

Phosphate-based acrylates are generally preferred for the preparation of photocurable materials over other organophosphorus compounds. The attraction of using phosphate-based acrylates is the ease of the synthetic procedures when compared to other compounds such as phosphine oxides. However, the hydrolytic stability of the latter class of compounds is higher

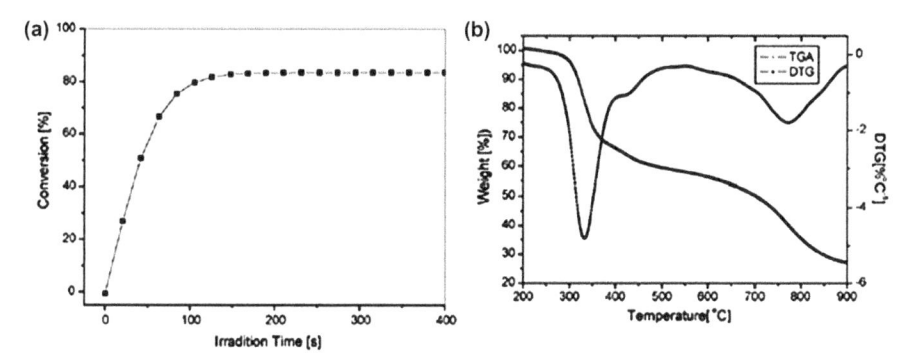

**Figure 9.4** (a) Double-bond conversion and (b) TGA of DMPE-based photocured coatings.

**Scheme 9.9** Preparation of TGMAP.

than these widely used acrylated phosphates. For instance, Kayaman-Apohan et al., prepared UV-curable coating materials by using an acrylated phenylphosphine oxide oligomer (APPO) that was synthesized in four steps.[31] APPO, a triphenylphosphine oxide containing diacrylate was added in different amounts to a mixture of silane precursor and urethane/polyester acrylates. The obtained formulations were poured into Teflon molds and cured under UV light. APPO containing organic–inorganic hybrid coatings exhibited high conversions and high gel contents. According to Si-NMR results the incorporation of APPO increased the condensation of the silica network. Gloss and pendulum hardness values of the hybrid coatings increased as the APPO concentration was increased. Interestingly, as was evident from SEM images, fibrillar structures were formed when APPO was added to the sol–gel containing acrylate mixture. These fibrillar structures at the nanoscale reinforced the coating material and resulted in an increase in the modulus of the coatings. The thermal properties of this phosphine-oxide-containing photocured coatings were investigated by means of TGA. The thermal stability of the coating materials increased with increasing sol–gel and APPO resin percentages. These coating materials exhibited higher experimental char yields than the theoretical values due to the strong interactions between the silica network and the organic matrix.

One year later, the same group synthesized another phosphine oxide derivative and prepared organosilica containing hybrid coatings by photopolymerization and sol–gel process.[32] Diallylphenylphosphine oxide monomer (DAPPO) was synthesized *via* Grignard reaction with allylmagnesium bromide and dichlorophenlyphosphine oxide at a yield of 55% (Scheme 9.10). A sol–gel precursor that constitutes the inorganic part was prepared by tetraethoxysilane (TEOS) and 3-methacryloxypropyl trimethoxysilane (MAPTMS) in the presence of ethyl alcohol, water and catalyst (*para*-toluene sulfonic acid). The organic part consisted of a mixture of an aliphatic urethane acrylate, reactive diluents, photoinitiator and DAPPO. Then, different amounts of DAPPO-containing hybrid coatings were applied on corona-treated Plexiglas materials. The gloss of the coatings did not change much with the addition of DAPPO. On the other hand, DAPPO enhanced the mechanical properties by increasing the crosslinking density and chain–chain interactions due to the presence of the polar phosphine

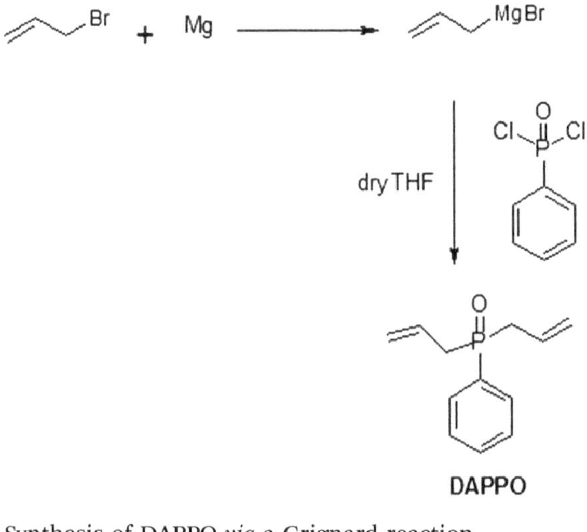

**Scheme 9.10** Synthesis of DAPPO *via* a Grignard reaction.

oxide group. The polar phosphine oxide moiety also enhanced the adhesion of the coatings to Plexiglas substrates. TGA under an air atmosphere showed that the incorporation of the thermally phosphine-oxide monomer increased the thermal stability and the char yield of the coatings. High char yields of the DAPPO and silica-containing coatings were correlated to the flame retardancy. All coating materials exhibited a single-step degradation.

Another Si-containing coating material was prepared by Xing et al.[33] First, a silicon-containing diacrylate (SHEA) was prepared from the reaction of HEA and dimethyldichlorosilane. SHEA was then mixed with the popular phosphate-based triacrylate; TAEP, at different ratios and the resulting mixtures were cured under UV irradiation in the presence of 3 wt% Darocur 1173 (photoinitiator). Thermal stability and the flame-retardant behavior of the resulting coatings were investigated by TGA, TGA-FTIR and microscale combustion calorimetry (MCC). All formulations exhibited 50 wt% weight loss at temperatures around 420–450 °C. The increase of TAEP concentration increased the maximum weight loss temperatures (50 wt% weight-loss temperatures). The formulation with 1:1 mass ratio of TEAP:SHEA exhibited the highest thermal stability among other formulations. This coating material was found to produce higher char yield than TEAP or SHEA used alone and it displayed the highest onset degradation temperature value. The authors attributed the enhanced thermal properties of this composition to the formation of a Si–O–Si network and silicon dioxide formation along with the synergistic effect between P and Si. In this system the early decomposition of the P–O–C and Si–O–C bonds form high char material at the beginning of combustion and thus increase the thermal stability of the system. On the other hand, the increase in the concentration of SHEA increases the Si–O–C bonds that have relatively low stability and result in a decrease in the thermal stability of the coatings. According to MCC results, while the HRR of SHEA was 280 W g$^{-1}$, it decreased remarkably with the addition of TAEP and the best formulation displayed an HRR value of 65 W g$^{-1}$, indicating a synergistic effect between Si and P. Moreover, the authors investigated the type of evolved gases by using TGA-FTIR technique (Figure 9.5(a)). They found that only SHEA-containing films and the films with the 1:1 mass ratio of SHEA:TAEP displayed different thermal degradation mechanisms evidenced from the TGA-FTIR results. They found that the gas products of these two systems were different. SHEA released $CO_2$ and gaseous alkane compounds at around 500 °C (Figure 9.5(c)), meanwhile the release of the volatile compounds in SHEA:TAEP-containing films (Figure 9.5(b)), were found to occur at approximately 100 °C lower temperatures due to the early decomposition of TAEP that generated poly(phosphoric acid) followed by the formation of high amounts of carbonaceous char layer.

Fluorinated materials are known to exhibit hydrophobic surface properties due to their low surface energies and fluorinated monomers are widely utilized for the preparation of photocurable hydrophobic coatings.[34] Miao et al., combined the low surface energy properties of the fluorinated

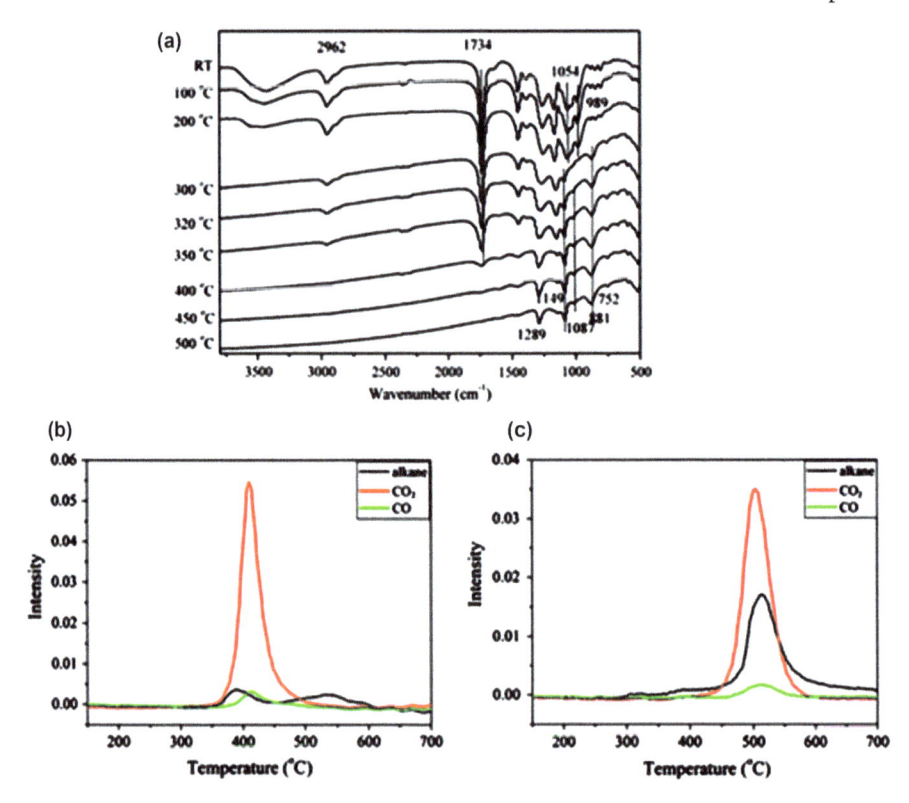

**Figure 9.5** (a) Real-time FTIR spectra of SHEA : TAEP (1 : 1) mixture and the relationship between intensity of characteristic peak and temperature for evolved gases for (b) SHEA : TAEP (1 : 1) and (c) SHEA alone.

monomers with the flame-retardant properties of phosphorus compounds and synthesized acrylated perfluoroalkyl phosphate oligomers.[35] Scheme 9.11 displays the two different strategies employed for the synthesis of these oligomers. The structures of these both fluorine and phosphorus-containing acrylates were characterized by FTIR and NMR techniques. UV-cured coatings containing these acrylates showed moderate double-bond conversions. The flame retardancy of these acrylates was investigated by LOI measurements. The LOI values of the coatings were found to change between 36 and 38. These values represent the limiting oxygen index values for the coatings containing 100% acrylated perfluoroalkyl phosphate oligomers. When the authors tried to prepare photocurable compositions with an aliphatic urethane acrylate, a reactive diluent and these novel acrylated species at high concentrations, they observed phase separation and thus they succeeded in only preparing photocurable formulation containing low levels of fluorine. However, these low levels of fluorine-containing coatings exhibited high water contact angles. When only 1 wt% acrylated perfluoroalkyl phosphate oligomer was added to the urethane acrylate–reactive diluent mixture, the contact angle of the coating

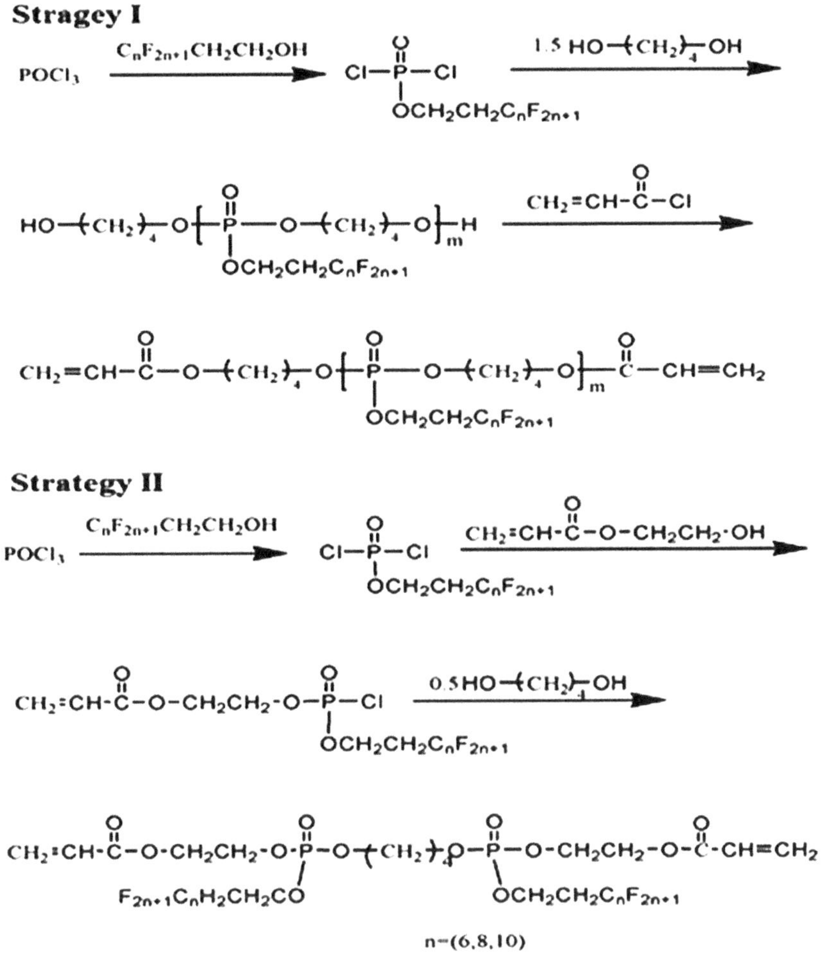

**Stragey I**

$$POCl_3 \xrightarrow{C_nF_{2n+1}CH_2CH_2OH} \underset{OCH_2CH_2C_nF_{2n+1}}{Cl-\overset{\overset{O}{\|}}{P}-Cl} \xrightarrow{1.5\ HO-(CH_2)_4-OH}$$

$$HO-(CH_2)_4-O{\left[-\underset{OCH_2CH_2C_nF_{2n+1}}{\overset{\overset{O}{\|}}{P}}-O-(CH_2)_4-O\right]}_m H \xrightarrow{CH_2=CH-\overset{\overset{O}{\|}}{C}-Cl}$$

$$CH_2=CH-\overset{\overset{O}{\|}}{C}-O-(CH_2)_4-O{\left[-\underset{OCH_2CH_2C_nF_{2n+1}}{\overset{\overset{O}{\|}}{P}}-O-(CH_2)_4-O\right]}_m \overset{\overset{O}{\|}}{C}-CH=CH_2$$

**Strategy II**

$$POCl_3 \xrightarrow{C_nF_{2n+1}CH_2CH_2OH} \underset{OCH_2CH_2C_nF_{2n+1}}{Cl-\overset{\overset{O}{\|}}{P}-Cl} \xrightarrow{CH_2=CH-\overset{\overset{O}{\|}}{C}-O-CH_2CH_2\cdot OH}$$

$$CH_2=CH-\overset{\overset{O}{\|}}{C}-O-CH_2CH_2-O-\underset{OCH_2CH_2C_nF_{2n+1}}{\overset{\overset{O}{\|}}{P}}-Cl \xrightarrow{0.5\ HO-(CH_2)_4-OH}$$

$$CH_2=CH-\overset{\overset{O}{\|}}{C}-O-CH_2CH_2-O-\underset{F_{2n+1}C_nH_2CH_2CO}{\overset{\overset{O}{\|}}{P}}-O-(CH_2)_4-O-\underset{OCH_2CH_2C_nF_{2n+1}}{\overset{\overset{O}{\|}}{P}}-O-CH_2CH_2-O-\overset{\overset{O}{\|}}{C}-CH=CH_2$$

n~(6,8,10)

Scheme 9.11 Preparation of acrylated perfluoroalkyl phosphate oligomers.

was around 110–115°. This value is remarkable when it is considered that a flat surface (no roughening) covered with only $-CF_3$ groups can only exhibit a maximum contact angle value of 120.[36]

In the previous example, the authors prepared fluorine- and phosphorus-containing oligomers *via* an end-capping strategy. Either fluorine and acrylate groups containing phosphate derivative was first prepared and reacted with butanediol or a fluorinated phosphate derivative was first reacted with butanediol and then end capped with acryloyl chloride.[35] A similar strategy was also applied by other authors.[37,38] For instance, a tetra acrylate (BDEEP) was synthesized using the same strategy and controlling the stoichiometry of the reactants.[37] 2 moles of phosphorus oxychloride was first reacted with a mole of butanediol and then it was reacted with 4 moles of HEA. The resulting tetra acrylate, which has a relatively high P content (10.4%),

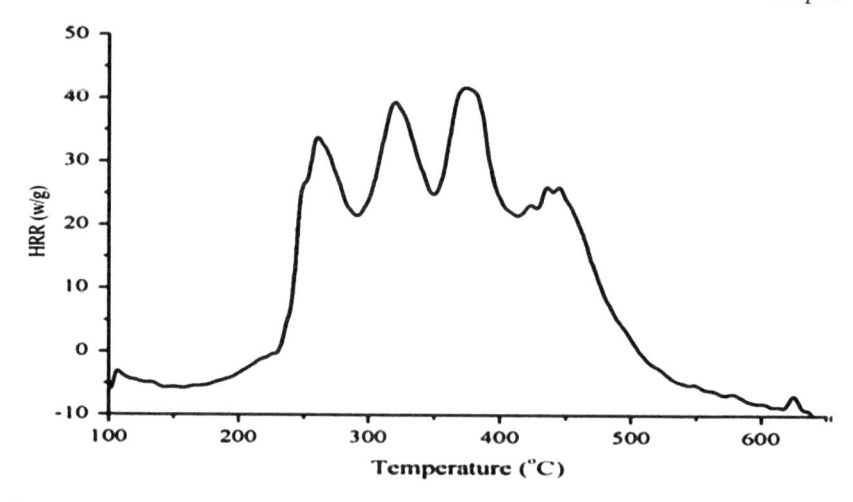

**Figure 9.6** The curve of specific heat-release rate *versus* temperature for BDEEP.

displayed a high gel content (96%) and high double bond conversion (84%) values. The fire-retardant properties of the coating material were investigated by MCC, TGA and TGA-FTIR. The HRR of this coating was found to be 42.1 W g$^{-1}$ (Figure 9.6). According to TGA and MCC results, BDEEP displayed a fast decomposition at relatively low temperatures around 200–250 °C. On the other hand, it exhibited a high maximum weight loss temperature (357 °C) and high char yield (26%) at 800 °C under an air atmosphere. TGA-FTIR results revealed that BDEEP has similar degradation characteristics to TAEP and DAEEP.

In another work, piperazine was end capped with an acrylated phosphate and a novel intumescent FR diacrylate; piperazine-*N*,*N*′-bis(acryloxyethylaryl-phosphoramidate (NPBAAP) was synthesized.[38] The structure of NPBBAP was characterized by FTIR, $^1$H-NMR and $^{31}$P-NMR measurements. This P- and N-containing monomer was mixed with an epoxy acrylate at different ratios. NPBAAP displayed a three-stage degradation profile in the presence of oxygen: onset degradation, a maximum weight loss between 250–450 °C and a final weight loss above 650 °C.

On the other hand, under a nitrogen atmosphere it displayed a similar thermal behavior without the final weight loss observed at high temperatures under an air atmosphere. This novel coating material highly expanded upon combustion and honeycomb-like structures and bubbles were observed from the SEM images of the char of NPBAAP, which all displayed the characteristics of an intumescent behavior (Scheme 9.12). The thermal degradation of this monomer was investigated by real-time infrared studies. First, the peak intensity of the aliphatic groups decreased with increasing temperature. Then, at above 200 °C, P–N–C bonds started to decompose and P/N-containing species were formed. At higher temperatures P–O–P bonds containing compounds occurred. Moreover, the authors examined the photocured films of NPBAAP with TGA-FTIR in order to identify the volatile

**Scheme 9.12** Photograph and SEM micrograph of the intumescent char residue of NPBAAP.

species generated during the combustion process. This allowed them to make a projection of the degradation mechanism, as summarized in Figure 9.7.

The effect of NPBAAP on the thermal properties of epoxy acrylates was also investigated. Both under air and nitrogen atmospheres, the char yield of the coatings increased with increasing amount of NPBAAP. Interestingly, the char yield of the 30 wt% NPBAAP-containing epoxy acrylate films yielded a significant amount of char in the presence of oxygen, even higher than the pure NPBAAP. It was stated than this situation could be attributed to the interactions between the degradation products of NPBAAP and epoxy acrylate in an oxidative environment. Moreover, the HRC values decreased with the addition of NPBAAP.

In another study an acrylate oligomer, similar to NPBAAP was prepared and also utilized as a FR monomer for the epoxy acrylates.[39] POCl$_3$ was reacted with HEA and then with piperazine in order to obtain the P- and N-containing FR monomer, POPHA. 10%, 20% and 30% POPHA-containing epoxy-acrylate-based coatings were prepared by UV-induced free-radical polymerization in the presence of 3 wt% photoinitiator (Dorocure 1173). The fire resistance of the photocured coating material was characterized by LOI measurements. The thermal behavior of the coatings was analyzed by TGA, TGA-FTIR and pyrolysis/mass (DP-MS). The LOI of the coatings increased as the concentration of POPHA increased in the coatings. The LOI of the pristine epoxy acrylate coatings was 21 while it increased to 29 when the POPHA loading was 20 wt%. This composition contained 2.4% P and 1.1% N. With further increase in POPHA concentration (30 wt%) the LOI only increased to 29.5 although the P and N percentages were increased to 3.6 and 1.65%, respectively. According to TGA, char yields increased with POPHA loadings, while 10 wt% loss temperatures and the maximum weight loss temperatures decreased due to the early decomposition of POPHA. As observed in previous studies where phosphorus-containing acrylates were used, as a result of the degradation of POPHA poly(phosphoric acid) was generated and it catalyzed the degradation of the epoxy acrylate that in turn decreased maximum weight loss temperature. The RTIR and TGA-FTIR results were consistent with the previous studies. DP-MS is a

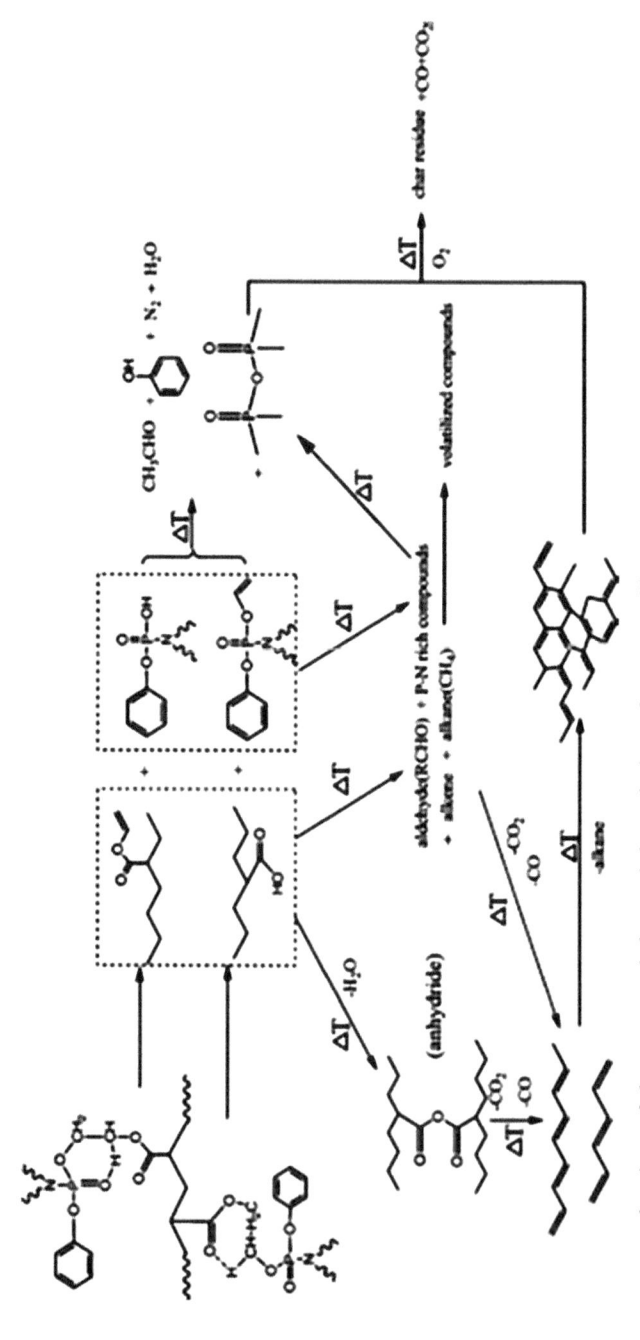

**Figure 9.7** Mechanism of the proposed thermal degradation of NPBAAP film.

technique where the polymeric material is subjected to pyrolysis under high vacuum. Then, the ionized volatile compounds are collected by the detector. The advantage of this technique is that primary pyrolysis products can be detected.[40] According to DP-MS results at 190 °C, piperazine and its derivatives were released from the 30 wt% POPHA-containing coatings. On the other hand, at the same temperature, the intensity of the peaks for the piperazine derivatives was found to be low for the 20 wt% POPHA containing coatings. This situation was attributed by the authors to the interactions between the epoxy acrylate and POPHA when the concentration of POPHA was 20%.

TAEP kept its popularity among researchers working in the area of flame-retardant photocurable materials in the last few years and it was mixed with another monomer, triglycidyl isocyanurate acrylate (TGICA), which has also recently gained popularity.[41-43] TGICA is a nitrogen-containing triacrylate monomer that is synthesized by reacting triglycidyl isocyanurate with acrylic acid. In one study a 1:1 mixture of TAEP and TGICA was used to prepare different amounts of P- and N-containing solutions in acetone.[41] Cotton fabrics were impregnated with these flame-retardant acrylic solutions and then cured with UV irradiation. Modification of the cotton fabrics was proved by FTIR and SEM analysis. ICP-AES was used to determine the phosphorus content of the coated cotton fabrics. Nitrogen content was measured by elemental analysis. The thermal properties of the modified fabrics were investigated by TGA and the flame retardancy of the fabrics was determined by TGA-IR, LOI and MCC measurements. The amount of P and N contents on the fabrics increased with the increase in the concentration of TAEP–TGICA in acetone solutions. The LOI value of the uncoated fabric was found to be 21. It was increased to 24.5 when the P and N concentrations were 0.72% and 1.8%, respectively. As expected, the char yields of the coated fabrics increased and HRR values were decreased. The same research group also investigated the effect of TAEP–TGICA mixture on the flame retardancy of epoxy acrylates.[42] In addition to these monomers, alpha-zirconium phosphate ($\alpha$-Zr(HPO$_4$)$_2$H$_2$O, $\alpha$-ZrP) was incorporated into this epoxy-acrylate-based UV-curable coating. $\alpha$-ZrP is a clay-like, layered inorganic additive. XRD results displayed that epoxy acrylate was intercalated and/or exfoliated. The flame retardancy of $\alpha$-ZrP-containing epoxy-acrylate system was studied by means of MCC. As a result, the HRR value of the only TEAP–TGICA-containing epoxy acrylate was decreased. When only 1 wt% $\alpha$-ZrP was added, the HRR value was further decreased slightly. However, with further addition of $\alpha$-ZrP, HRR values increased. Char residues increased with increasing $\alpha$-ZrP content. The authors concluded that $\alpha$-ZrP increased the flame retardancy of these UV-curable coatings by decreasing the heat transfer between polymer chains. Thus, the decomposition of the polymeric chains intercalated inside the layers of $\alpha$-ZrP was retarded. Furthermore, the release of water from the $\alpha$-ZrP containing films was higher than the $\alpha$-ZrP free films upon combustion. Also the release of $CO_2$ was found to be less in $\alpha$-ZrP-containing coatings.

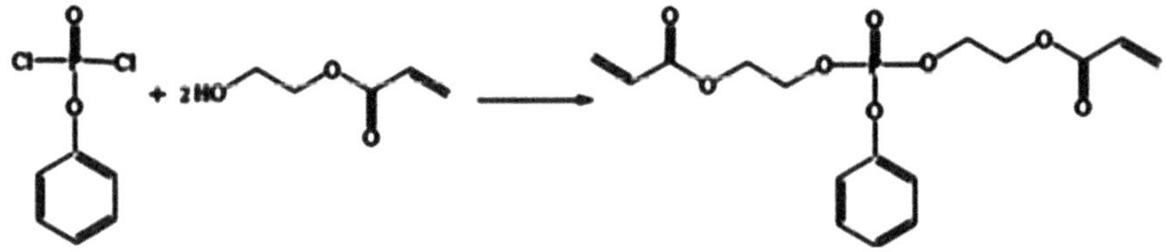

**Scheme 9.13** Preparation of PDHA.

Textile materials constitute an important part of our daily life, but unfortunately these materials are generally flammable. Therefore, preparing flame-retardant textile materials is of great importance. In another study a novel phosphorous monomer, PDHA, was grafted on cotton fabrics by UV-induced graft polymerization.[43] PDHA was prepared from the reaction of phenyl dichlorophosphate and HEA (Scheme 9.13). Then, cotton fabrics were treated similar to ref. 41. PDHA-grafted fabrics resulted in higher LOI values than the previous study (ref. 41) that could be related to the increase in the phosphorus percentage and the presence of aromatic moiety in PDHA. PDHA was able to increase the LOI of the cotton fabrics up to 27. The durability of the flame-retardant coating on the cotton fibers was investigated by comparing the MCC results before and after washing the cotton fabrics. The HRR values of the PDHA-coated fabrics slightly increased after washing.

PDHA was also used to prepare nanocomposites.[44] It was blended with TGICA, 2-phenoxyethyl acrylate (PHEA) and α-ZrP. Zirconium phosphate was randomly distributed in the polymer matrix according to XRD results. The thermal and flame-retardant properties of the coatings were improved by the addition of 0.5% α-ZrP. However, higher α-ZrP concentrations resulted in slight decreases in thermal properties when compared 0.5% α-ZrP containing coatings. This situation was attributed to the high phosphate concentration that restrained the expansion of intumescent flame-retardant system. Also, it was found that the addition of relatively low amounts of zirconium phosphates did not cause a dramatic decrease in the optical transmittance properties of the coatings.

Our popular monomer TGICA was also utilized in another study where another novel monomer was synthesized.[45] First, 2,2-dimethyl-1,3-propanediol phosphoryl chloride was prepared by reacting neopentyl glycol with POCl$_3$. The resulting phosphorus chloride was then reacted with HEA to obtain 2,2-dimethyl-1,3-propanediol acryloyloxyethyl phosphate (DPHA) monomer. DPHA- and TGICA-containing photocured coatings were prepared. The structure of DPHA and its synthetic reactions can be seen in Scheme 9.14. DPHA was found to exhibit an HRR value as high as 476 W g$^{-1}$. On the other hand, DPHA when used in combination with TGICA resulted in coatings with low HRR values. This situation can be attributed to the synergistic effect between these monomers and their intumescent flame-retardant characteristics as supported by previous studies. CO, CO$_2$, water, ammonia, some carbonyl-containing compounds, phosphorus oxides and

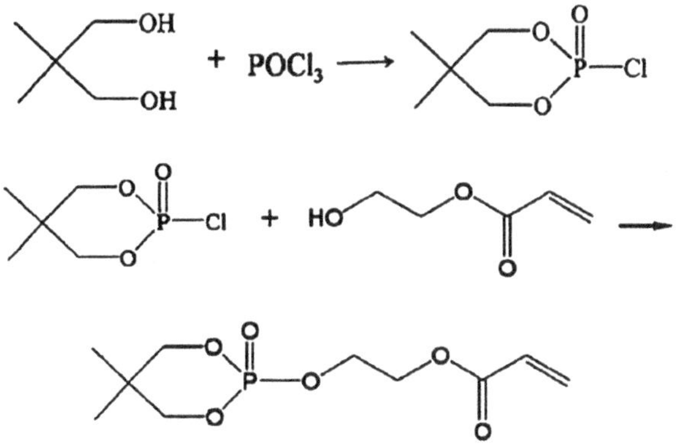

**Scheme 9.14** Synthesis of DPHA.

aromatic compounds were formed during combustion, as evidenced by TGA-FTIR results.

Up to now we have tried to summarize some developments in the area of FR photocurable coatings. Although we mentioned some of the studies that focus on the enhancement of the flame retardancy of especially epoxy-acrylate-based coatings, we now want to give other examples of flame-retardant epoxy acrylate systems that were prepared in the last decade.

While phosphorus-based flame retardants are being discussed, we cannot leave without saying a few words about another class of polymers: phosphazenes. Due to the presence of both P and N in their structures, they inherently display intumescent FR characteristics.

Ding and Shi, prepared an acrylated phosphazene derivative (HACP) by simply reacting HEA with hexachlorocyclotriphosphazene.[46] They blended an epoxy acrylate oligomer, photoinitiator and HACP and then cured under UV irradiation. They also prepared another phosphazene FR where they used ethyl alcohol for the substitution of the chlorine units and named the product as HECP. HACP and HECP mixed well with the epoxy acrylate and transparent films were obtained. HACP- or HECP-containing epoxy acrylates produced more char than the neat epoxy according to TGA results. HECP-containing films even produced higher char yields than the same amount of HACP containing films due to an increase in P content. On the other hand, the LOI value of the HECP-epoxy acrylate systems failed to reach the LOI values of HACP-containing epoxy acrylate systems. The highest LOI value was found to be 28.5 for the 40 : 60 HACP : epoxy-acrylate-containing coatings. The lower flame retardancy of the HECP-containing films was attributed by the authors to the presence of weakly bonded ethyl groups and their flammable volatile degradation products.

Wang and Shi prepared acrylated benzenephosphonates (Scheme 9.15) in an effort to achieve high flame retardancy in epoxy acrylates.[47] Phenyl-phosphonic dichloride was reacted with two moles of HEA or first reacted with 1 mole of phenol and then with 1 mole of HEA in order to obtain

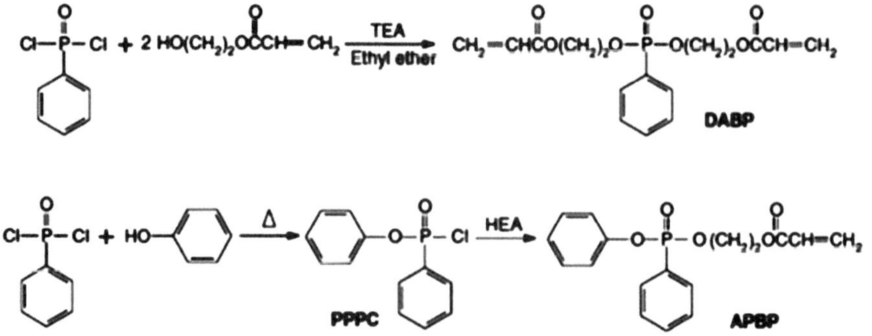

**Scheme 9.15** Synthesis of acrylated benzene phosphonates.

di(acryloyloxyethyl) benzenephosphonate (DABP) and acryloyloxyethyl phenyl benzenephosphonate (APBP). These benzene phosphonates were then blended with an epoxy acrylate and photocurable coatings were prepared. Due to the decrease in the viscosity of the blends compared to neat epoxy, the photopolymerization rate was increased. DABP-containing epoxy acrylates behaved almost the same as APBP-containing ones. No significant differences in wither the thermal degradation behaviors or in LOI values, were observed. The highest LOI value (30) was obtained with the films containing 50 wt% DABP, which alone had a LOI value of 32.

In one study[48] an epoxy acrylate was directly modified with a phosphorus-containing isocyanate in contrast to other studies that generally blend epoxy acrylate with another acrylic precursor. Xilei Chen and Chuanmei Jiao first prepared an isophorone diisocyanate (IPDI) adduct of 1-oxo-4-hydroxymethyl-2,6,7-trioxa-1-phosphabicyclo[2.2.2]octane (PEPA). Then, the epoxy acrylate was modified with this adduct by the formation of urethane linkages from the reaction of isocyanate with hydroxyl groups. 30% PEPA-IPDI adduct containing films displayed a LOI value of 29.5 and were able to pass the UL94 test. All modified epoxy acrylates exhibited decreased onset degradation temperatures but increased char yields. On the other hand, it was found that there was no significant difference in the char yields of the different amounts of PEPA-containing epoxy acrylates. This situation led the authors to conclude that the increased flame retardancy was also due to the gas-phase flame-retardant mechanism in addition to the condensed-phase flame retardancy. HRR values decreased in the modified epoxy-acrylate-based coatings, while the combustion times were prolonged.

Recently, an interesting study was performed by Hu et al.[49] They studied the effect of modified chitosan (Scheme 9.16) on the flame retardancy of epoxy acrylates. Chitosan was first phosphorylated then further functionalized with GA. 5, 10 and 20 wt% modified chitosan-containing coatings were prepared by photoinitiated free-radical polymerization. The LOI value was increased gradually from 21 to 26. HRR values of the modified

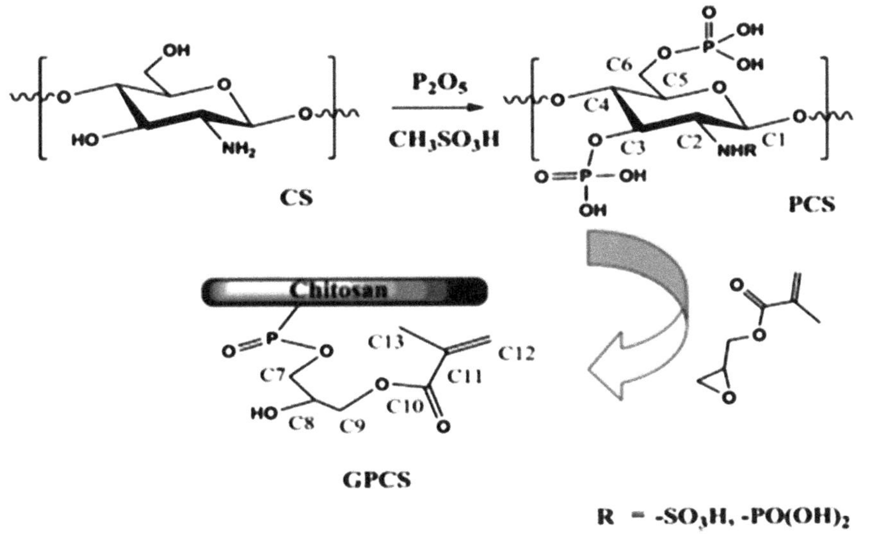

**Scheme 9.16** Preparation of acrylated and phosphorylated chitosan.

chitosan-containing coatings decreased as the P content was increased. The incorporation of the modified chitosan into the epoxy acrylate caused a decrease in the early stages of thermogravimetric analysis due to the relatively low stability of chitosan. On the other hand, when the temperature exceeded 450 °C, chitosan-containing films exhibited higher thermal stability and produced higher char yields than the neat epoxy. According to the dynamic mechanical analysis (DMA) results the modulus of the coatings first increased with the addition of chitosan but decreased with the further addition due to the decrease in crosslinking density. The decrease in crosslinking density also influenced the $T_g$ of the coatings and a continuous decrease in $T_g$ was observed with the addition of the modified chitosan.

In 2012, we saw a familiar phosphorus compound, DOPO once again.[50] In this study sol–gel and UV-curing technologies were combined to prepare FR epoxy acrylate coatings. DOPO was reacted with vinyltrimethoxysilane. A commercial epoxy acrylate was then modified with (3-isocyanatopropyl)triethoxysilane. Then both components were mixed at different ratios and stirred for 5 h under acidic conditions in order to form a silicon network. After that samples were cured under UV illumination. Si-NMR displayed that the system was efficiently crosslinked by the condensation of the siloxane units. As the P and Si contents were increased the LOI value of the epoxy acrylate was increased and when 30 wt% DOPO-vinyltrimethoxysilane was used, the LOI value reached 31.5. According to TGA and DP-MS results, first phosphate and silicon groups decomposed than the main epoxy acrylate chains started to degrade. Also, the pencil hardness values increased with the increase in the siloxane content.

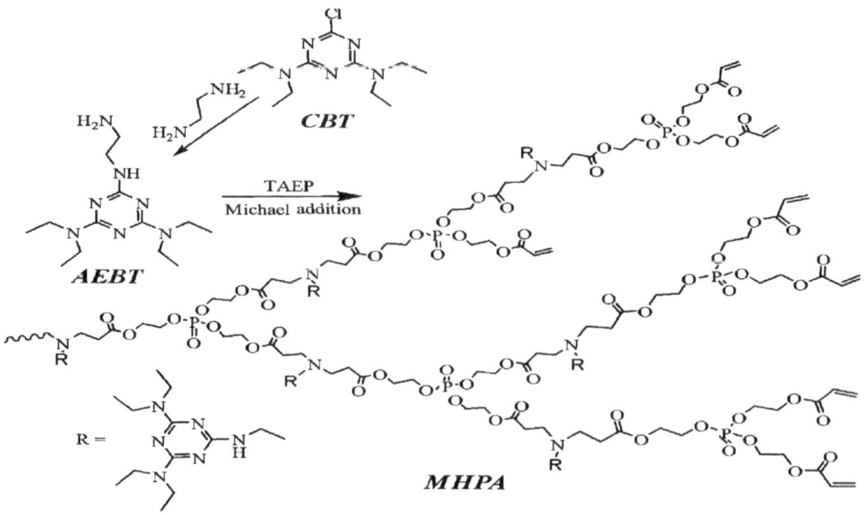

**Scheme 9.17** Preparation of hyperbranced polyphosphate acrylates by Michael addition (CBT: 2-chloro-4,6-bis(diethylamino)-s-triazine, AEBT: 2-(2-amino-ethylamino)-4,6-bisethylamino-1,3,5-triazine and MHPA: melamine-based hyperbranched polyphosphonate acrylate.)

We now want to discuss the studies on flame-retardant epoxy acrylates and continue with an interesting strategy to prepare flame-retardant coatings.

The group of Wenfang Shi from China and also some other authors prepared phosphorus–nitrogen-containing hyperbranched polyphosphate acrylates by the Michael addition of amine groups bearing compounds like piperazine or butyl amine to phosphorus acrylates like TAEP.[51–56] This methodology allowed researchers to prepare FR coatings with LOI values above 40.[51,52]

This strategy was also applied to epoxy acrylates.[55,56] In one study, TAEP was reacted with butyl amine[55] and in another one it was reacted with a melamine derivative.[56] An example for this strategy is given in Scheme 9.17. The addition of these hyperbranched materials into epoxy acrylate improved the flame retardancy and the LOI values were found to change to around 24.5–28 with respect to the type of the hyperbranced polyphosphate acrylate used.

Finally, we want to discuss another study where Qian et al., prepared a novel phosphorus- and silicon-containing monomer (KHPDH).[57] This monomer was prepared by reacting one mole of phenyl dichloro-phosphate with a mole of HEA and subsequently with a mole of 3-aminopropyl triethoxysilane in the presence of triethylamine. As described in the previous similar monomers, first KHPDH was mixed with an epoxy acrylate in different ratios then the siloxane groups of KHPDH were hydrolyzed and condensed. After that, the resulting hybrid materials were cured under UV light. 10 to 30 wt% KHPDH-containing coatings were prepared. The chemical structure of the KHPDH was characterized by $^1$H NMR, $^{31}$P NMR and $^{29}$Si

NMR techniques. DSC results revealed that the $T_g$ of the coatings decreased as the amount of KHPDH increased. This was attributed to the increase in the free volume of the coatings. According to TGA results neat epoxy acrylate coatings had better thermal stability at low temperatures when compared to the hybrid coatings. This situation was once again attributed to the relatively low stability of the P–O and P–N bonds. On the other hand, the thermal stability of the KHPDH-containing hybrids exhibited enhanced thermo-oxidative stability at temperatures above 400 °C. The LOI value of the 30 wt% KHPDH containing coatings was found to be 28. In contrast to previous studies, the char was investigated by XPS analysis in this work. It was found that the surface of the char residue was richer in silicon than the interior part of the char. Thus, the authors concluded that both the phosphorus and the silicon layer formed in the char enhanced the flame retardancy of the epoxy acrylate.

One year after the year of publication of the above article, we prepared a similar Si- and P-containing compound.[58] This compound, bis-(triethoxysilylpropyl) phenyl phosphamide (BESPPA) was synthesized by reacting dichlorophenyl phosphine oxide and 3-aminopropyltriethoxysilane (Scheme 9.18). BESPPA was then mixed with tetra (3-mercaptopropionate), glyoxal bis(diallyl acetal) and with different amounts of MAPTMS based sol–gel precursor. These compositions were first cured with UV light then postcured thermally at 100 °C for 24 h. Onset degradation temperatures

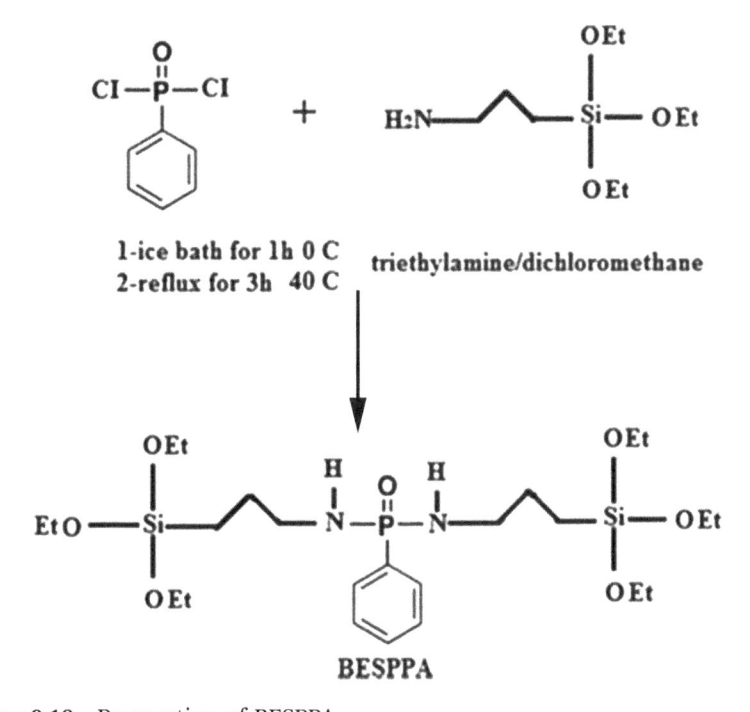

Scheme 9.18 Preparation of BESPPA.

slightly decreased with increasing amount of sol–gel component and BESPPA. On the other hand, both the maximum weight loss temperatures and the char yield of the hybrid thiol–ene photocured coatings increased. Water contact-angle values also increased when the sol–gel precursor was increased. The LOI of the thiol–ene photocured hybrid coatings increased from 20.9 to 26.5.

## 9.3 Boron-Containing Flame-Retardant Photocurable Coatings

With a share of 72.2%, Turkey has the largest boron reserves in the world.[59] Boron-containing flame retardants such as boric acid and its derivatives, borates and borax have been in use for a long time. Similar to phosphorous-based FRs, boron-containing flame retardants also mostly act in the condensed phase.

When compared to the studies in which phosphorous compounds were used, we see that there are only a few studies (patents not included) on boron-containing flame-retardant photocurable systems. In spite of the rich possibilities in organoboron chemistry[60] and there are some commercially available boron-containing vinyl and allyl compounds, the low number of published articles in this area may be due to the relatively higher cost of boron compounds.

We want to start with a B- and P-containing FR additive, boron phosphate ($BPO_4$) whose effect on flame retardancy was investigated by several studies for several polymers.[61] Petric et al. prepared $BPO_4$-containing photocured urethane-acrylate- and urethane/epoxy-acrylate-based FR coatings.[62] Boron phosphate is a low-cost additive that is generally prepared by reacting equimolar amounts of boric acid and phosphoric acid. They were able to prepare flexible films with $BPO_4$ loadings as high as 50 wt%. When urethane acrylate was mixed with 50 wt% $BPO_4$, the LOI value increased to 38. The LOI value of the 1 : 1 mixture of urethane/epoxy acrylate containing films was found to be 16 and the incorporation of 50% $BPO_4$ doubled this LOI value. The authors also stated that the smoke generation was less when $BPO_4$ was used. Thus, it can be concluded that $BPO_4$ is a good FR additive for acrylic coatings. Also, it can be said that $BPO_4$ follows the general rule that high loadings of FR additives are necessary to reach high LOI values, On the other hand, its effects on the mechanical, thermal and also optical properties were not investigated.

Boron-containing compounds have special features like neutron shielding beside their flame-retardant properties.[63] Mülazim et al., prepared $^{10}$B-containing organic–inorganic hybrid coating materials via a sol–gel route. In order to prepare the sol–gel component, methacryloxymethy-triethoxysilane was hydrolyzed with $H_3{}^{10}BO_3$. The resulting mixture consisted of hydrolyzed MEMO and a borate ester, $^{10}B(OEt)_3$. They mixed this inorganic component with the organic part of the hybrid system that was prepared by mixing a polyester acrylate oligomer, an epoxy acrylate oligomer, a crosslinker, a reactive diluent and photoinitiator. Also, they added diphenylsilane diol and boron-containing commercial vinyl compound, 4-vinylphenyl boronic acid (4-VPBA) to this mixture. Hybrid coating materials were applied on corona-treated PMMA substrates and coatings were cured under UV light and subsequently subjected to thermal treatment as follows: 70 °C for 1 h, 80 °C for 1 h and then 60 °C for 1 day. As a result, hybrid coatings were obtained by the formation of borosilicate linkages (–B–O–Si–).

The gel content of these hybrid coatings was found to be above 95%. These high gel content values indicated that a highly crosslinked polymeric network was formed. Si- and B-containing hybrid coatings displayed enhanced gloss and pendulum hardness values. These coatings were also found to be highly resistant to MEK rubbing. Moreover, the presence of both Si and B greatly improved the abrasion resistance of the coatings. The tensile modulus and tensile strength of the hybrid films increased as the inorganic content was increased. Char yields at 750 °C under an air atmosphere increased remarkably and the coating material with the highest inorganic percentage (5.79%) displayed a LOI value of 27 and a char residue of 25 wt%.

Later, we used this useful monomer, 4-vinylphenyl boronic acid, and prepared flame-retardant coatings by using photoinitiated thiol–ene polymerization.[64] We mixed a commercial modified aliphatic urethane triacrylate, a reactive diluent; N-vinyl pyrrolidone (NVP), tris (3-mercaptopropionate) (TMPMP), 4-VPBA and photoinitiator. Preparation of these coatings is illustrated in Scheme 9.19. Photopolymerization kinetics was investigated by real-time infrared spectroscopy. This hybrid system displayed high –SH and double-bond conversions. Due to interactions between the urethane acrylate and the boronic acid sites, the modulus of the coatings increased and the elongation at break values decreased. According to TGA results, maximum weight loss temperatures were increased as the amount of 4-VPBA was increased. 1.44 wt% B-containing thiol–ene photocured coatings produced a high amount of char (5%). This high char yield could be due to the formation of a boroxine network. It was also found that these B-containing coatings were flame retardant to some extent. The LOI value of the base coating was increased from 19.1 to 24.2 as the B percentage was increased. This high LOI value could be due to a synergistic effect of boron, sulfur and also nitrogen atoms.

Another B-containing flame-retardant photocured coating was prepared by Xu et al. via sol–gel.[65] They prepared a $SiO_2$–$P_2O_5$–$B_2O_3$ sol–gel mixture by mixing trimethyl phosphate (TMP), trimethyl borate (TMB), TEOS and MAPTMS. These sol–gel precursors were hydrolyzed in the presence of ethyl alcohol–water mixture under acidic conditions. The prepared sol was added to an epoxy acrylate at different ratios and the coating materials were applied on PMMA substrates. The pendulum hardness of the coatings increased as the amount of the sol was increased. The coatings exhibited good adhesion to PMMA plates according to crosscut test results. The addition of the sol–gel component did not significantly affect the

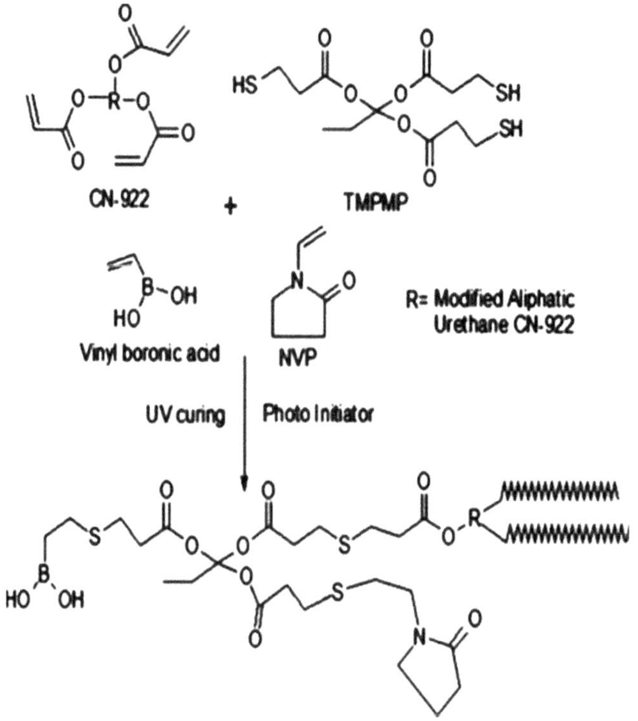

**Scheme 9.19** Preparation of VPBA-containing thiol–ene cured coatings.

transmittance of the coatings. The thermal properties of coatings were characterized by TGA under nitrogen atmosphere. The onset weight loss temperatures decreased while the maximum weight loss and the char yields of the coatings improved with the addition of the prepared $SiO_2$–$P_2O_5$–$B_2O_3$ sol–gel mixture. Moreover, the authors investigated the flame retardancy of these hybrid coatings *via* MCC. According to MCC results the HRR value of the hybrid coatings decreased when compared to the neat epoxy acrylate. Thus, it was concluded that these $SiO_2$–$P_2O_5$–$B_2O_3$ sol-containing hybrid coatings were more thermally stable and flame retardant than the neat epoxy acrylate-based coatings.

Kübra Kaya from Istanbul Technical University, Turkey prepared her doctor's dissertation on boron-containing epoxy acrylate coatings.[66] She synthesized a phosphine oxide diacrylate oligomer; APPO that we mentioned in the previous section and two boron-containing acrylic monomers. The first monomer was 2-(5,5-dimethyl-1,3,2-dioxaborinan-2-yloxy)ethylmethacrylate and it was denoted as BM-M (boron methacrylate monomer) and the other one was an oligomer (BM-O); 4,4′-(2,2′-oxybis(ethane-2,1-diyl)bis(oxy))bis(10-methyl-9-oxo-3,5,8-trioxa-4-boraundec-10-ene-4,1-diyl)bis(2-methylacrylate). BM-M was prepared according to a patented work in 1961[67] and BM-O was prepared according to a report by Cytec Industries.[68]

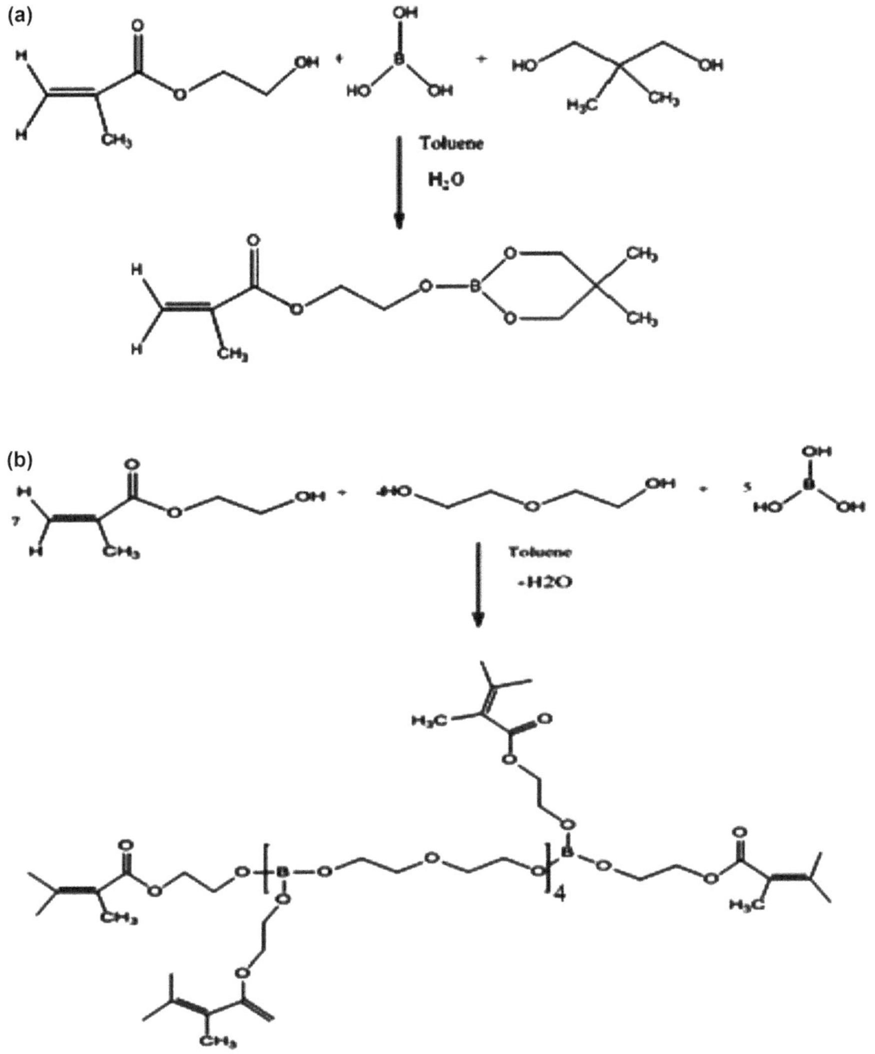

Boron/Phosphorus-Containing Flame-Retardant Photocurable Coatings 181

**Scheme 9.20** (a) Preparation of BM-M and (b) BM-O.

The synthetic procedures for BM-M and BM-O are shown in Scheme 9.20. They tried to improve the properties of epoxy acrylates. They prepared several formulations that contained BM-M or BM-O in combination with APPO and a sol–gel mixture. Boron-containing coatings exhibited high solvent and chemical resistance. The addition of BM-M or BM-O improved the thermal properties and the hardness of the coatings. According to thermogravimetric analysis under nitrogen purge boron-containing coatings provided char residues above 10 wt% at 800 °C.

Zeytuncu et al., prepared boron-, fluorine- and silica-containing UV-cured hybrid coatings (Scheme 9.21).[69] They synthesized a boron-containing

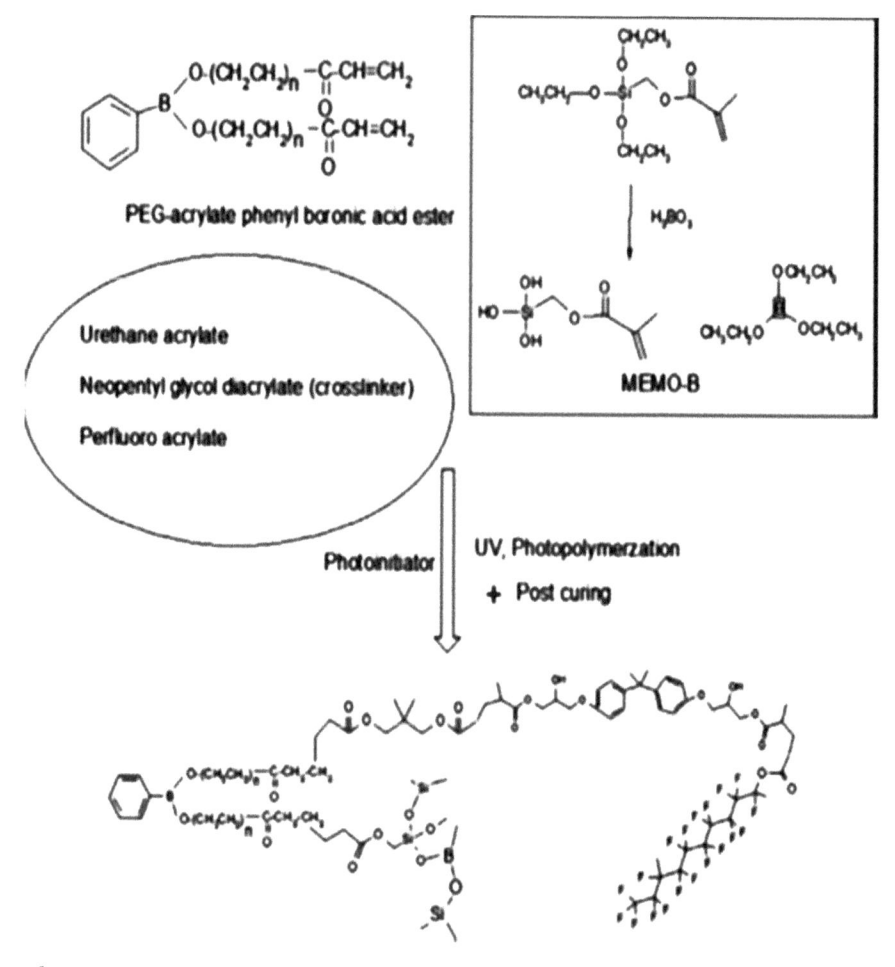

**Scheme 9.21** Preparation of B/Si/F-containing hybrid flame-retardant coatings.

acrylate monomer by reacting phenyl boronic acid with poly(ethylene glycol) acrylate. They introduced silica and fluorine to a UV-curable epoxy/urethane acrylate system by adding hydrolyzed MEMO and a perfluoroalkyl ethyl acrylate monomer, respectively. These B/F/Si containing coatings were applied on polycarbonate substrates. As evidenced by pendulum and pencil hardness tests, the surface hardness of the coatings was increased while the mass loss upon abrasion decreased. The thermal properties of these hybrid flame-retardant coatings were investigated by TGA. Interestingly, it was discovered that the addition of B, S, and F did not cause a remarkable improvement on the first and the maximum weight loss temperatures, which could be due to the low amount of these inorganic atoms in the polymer matrix. On the other hand, char yields increased as the inorganic domains were increased. Furthermore the flame retardancy of these coatings was investigated by means of the LOI test and the LOI value was increased from 20.4 to 23.1. Again, this small increase in LOI value can be attributed to the low level of the inorganic moieties used.

## 9.4 Conclusions

Finally, it can be said that both the organophosphorus or organoboron chemistries offer a wide range of possibilities in order to prepare novel fire-resistant photocurable monomers and additives. Generally, phosphorus-based monomers act in the condensed phase and contribute to char formation by either forming phosphoric anhydride linkages or by forming poly(phosphoric acid) that has catalytic effects on char formation. Among several studies summarized in this text, it can be seen that VPA is a highly efficient flame-retardant monomer. However, it has the drawback that it has low double-bond conversion percentage. DOPO and TAEP are also very efficient flame-retardant monomers, especially when used in combination with nitrogen- or silica-containing comonomers or groups due to the synergism between these atoms.

Moreover, we see that very effective intumescent flame retardants were prepared and successfully applied to urethane or epoxy-based photocurable coatings. A comparison of the LOI values of some of the flame-retardant phosphorus-containing monomers is given in Figure 9.8. It can be seen that different LOI values were reported for TAEP,[23,28,51] DABP[47,54] and DGTH.[26,27] All these phosphorus-based monomers exhibit high LOI values and they all can be classified as self-extinguishing materials when used alone.

The highest LOI value for epoxy acrylate systems among the studies summarized here was obtained for the coatings that were prepared by the addition of 50 wt% DABP and found to be 30. The effect of the concentration of some selected monomers on the LOI of epoxy acrylates can be seen in Figure 9.9. When we compare the studies mentioned in this text, for urethane-acrylate-based coatings, VPA-containing ones are by far better than

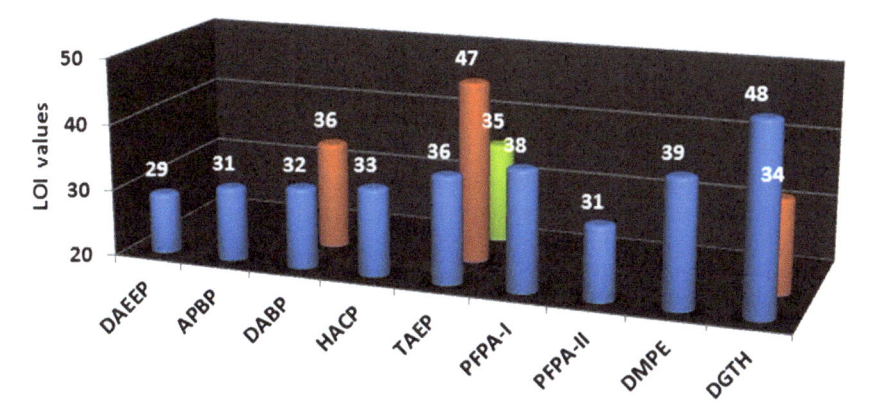

**Figure 9.8** Comparison of the LOI values of selected monomers.

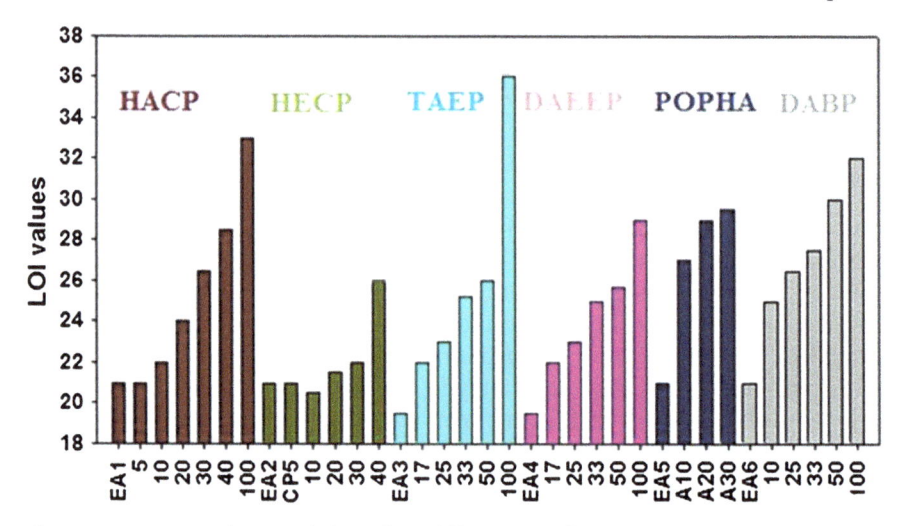

**Figure 9.9** Comparison and the effect of flame-retardant monomer concentration of the LOI values for epoxy-acrylate-based coatings prepared in recent years. (Neat epoxy acrylates were denoted as EA1, EA2, *etc*. Numbers on the X-axis indicate the weight percentages of the flame-retardant monomers.)

the other samples. While 1 wt% VPA can increase the LOI value of urethane acrylates to 42, the addition of 50 wt% TAEP can only reach to a LOI value of 31.5.

As is evident from the studies above, phosphate-based monomers are highly dominant among other phosphorus-containing monomers like phosphine oxides. Generally the synthesis of these phosphate-based acrylic precursors is much easier and therefore they are preferred, but the hydrolytic stability and the chemical resistance of these phosphates should be questioned. Although in recent years there have been only a few studies that investigated the thermal properties of boron-containing flame-retardant photocurable coatings, it can be said that boron-based UV-curable materials have good flame-retardant properties in addition to their characteristic abrasion resistance and enhanced hardness properties. Furthermore flame-retardant monomers were also successfully applied in thiol–ene photocurable systems. It is hoped that these flame-retardant thiol–ene curable coatings will find use in industrial applications.

From the studies that we reviewed here we see that most of the time the authors ignore investigation of the coating properties such as hardness, gloss, adhesion, gel content, chemical resistance, *etc*. However along with the flame retardancy it is also desired to prepare coatings with good mechanical and optical properties. Monomers designed for flame-retardant photocurable coatings should not only contribute to the flame retardancy of the coatings without sacrificing product quality, durability and performance but they should also have features like low smoke generation, low toxicity, low cost and they must be suitable for industrial applications.

# References

1. http://www.nfpa.org/research/fire-statistics/the-us-fire-problem
2. G. L. Nelson, in *Fire and Polymers II*, ed. G. L. Nelson, ACS Symposium Series, Washington, 1995, ch. 1, pp. 1–2.
3. A. B. Morgan and J. W. Gilman, *Fire Mater.*, 2013, **37**, 259.
4. S. Hussain, in *History of Polymeric Composites*, ed. R. B. Seymour and R. D. Deanin, VNU Science Press, Utrecht, 1987, pp. 272.
5. J. Davis and M. Huggard, *J. Vinyl Addit. Technol.*, 1996, **2**, 69.
6. P. Kiliaris and C. D. Papaspyrides, *Prog. Polym. Sci.*, 2010, **35**, 902.
7. I. van der Veen and J. de Boer, *Chemosphere*, 2012, **88**, 1119.
8. F. Laoutid, L. Bonnaud, M. Alexandre, J.-M. Lopez-Cuesta and Ph. Dubois, *Mater. Sci. Eng., R*, 2009, **63**, 100.
9. S.-Y. Lu and I. Hamerton, *Prog. Polym. Sci.*, 2002, **27**, 1661.
10. S. Bourbigot and S. Duquesne, *J. Mater. Chem.*, 2007, **17**, 2283.
11. B. Schartel, *Materials*, 2010, **3**, 4710.
12. C. E. Hoyle, in *Radiation Curing of Polymeric Materials*, ed. C. E. Hoyle and J. F. Kinstle, ACS Symposium Series, Washington, 1990, ch. 1, p. 1.
13. Th. Randoux, J.-Cl. Vanovervelt, H. Van den Bergen and G. Camino, *Prog. Org. Coat.*, 2002, **45**, 281.
14. S. Liang, N. M. Neisius and S. Gaan, *Prog. Org. Coat.*, 2013, **76**, 1642.
15. T. Imamoglu and Y. Yagci, *Turk. J. Chem.*, 2001, **25**, 1.
16. E. Basturk, T. Inan and A. Gungor, *Prog. Org. Coat.*, 2013, **76**, 985.
17. S.-W. Zhu and W.-F. Shi, *Polym. Degrad. Stab.*, 2002, **75**, 543.
18. S.-W. Zhu and W.-F. Shi, *Polym. Inter.*, 2002, **51**, 223.
19. S.-W. Zhu and W.-F. Shi, *Polym. Degrad. Stab.*, 2003, **82**, 435.
20. S.-W. Zhu and W.-F. Shi, *Polym. Degrad. Stab.*, 2003, **81**, 233.
21. M. V. Kahraman, N. Kayaman-Apohan, N. Arsu and A. Gungor, *Prog. Org. Coat.*, 2004, **51**, 213.
22. E. Cakmakci, Y. Mulazim, M. V. Kahraman and N. Kayaman-Apohan, *React. Funct. Polym.*, 2011, **71**, 36.
23. H. Liang and W.-F. Shi, *Polym. Degrad. Stab.*, 2004, **84**, 525.
24. H. Liang, A. Asif and W.-F. Shi, *J. Appl. Polym. Sci.*, 2005, **97**, 185.
25. H. Liang, W.-F. Shi and M. Gong, *Polym. Degrad. Stab.*, 2005, **90**, 1.
26. X. Chen, Y. Hu, C. Jiao and L. Song, *Polym. Degrad. Stab.*, 2007, **92**, 1141.
27. X. Chen, Y. Hu and L. Song, *Polym. Eng. Sci.*, 2008, **48**, 116.
28. X. Chen, Y. Hu, L. Song and W. Xing, *Polym. Adv. Technol.*, 2008, **19**, 393.
29. X. Chen, Y. Hu, C. Jiao and L. Song, *Prog. Org. Coat.*, 2007, **59**, 318.
30. X. Chen and C. Jiao, *Polym. Degrad. Stab.*, 2008, **93**, 2222.
31. N. Kayaman-Apohan, S. Karatas, B. Bilen and A. Gungor, *J. Sol-Gel Sci. Technol.*, 2008, **46**, 87.
32. S. Karatas, Z. Hosgor, N. Kayaman-Apohan and A. Gungor, *Prog. Org. Coat.*, 2009, **65**, 49.
33. W. Xing, L. Song, Y. Hu, S. Zhou, K. Wu and l. Chen, *Polym. Degrad. Stab.*, 2009, **94**, 1503.

34. Y. Mulazim, E. Cakmakci and M. V. Kahraman, *J. Vinyl Addit. Technol.*, 2013, **19**, 31.
35. H. Miao, Z. Huang, L. Cheng and W. Shi, *Prog. Org. Coat.*, 2009, **64**, 365.
36. T. Nishino, M. Meguro, K. Nakamae, M. Matsushita and Y. Ueda, *Langmuir*, 1999, **15**, 4321.
37. W. Xing, Y. Hu, L. Song, X. Chen, P. Zhang and J. Ni, *Polym. Degrad. Stab.*, 2009, **94**, 1176.
38. L. Chen, L. Song, P. Lv, G. Jie, Q. Tai, W. Xing and Y. Hu, *Prog. Org. Coat.*, 2011, **70**, 59.
39. X. Qian, L. Song, Y. Hu, R. K. K. Yuen, L. Chen, Y. Guo, N. Hong and S. Jiang, *Ind. Eng. Chem. Res.*, 2011, **50**, 1881.
40. G. Montaudo and C. Puglisi, in *Mass Spectrometry of Polymers*, ed. G. Montaudo and R. P. Lattimer, CRC Press, Boca Raton, 2001, ch. 5, pp. 181–183.
41. W. Xing, G. Jie, L. Song, S. Hu, X. Lv, X. Wang and Y. Hu, *Thermochim. Acta*, 2011, **513**, 75.
42. W. Xing, G. Jie, L. Song, X. Wang, X. Lv and Y. Hu, *Mater. Chem. Phys.*, 2011, **125**, 196.
43. H. Yuan, W. Xing, P. Zhang, L. Song and Y. Hu, *Ind. Eng. Chem. Res.*, 2012, **51**, 5394.
44. W. Xing, P. Zhang, L. Song, X. Wang and Y. Hu, *Mater. Chem. Phys.*, 2014, **49**, 1.
45. W. Xing, L. Song, P. Lv, G. Jie, X. Wang, X. Lv and Y. Hu, *Mater. Chem. Phys.*, 2010, **123**, 481.
46. J. Ding and W.-F. Shi, *Polym. Degrad. Stab.*, 2004, **84**, 159.
47. Q. Wang and W.-F. Shi, *Eur. Polym. J.*, 2006, **42**, 2261.
48. X. Chen and C. Jiao, *Polym. Adv. Technol.*, 2010, **21**, 490.
49. S. Hu, L. Song, H. Pan, Y. Hu and X. Gong, *J. Anal. Appl. Pyrolysis*, 2012, **97**, 109.
50. X. Qian, H. Pan, W. Y. Xing, L. Song, R. K. K. Yuen and Y. Hu, *Ind. Eng. Chem. Res.*, 2012, **51**, 85.
51. Z. Huang and W.-F. Shi, *Eur. Polym. J.*, 2007, **43**, 1302.
52. Z. Huang and W.-F. Shi, *Polym. Degrad. Stab.*, 2007, **92**, 1193.
53. Z. Huang and W.-F. Shi, *Prog. Org. Coat.*, 2007, **59**, 312.
54. H. Wang, S. Xu and W.-F. Shi, *Prog. Org. Coat.*, 2009, **65**, 417.
55. X. Wang, W. Xing, L. Song, B. Yu, Y. Shi, W. Yang and Y. Hu, *J. Polym. Res.*, 2013, **20**, 165.
56. X. Wang, B. Wang, W. Xing, G. Tang, J. Zhan, W. Yang, L. Song and Y. Hu, *Prog. Org. Coat.*, 2014, 77, 94.
57. X. Qian, L. Song, Y Hu and R. K. K. Yuen, *J. Polym. Res.*, 2012, **19**, 9890.
58. E. Basturk, B. Oktay, M. V. Kahraman and N. Kayaman-Apohan, *Prog. Org. Coat.*, 2013, **76**, 936.
59. http://en.etimaden.gov.tr
60. *Boronic Acids: Preparation and Applications in Organic Synthesis, Medicine and Materials*, ed. D. G. Hall, Wiley-VCH, Weinheim, 2011.
61. E. Cakmakci and A. Gungor, *Polym. Degrad. Stab.*, 2013, **98**, 927.

62. M. Petric, M. Grozav and G. Ilia, *Rev. Chim.*, 2010, **61**, 1183.
63. Y. Mulazim, M. V. Kahraman and N. Kayaman-Apohan, *J. Sol-Gel Sci. Technol.*, 2011, **59**, 613.
64. E. Cakmakci, Y. Mulazim, M. V. Kahraman and N. Kayaman-Apohan, *Prog. Org. Coat.*, 2012, **75**, 28.
65. F. Xu, X. Sun, J. Hang, D. Shang, L. Shi, W. Sun and L. Wang, *J. Sol-Gel Sci. Technol.*, 2012, **63**, 382.
66. K. Kaya, PhD Thesis, Istanbul Technical University, 2013.
67. C. A. Lane, *US Pat.* 2994713, 1961.
68. H. Van den Bergen and P. Roose, presented at *Flame Retardancy in Polymers, Present Status and New Developments*, Elewijt, Belgium, April, 2008.
69. B. Zeytuncu, M. V. Kahraman and O. Yucel, *J. Vinyl Addit. Technol.*, 2013, **19**, 39.

CHAPTER 10

# Lamellar and Circular Constructs Containing Self-Aligned Liquid Crystals

LUCIANO DE SIO,*[a,b] NELSON TABIRYAN[a] AND TIMOTHY BUNNING[c]

[a] Beam Engineering for Advanced Measurements Company, Winter Park, Florida 32789, USA; [b] Department of Physics and Centre of Excellence for the Study of Innovative Functional Materials, CEMIF-CAL, University of Calabria, 87036, Arcavacata di Rende, Italy; [c] Air Force Research Laboratory, Wright-Patterson Air Force Base, Ohio 45433-7707, USA
*Email: luciano.desio@fis.unical.it

## 10.1 Introduction

Liquid-crystal materials (LCs) were discovered in 1888 by an Austrian botanist, F. Renitzer, but only in the last 30 years have these materials been developed sufficiently to be used in optics,[1] electronics,[2] plasmonics[3] and optofluidics.[4] LCs are a state of matter between a conventional liquid and a crystalline solid and temperature is most commonly used to induce various phase changes. There are many different types of LC phases (known as mesophases) characterized by both positional order (molecules arranged in an ordered lattice) and orientational order (molecules pointing in the same direction). LC molecules can be either elongated (calamitic molecules) or disk-like (discotic molecules). In general, the inner part of a mesogenic molecule is rigid (phenyl groups) while the outer part is flexible (aliphatic chains) causing steric interactions between the centralized core regions resulting in orientational order within the fluid. The main types of LC phases are:

- *Nematic phase* – molecules have no positional order but tend to point in the same direction (along the molecular director).
- *Cholesteric phase* – molecules contain a chiral center that produces intermolecular forces that favor alignment between molecules at a slight angle to one another. This leads to the formation of a helical mesoscopic structure (orthogonal to the average director orientation) that can be visualized as a stack of very thin 2-D nematic-like layers with the director in each layer twisted with respect to those above and below.
- *Smectic phase* – molecules maintain the general orientation order of a nematic phase, but also tend to align themselves in layers or planes.
- Among the many different experimental techniques available to investigate the structure and physical properties of LC phases,[5,6] polarized optical microscopy (POM) observations often are sufficient for determining the type and structure of LCs.

A POM image of a typical texture of a *nematic* phase is shown in Figure 10.1(a). In a planar aligned state, the molecular director aligns parallel to the glass surface. With no treatment of the substrates inducing macroscopic alignment, the nematic textures form disclination lines,[7] which are easily identifiable under the microscope. Figure 10.1(b) shows a POM image of a short-pitch *cholesteric* phase in the Grandjean[8] or standing helix texture. The bright colors are due to the difference in rotating power of domains with varying cholesteric pitch. Figure 10.1(c) shows a POM image of a layered structure of a *smectic* phase[9] where the layers preserve their thickness and can slide over one another in order to adjust to the surface conditions. The optical properties (focal conic texture) of the *smectic* state arise from the layer distortions.

## 10.2 Polymeric Template

Over the last couple of decades, great attention has been devoted to the exploitation of LCs in confined geometries for the realization of switchable

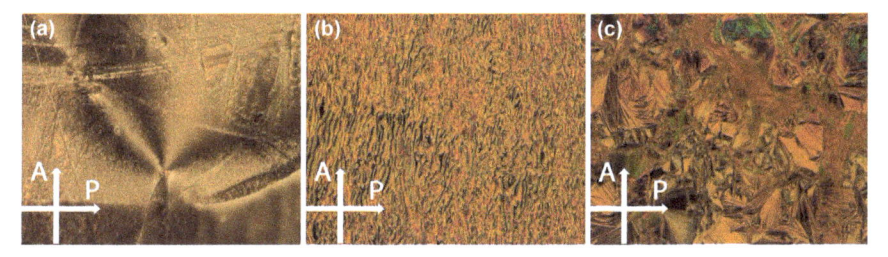

**Figure 10.1** POM view of a random aligned nematic (a), cholesteric (b) and smectic (c) liquid-crystal phases.

structures, with particular attention to switchable holographic diffraction gratings.[10–15] Gratings with large feature sizes have been explored through the use of classic two-step lithography procedures, *etc*. To make gratings on the hundreds of nanometer to several micrometer length scale, one-step photopolymerization techniques have been explored extensively. This work built on the previous work of photopolymerization-based polymer dispersed liquid crystals (PDLC) that resulted in the random formation of small LC droplets in a polymer binder. Two different kinds of switchable gratings grew from this work including holographic PDLCs (HPDLCs), wherein the distribution of LC nano/microdomains (droplets) is modulated in space and POLICRYPS structures that are composed of slices of almost pure polymer separated by NLC layers whose molecular director is uniformly aligned perpendicularly to the polymeric walls.[16–18] Typical fields needed to switch PDLCs (HPDLCs) films are 5–10 V $\mu m^{-1}$, which is the number one drawback of such systems. Improvements to the field strengths can be obtained by introducing specialty monomers and additives into the system, but in general a large surface to volume ratio is introduced by such small size droplets.[19] Fluorinated monomers that reduce the surface energy and thus the interaction of the polymer and LC molecules at the boundaries and the addition of inert surfactant molecules[20] that act to minimize this surface interaction have been moderately successful in reducing these fields. POLICRYPS gratings are realized by photocuring a prepolymer mixture in a stabilized holographically formed interference pattern. The mixture is obtained by diluting 26% in weight of NLC BL-001 (Merck) with a UV-sensitive prepolymer system NOA-61 (Norland). The curing process is performed at high temperature (65 °C) to exploit the high diffusivity of NLC molecules in the isotropic state, thus avoiding the formation and separation of the nematic phase as droplets during the curing step. The temperature is controlled by a hot-stage that is also used to slowly cool (0.2 °C min$^{-1}$) the fabricated grating to room temperature by means of a controlled ramp once the curing process has come to an end. The absence of individual NLC droplets (present in both PDLC and H-PDLC) prevents light scattering and allows fabrication of structures with a high optical quality as diffraction efficiencies as high as 98% have been measured. Furthermore, the confinement and subsequent alignment of NLC molecules in these uniform channels enable fields of a few V $\mu m^{-1}$ to reorient the NLC director on a millisecond timescale. Control of the polymer refractive index (RI) and the ordinary/extraordinary refractive indices of the NLC is obtained by selecting the right materials and the curing conditions. As a result, the refractive index contrast of the grating can be switched on/off by applying external perturbations such as electric field, temperature, *etc*.

Figure 10.2(a) is a typical POM image of a POLICRYPS sample and reflects the almost complete phase separation that is usually achieved in this kind of system. Most importantly, the NLC alignment is obtained spontaneously without any kind of surface treatments. These structures, originally designed and exploited as high-quality, switchable, diffraction grating architectures,

**Figure 10.2** POM view of the POLICRYPS structure before (a) and after (b) the microfluidic etching process (a).

also represent an excellent candidate as a template for the introduction of other materials after initial formation. We have recently shown[21,22] that it is possible to remove the LC from the polymeric structure by dipping the sample in a mixture of water and tetrahydrofuran (THF); Capillary forces remove the LC from the microchannels (microfluidic etching) without affecting the morphology of the polymeric structure. Figure 10.2(b) is a POM image of the empty polymeric template between parallel polarizers showing sharp polymer slices separated by empty channels.

### 10.2.1 Lamellar Polymer Templated Gratings

The performance of LC-based optical devices is determined by the director orientation of the LC as well as the confinement topology. These microstructures (Figure 10.2(b)) have been exploited for realizing the micro/nanoconfinement of a wide variety of organic materials including LCs, stabilized through self-organization processes at the nanoscale without the need for any chemical or mechanical treatment.

We overview several examples of this using a variety of different fluids:

- *Nematic Liquid Crystals (NLC)* – We have backfilled the empty POLICRYPS template with the same NLC used during the curing process. The sample was infiltrated with NLC BL-001 at elevated temperature (70 °C) during the filling process to ensure that a complete transition to the isotropic state ($T_{N-I} = 67$ °C) had occurred. The self-organization process giving rise to uniform and stable alignment of the NLC within the microchannels is induced after the filling process by slowly (0.5 °C min$^{-1}$) cooling the sample to room temperature. The excellent optical quality of the sample is evident in the POM image of Figure 10.3(a) and reflects the good NLC alignment.
- *Photoresponsive Liquid Crystals (Photo-LCs)* – Doping NLCs with photochromic molecules, such as azo-compounds, offers the possibility of cheap, clean, and wireless control of the optical properties of the grating. In this framework, the development of *Photo-LCs* has provided outstanding solutions for all-optical switching applications.[23,24] PLCs

are most commonly LCs that contain an azo group in their structure, thus exhibiting both the photosensitivity of azobenzene compounds and the birefringence of LCs. When acted on by UV or visible light, these materials undergo a reversible isomerization from a thermodynamically stable *trans* to a *cis* conformation, followed by a dramatic change in the macroscopic optical properties of the material.[25] A mesogenic azo dye mixture CPND-57 (provided by Beam Engineering),[26] a eutectic mixture of CPND-5 and CPND-7 (1:3 ratio) azo dyes, 1-(2-Chloro 4-N-n-alkylpiperazinylphenyl)-2-(4-nitrophenyl)diazenes, was dissolved in BL-001 NLC (by Merck) at a concentration of 15%. The empty POLICRYPS template has been filled by capillarity with CPND-57–BL-001 mixture according to the previously described method.[21] Excellent optical quality of the sample is evident in the polarizing optical microscopy (POM) photo shown in Figure 10.3(b).

- *Cholesteric Liquid Crystals (CLCs)* – A few years ago, Patel and Meyer[27] showed that the flexoelectro-optic effect in CLCs can be used to obtain switching effects with submillisecond response times, although with weak optical phase modulation and the need for relatively high electric fields. In this framework, short-pitch CLCs with uniform lying helixes (ULH) oriented in the plane of the layer have been studied in detail.[28]

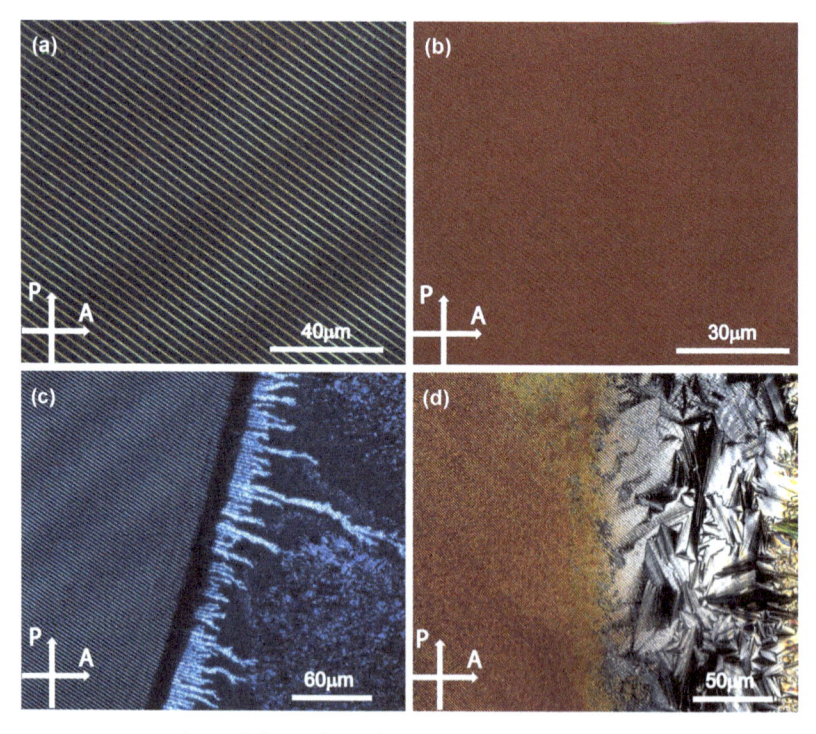

**Figure 10.3** POM view of the polymeric template backfilled with NLC (a), PLC (b), CLC (c) and FLC (d).

The flexoelectric effect in CLCs, aligned in a ULH geometry, induces a tilt of the optical axis in the plane of the layer, with a tilt angle that linearly depends on the amplitude of the applied electric field. Despite the ULH texture exhibiting unique features, it usually needs to be stabilized by an external electric field since the texture tends to relax into a Grandjean configuration over time.[29] To overcome this drawback, several approaches for achieving a stable ULH texture have been studied, which include the realization of periodic anchoring conditions, or polymer stabilized network structures.[30] We utilized the periodic presence of the polymeric slices in our empty template to break the in-plane translational symmetry while imposing to the LC director an alignment perpendicular to the confining surfaces.[31] In this geometry, the CLC helix can only orient along the polymeric channels. We utilized an indium tin oxide (ITO)-coated cell of 10 μm thickness. A drained template of 5 micrometer periodicity was filled by capillary action with the BL088 CLC (Merck) (helix pitch ≈ 400 nm) according to the previously described method.[21] Figure 10.3(c) is a POM image of the sample at the edge of the grating region showing to the right the existence of a standard focal conic texture induced by a random distribution of the CLC helical axes, and on the left, the ULH geometry induced by the polymeric structure.

- *Ferroelectric Liquid Crystals (FLCs)* – FLC materials can be exploited for the realization of fast electro-optical devices.[32] They are characterized by both orientational and positional order of the molecules that changes layer by layer. Molecules are organized in a layered structure, with a molecular director axis tilted at a given angle ($\theta$) with respect to the layer normal ($\hat{K}$). The existence of spontaneous polarization is a microscopic property of a thin FLC film. Bulk samples do not show, in general, any spontaneous polarization because of the random helicoidal organization of the FLC molecules in the bulk. In order to realize bistable switching between two different polarized states, the helicoidal structure has to be unwound and stabilized. This is typically accomplished by confining the FLC sample between two glass plates that is a new sample geometry called the surface-stabilized FLC (SSFLC).[33] In this configuration, application across the cell of an external electric field causes a switching of the molecular director ($n$) around a cone, so that the spontaneous polarization ($P$) is brought to align to the applied field. This configuration exhibits some intrinsic drawbacks such as the need for using thin cells, defects formation and thickness variation under the influence of the electric field. However, we have utilized the polymeric template to accomplish the same thing by infiltrating the drained template with the LC CS-1024 FLC (by Chisso). The best results in terms of uniformity (verified by POM analysis) were obtained in a sample with $\Lambda = 3$ μm and a corresponding channel width of about 1.0 μm. Figure 10.3(d) is a POM image of the sample at the edge of the grating area showing the typical conic texture of random aligned FLC molecules outside the grating area (right view) and the very well-aligned SSFLC geometry inside the grating area (left view).

## 10.3 Circular Polymer Templated Gratings

The confinement of LCs imposed by surface boundary conditions is at the heart of most LC device applications. This confinement can come in the form of boundary conditions enforced by planar substrates[34] or curved boundary conditions.[35] Interest in liquid-crystalline materials confined to curved geometries has expanded greatly in recent years because of the important role in new and emerging electro-optic technologies and the richness in physical phenomena. Many approaches have been explored for nonplanar confinement of LC materials including phase separation techniques,[36] encapsulation methodologies[37] or curved boundary conditions.[38] To date, all methodologies require surface-chemistry treatments. Alignment of the LC molecular director represents a key issue for nonplanar LC-based photonic devices. We show here an alternative approach, one not needing extrinsic surface chemistry treatments, enabled by utilizing curved POLYCRYPS templates. The initial gratings are photochemically formed using a single beam and a commercially available Fresnel mask[38] on a glass substrate. To reduce the attenuation of the glass substrate as they absorb in the UV range, we utilized different photochemistry enabling visible photocuring of the "circular-like" POLICRYPS.

Figure 10.4 depicts the experimental setup used to record the circular POLICRYPS structures. A single-mode beam from an Ar-ion laser operating at $\lambda = 488$ nm is broadened by a spatial filter (composed by two lenses L1, L2, and a small aperture) up to a diameter of about 1 cm. The light beam passes through a patterned Fresnel photomask ($I \approx 30$ mW cm$^{-2}$) and then onto the photosensitive mixture. We have slightly modified the standard mixture (NOA 61 + BL-001) used for POLICRYPS fabrication by adding a small amount (1% in weight) of the visible photoinitiator (Irgacure 784). ITO-covered glass cells (10 μm thick) were filled by capillary action with the modified POLICRYPS mixture. The sample was kept in the dark during the whole filling process. The curing process was performed at high temperature

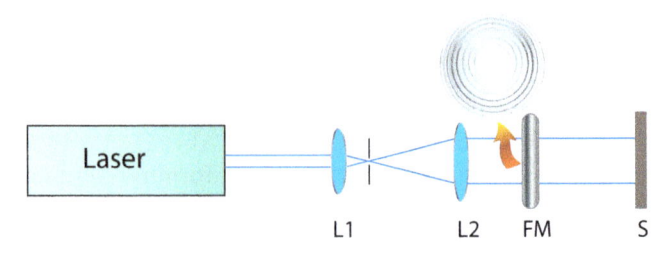

**Figure 10.4** Optical recording setup: L1 and L2 lenses, FM Fresnel Mask, S sample.

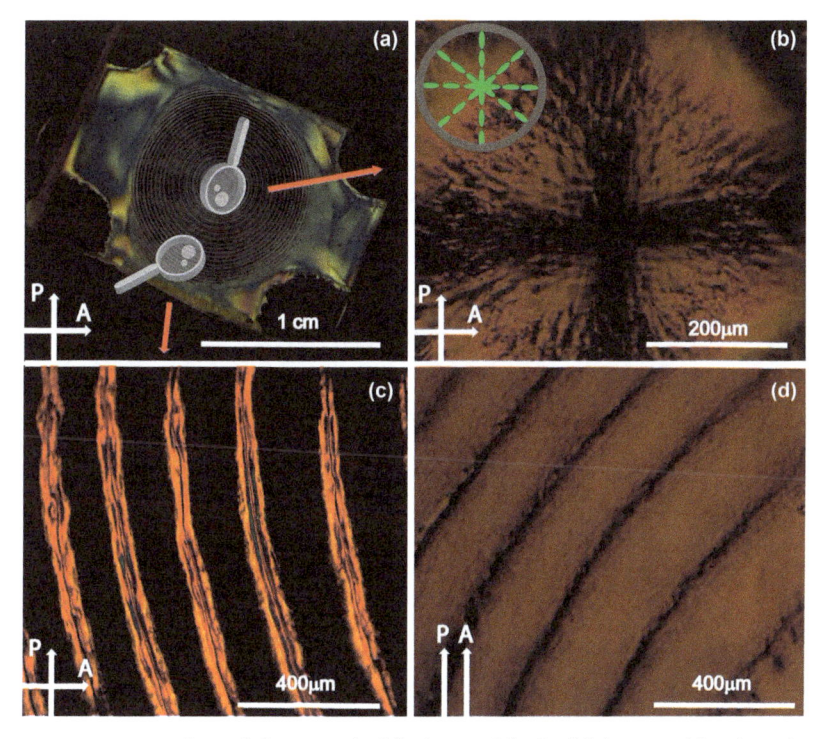

Lamellar and Circular Constructs Containing Self-Aligned Liquid Crystals 195

**Figure 10.5** POM view of the sample (a) along with the high magnification view of the central (b) and external area (c) POM view of the curved polymeric walls between parallel polarizers (d).

($\approx 65$ °C) in order to exploit the high diffusivity of NLC molecules in the isotropic state, thus avoiding the formation and separation of nematic droplets. The temperature was controlled by a hot stage and was used to slowly cool (0.2 °C min$^{-1}$) the fabricated structure to room temperature after the curing process had finished.

Figure 10.5(a) shows the polarized image of the circular POLICRYPS structure indicating concentric rings with a varying interdistance made of curved polymeric slices with alternative channels of well-aligned NLC whose the molecular director is radial aligned (see schematic in the inset of Figure 10.5(b)) to the polymeric slices. High magnification POM images reported in Figure 10.5(b) and (c) shows that the NLC is well aligned between the curved polymeric walls (see Figure 10.5(d)). In particular, Figure 10.5(b) shows the familiar "Maltese cross" pattern, which is a marker of NLC orientation with circular (or spherical) symmetry.

To better investigate the radial NLC alignment inside the structure, we have analyzed the central area reported in Figure 10.5(b) while rotating the analyzer axis. Figure 10.6(a)–(f), indicate that areas where molecules are aligned parallel/perpendicular to the analyzer appear darkest, whereas areas

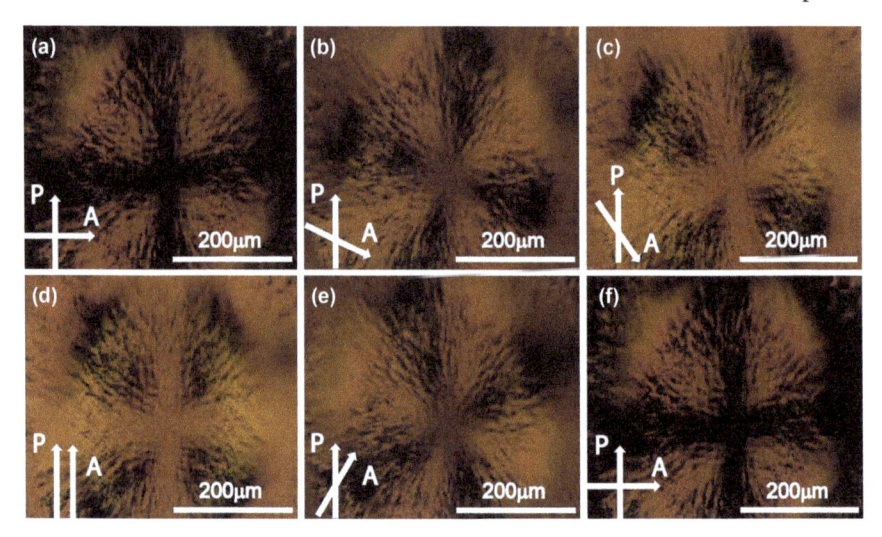

**Figure 10.6** POM view of the central area of the sample while rotating the analyzer axis.

where molecules are at 45° to the analyzer are the brightest. This behavior clearly confirms that the NLC exhibits a radial distribution within the curved polymeric slices and highlights the extraordinary capability of our POLICRYPS technique to induce long-range order in a circular geometry.

## 10.3.1 Electro-Optical Properties

A first study has been carried by probing the area shown in Figure 10.5(b) with white light and monitoring the POM image between crossed polarizers while increasing the applied external voltage.

Due to NLC molecular director reorientation, the "Maltese cross" gradually broadens (see Figure 10.7(a)–(d)). Because of strong anchoring forces at the polymeric walls, the NLC reorientation is inhibited at the interface, which explains the continued visibility of the "Maltese cross" at high electric fields $\sim 7$ V $\mu m^{-1}$ (Figure 10.7(d)). The dynamic properties were measured by placing the sample between crossed polarizers and monitoring the transmitted intensity of a monochromatic light source (He–Ne laser, $\lambda = 633$ nm) while applying an a.c. external voltage (square wave, 1 kHz). The switching cycles of the transmitted intensity are shown in Figure 10.7(e). The switching is fully reversible, indicating no dynamic instabilities of the NLC. The switching voltages needed (7 V $\mu m^{-1}$) are lower than standard H-PDLC gratings but higher than linear POLICRYPS structures (4–5 V $\mu m^{-1}$). The response times are in the milliseconds range. A qualitative explanation can be found in the curved confinement effect that increases the anchoring energy of the NLC to the polymeric surfaces.

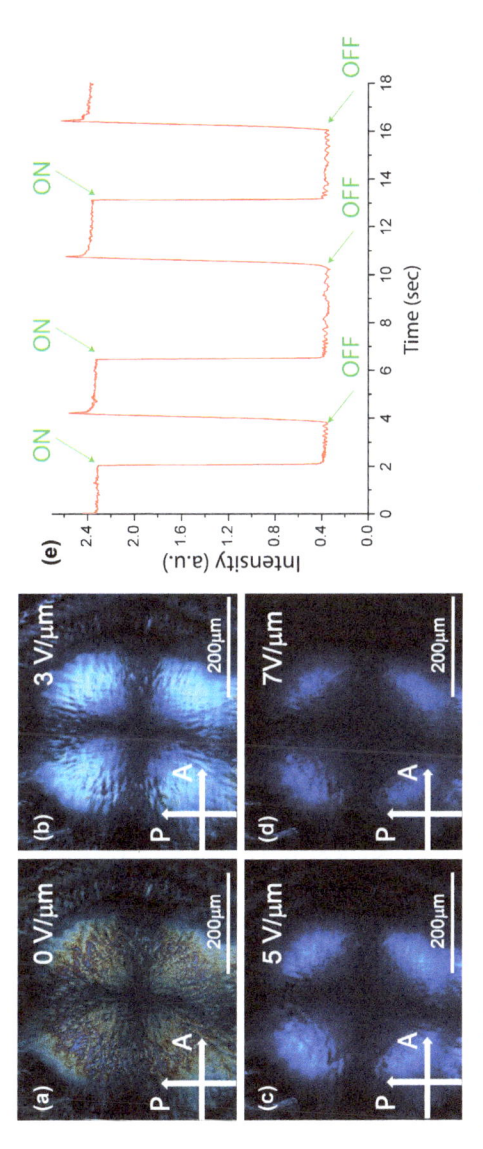

**Figure 10.7** POM view of the central area of the sample *versus* the external electric field (a)–(d) along with its electro-optical response (e).

## 10.4 Conclusion

We report the fabrication and characterization of two different kinds of switchable structures in composite mixtures of polymer and liquid crystals. The combination of intensity holography and photopolymerization, allows the fabrication of periodic lamellar structures called POLICRYPS made of slices of almost pure polymer alternated with well-aligned NLC. These structures act as templates after selectively removing one of the components and can be utilized to self-organize a wide range of materials including nematic, cholesteric and ferroelectric LCs. Circular POLYCRYPS gratings can be realized by utilizing a Fresnel mask and a single-beam curing process where no extrinsic surface treatment is needed to induce bulk alignment. Both approaches reveal the extraordinary capability of the POLICRYPS technique to induce long-range organization in liquid-crystalline compounds on flat and curved surfaces without employing surface-chemistry functionalization.

## Acknowledgements

The research was partially supported by the Air Force Office of Scientific Research (AFOSR), Air Force Research Laboratory (AFRL), U.S. Air Force, under grant FA8655-12-1-003 (P. I. L. De Sio) and the Materials and Manufacturing Directorate, AFRL.

## References

1. P. G. De Gennes and J. Prost, *The Physics of Liquid Crystals*, Oxford Science Publications, 1993.
2. D. Armitage, *Appl. Opt.*, 1980, **19**, 2235.
3. L. De Sio, A. Cunningham, V. Verrina, C. M. Tone, R. Caputo, T. Buergi and C. Umeton, *Nanoscale*, 2012, **4**, 7619.
4. J. G. Cuennet, A. E. Vasdekis, L. De Sio and D. Psaltis, *Nat. Photonics*, 2011, **5**, 234.
5. D. Armitage and F. P. Price, *J. Phys., Colloq.*, 1975, **36**, 133.
6. L. V. Azároff, *Mol. Cryst. Liq. Cryst.*, 1980, **60**, 73.
7. C. Denniston, *Phys. Rev. B: Condens. Matter Mater. Phys.*, 1996, **54**, 6272.
8. P. Boltenhagen, M. Kleman and D. Laventrovich, *J. Phys. II France*, 1991, **1**, 1233.
9. S. T. Lagerwall and B. Stebler, *J. Phys., Colloq.*, 1979, **40**, 53.
10. J. D. Margerum, A. M. Lackner, E. Ramos, G. W. Smith, N. A. Vaz, J. L. Kohler and C. R. Allison, *US Pat.* 5096282, 1992.
11. R. L. Sutherland, L. V. Natarajan, V. P. Tondiglia and T. J. Bunning, *Chem. Mater.*, 1993, **5**, 1533.
12. A. Y.-G. Fuh, C.-Y. Huang, M.-S. Tsai, J.-M. Chen and L.-C. Chien, *Jpn. J. Appl. Phys.*, 1996, **35**, 630.
13. M. Date, Y. Takeuchi, K. Tanaka and K. Kato, *J. Soc. Inf. Disp.*, 1999, 7, 17.

14. M. Jazbinsek, I. D. Olenik, M. Zgonik, A. K. Fontecchio and G. P. Crawford, *J. Appl. Phys.*, 2001, **90**, 3831.
15. X. Sun, X. Tao, T. Ye, P. Xue and Y.-S. Szeto, *Appl. Phys. B: Lasers Opt.*, 2007, **87**, 267.
16. L. De Sio, A. Veltri, R. Caputo, A. De Luca, G. Strangi, R. Bartolino and C. P. Umeton, *Riv. Nuovo Cimento*, 2012, **35**, 575.
17. L. De Sio, A. Veltri, R. Caputo, A. De Luca, G. Strangi, R. Bartolino and C. P. Umeton, *Liq. Cryst. Rev.*, 2013, **1**(1), 2.
18. L. De Sio and N. Tabiryan, *J. Polym. Sci., Part B: Polym. Phys.*, 2014, **52**(3), 158.
19. M. Schulte, S. Clarson, L Natarajan, D. Tomlin and T. Bunning, *Liq. Cryst.*, 2000, **27**, 467.
20. J. Klosterman, L. Natarajan, V. Tondiglia, R. Sutherland, T. White, C. Guymon and T. Bunning, *Polymer*, 2004, **45**, 7213.
21. L. De Sio, S. Ferjani, G. Strangi, C. Umeton and R Bartolino, *Soft Matter*, 2011, **7**, 3739.
22. L. De Sio, S. Ferjani, G. Strangi, C. Umeton and R. Bartolino, *J. Phys. Chem. B*, 2013, **117**, 1176.
23. H. K. Lee, A. Kanazawa, T. Shiono and T. Ikeda, *Chem. Mater.*, 1998, **10**, 1402.
24. A. Shishido, A. Kanazawa, T. Shiono, T. Ikeda and N. Tamai, *J. Mater. Chem.*, 1999, **9**, 2211.
25. Y. Yu, M. Nakano and T. Ikeda, *Nature*, 2003, **425**, 145.
26. U. A. Hrozhyk, S. V. Serak, N. V. Tabiryan, L. Hoke, D. M. Steeves and B. R. Kimball, *Opt. Express*, 2010, **18**, 8697.
27. J. S. Patel and R. B. Meyer, *Phys. Rev. Lett.*, 1987, **58**, 1538.
28. R. B. Meyer, *Phys. Rev. Lett.*, 1969, **22**, 918.
29. A. Sauper and G. Meyer, *Phys. Rev. A: At., Mol., Opt. Phys.*, 1983, **27**, 2196.
30. G. Carbone *et al.*, *Mol. Cryst. Liq. Cryst.*, 2011, **544**, 37.
31. G. Carbone, P. Salter, S. J. Elston, P. Raynes, L. De Sio, S. Ferjani, G. Strangi, C. Umeton and R. Bartolino, *Appl. Phys. Lett.*, 2009, **95**, 011102.
32. P. G. D. Gennes and J. Prost, *The Physics of Liquid Crystals*, Oxford University Press, UK, 2nd edn, 1995.
33. N. A. Clark and S. T. Lagerwall, *Appl. Phys. Lett.*, 1980, **36**, 899.
34. J. L. Janning, *Appl. Phys. Lett.*, 1972, **21**, 173.
35. G. P. Crawford, S. Žumer, *Liquid Crystals in Complex Geometry: Formed by Polymer and Porous Networks*, Taylor & Francis, 1996.
36. P. Sheng, *Phys. Rev. Lett.*, 1976, **37**, 1059.
37. D. Jayasri, T. Sairam, K. P. N. Murthy and V. S. S. Sastry, *Phys. A*, 2011, **390**, 4549.
38. Y. H. Fan, H. Ren and S. T. Wu, *Opt. Express*, 2003, **11**, 3080.

CHAPTER 11

# POLICRYPS: A Multipurpose, Application-Oriented Platform

R. CAPUTO,*[a] M. INFUSINO,[a,d] A. VELTRI,[a,d] L. DE SIO,[c] A. V. SUKHOV[b] AND C. P. UMETON[a]

[a] Department of Physics and Centre of Excellence for the Study of Innovative Functional Materials University of Calabria, CNR-IPCF UOS, Cosenza 87036 Arcavacata di Rende (CS), Italy; [b] Institute for Problems in Mechanics, Russian Academy of Science, Moscow 119526, Russia; [c] Beam Engineering for Advanced Measurements Company, Winter Park, Florida 32789, USA; [d] Colegio de Ciencias e Ingeniería, Universidad San Francisco de Quito, Quito, Ecuador
*Email: roberto.caputo@fis.unical.it

## 11.1 Introduction

The possibility of shaping matter by using light is very fascinating. After more than twenty years from the first experiments performed by Sutherland *et al.* for realizing microstructures in polymer composite materials, the topic is still quite up-to-date. In those experiments, the 1D laser interference pattern produced by a holographic setup was replicated in a polymer-dispersed-liquid-crystal (PDLC) material[1] obtaining a specific kind of diffraction gratings, afterwards known in the literature as HPDLCs.[2] The great advantage introduced by HPDLCs was the possibility of electrically tuning all optical properties of the structure. This was allowed by the presence of liquid-crystal droplets within the polymeric matrix. At that time, this achievement represented a giant technological breakthrough in the field and, because the experiment was quite simple to realize, from these first pioneering experiments many others followed.[3-5] Despite the extreme variety of structures allowed by this technology: periodic and aperiodic, different dimensionalities (1D, 2D and 3D)[6,7] and high optical performances in terms of diffraction efficiency (which almost reached in some cases the theoretical maximum in the Bragg condition),[8] the success of the HPDLC paradigm was dampened by some intrinsic drawbacks. Indeed, the presence of LC droplets in HPDLCs encouraged and limited its success. Namely, if the droplet size is comparable with the wavelength of the light diffracted, the sample is strongly scattering. For this reason, an idea to improve the performance of HPDLCs was to reduce the nematic droplets average size to almost the nanoscale. This allowed low scattering losses and higher diffraction efficiencies.[9] Unfortunately, some new issues emerged: indeed, the smaller the size of droplets, the higher their surface tension. This corresponds to an increase of the required switching voltages thus limiting the possibility of developing commercial, switchable holographic elements. The introduction of POLICRYPS technology[10,11] represented a win-win solution to HPDLCs issues: the system was still tunable for the presence of liquid crystals, but free of droplets. This technology permits confinement of a continuous and well-aligned nematic phase between polymeric walls, which results in low scattering losses and low switching voltages. As reported in the following, this noticeable result opens a completely new framework of applications, many of them envisioned and realized in the last ten years.

## 11.2 POLICRYPS Photocured Materials

Many of the features of POLICRYPS rely in its morphology: polarized optical microscope (POM) and scanning electronic microscope (SEM) investigations have shown that the structure is made of rigid slices of almost pure polymer alternated to films of almost pure NLC (see Figure 11.1).

The polymeric slices are well glued to two glasses that confine and contain the POLICRYPS. These slices represent a rigid frame that, somehow, "stabilizes" the NLC component and, therefore, the whole sample. Separation interfaces between polymer slices and NLC films are quite regular and sharp; furthermore, there is convincing evidence that, at these interfaces, the NLC director is everywhere perpendicular to them, thus inducing a good, uniform alignment of the director in the whole NLC film standing between two polymeric slices. This circumstance represents one of the main features that determine the overall characteristics of the whole structure. The uniform and regular alignment of the director in the NLC films of the structure determines the main optical and electro-optical properties of the POLICRYPS. From the optical point of view, losses due to the scattering of the visible light (which is eventually brought to impinge onto the POLICRYPS) are reduced to less than 2%, thanks to the absence of droplets, which exist in HPDLC samples, with an average size comparable to the light wavelength and an arbitrary director alignment. From an electro-optical point of view, the fact that the NLC molecules are confined (and well aligned) in a uniform

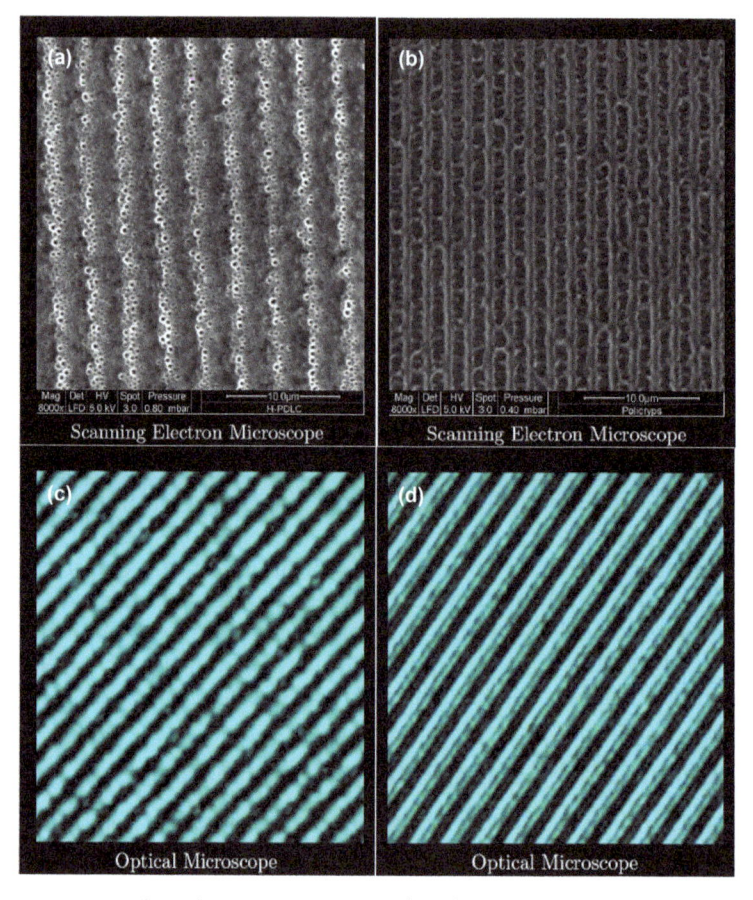

**Figure 11.1** Scanning electron microscope (SEM) of a HPDLC (a) and POLICRYPS (b) structure; Polarizing optical microscope (POM) micrographs of, respectively, a HPDLC (c) and POLICRYPS (d) structure.

film, rather than in a small droplet, allows a suitably oriented electric field of the order of a few V $\mu m^{-1}$ to uniformly "reorient" the NLC director in a millisecond timescale. Afterwards, by suitably choosing the values of the refractive index of the polymer and the ordinary/extraordinary refractive index of the NLC, this director reorientation can be exploited to vary the spatial modulation of the refractive index of the POLICRYPS.

## 11.2.1 POLICRYPS Photocuring Process

In order to fabricate a good POLICRYPS, a peculiar technique has to be used that exploits the high diffusivity of NLC molecules in the isotropic state and avoids the formation and separation of the nematic phase (as NLC droplets) during the curing process (see Chapter 5 of this book, ref. 12-14). The main fabrication steps can be illustrated as follows. By means of a hot stage, a syrup of NLC, monomer and photoinitiator is heated to a temperature that is above the nematic–isotropic transition point of the NLC component; the sample is then "cured" with the interference pattern of a UV radiation. After the curing process has come to an end, the sample is brought below the isotropic–nematic transition point by means of a controlled, very slow, linear cooling to room temperature. The experimental set-up exploits an active system for suppression of vibrations[15,16] and is presented in Figure 11.2. An Ar-ion laser is the source of a single-mode radiation at the wavelength $\lambda_B = 351$ nm. The beam is broadened up to a diameter of about 25 mm by the beam expander BE, and divided into two beams of almost equal intensity by the beam splitter BS. These two beams overlap and give rise to the "curing" interference pattern at the entrance plane of the sample cell S, whose temperature is controlled by the hot stage. Depending on the required nano/microscale dimensions of the structure, the spatial period of the interference pattern can be varied in the range $\Lambda = 0.2$–15 µm by adjusting the total interference angle $2\theta_{cur}$. A commercial, metal-coated, reflective diffraction grating (Edmund Optics) placed above the sample is used as a test element for the interferometric monitoring of vibrations. Part of each of the curing beams is reflected and diffracted by this grating. The setup is adjusted to make the reflected part of one beam spatially coincident with the diffracted part of the second one. These two radiations are wave coupled by the test

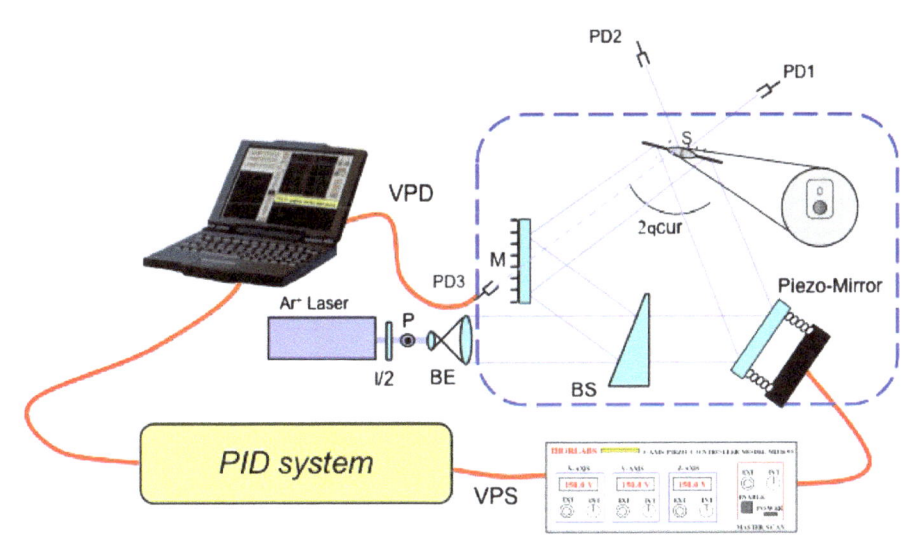

**Figure 11.2** Optical holographic setup for UV curing of gratings with stability check. P, polarizer; $\lambda/2$, half-wave plate; BE, beam expander; BS, beam splitter; $2\theta_{cur}$, total curing angle; M, mirrors; S, sample; PD1, first beam photodetector; PD2, second beam photodetector; PD3, diffracted/reflected beam photodetector. In the insertion the reference grating is shown (put immediately below the sample area) which enables the stability check.[16]
Reproduced by permission of the Optical Society of America.

grating and their interference pattern is detected by an additional photodiode PD3. The signal of this photodiode is sent to a computerized active-feedback system, which exploits a software that is based on a proportional–integral–derivative (PID) protocol; this drives a mirror holder whose position can be controlled by a piezoelectric mechanism, used in the feedback configuration. This control system has proved to be able to continuously compensate for changes in the optical path length due to vibrations as well as variations in environmental conditions such as room pressure, temperature or humidity; residual fluctuations are of the order of 6–7 nm, which correspond to the sensitivity of the piezo-system used.

## 11.3 Tunable Optical Switch

The basic device that can be realized by using electrically switchable holographic gratings in liquid-crystalline composite materials is an electro-optical switch.[17] Such a device should, in principle, completely diffract or transmit an impinging light beam, depending on the application of an external voltage (see Figure 11.3).

PDLCs have been actively utilized in the past in order to realize working prototypes of this kind of device; unfortunately, they still show issues that affect their performances. One of the main reasons that brought us to design POLICRYPS systems was the possibility of overcoming most of these issues. In the following, we report the results of an experimental comparison between an HPDLC and a POLICRYPS grating, in order to put into evidence

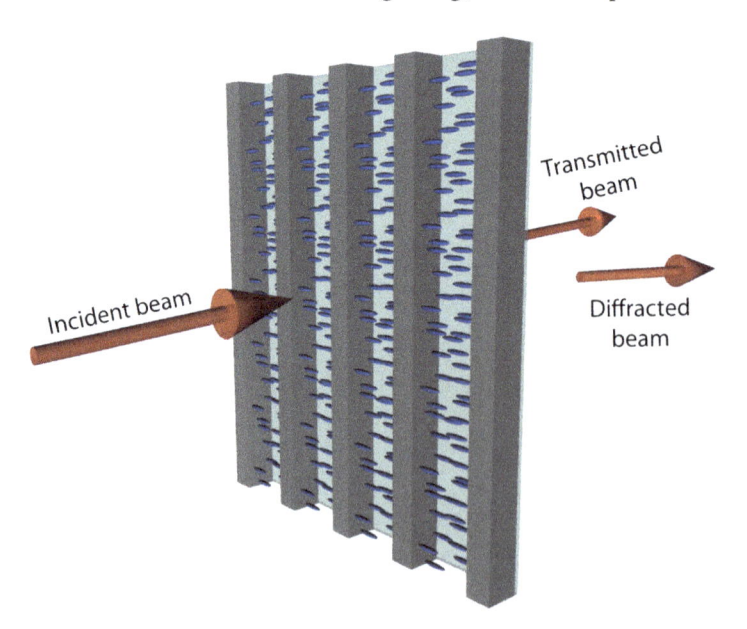

**Figure 11.3** Sketch of a POLICRYPS grating in transmission configuration.[11] Reproduced by permission of IOP Publishing. All Rights Reserved.

how microscopic features of the structure can influence the overall performance of the macroscopic device. We have realized a standard HPDLC and a POLICRYPS grating, both with a fringe spacing $\Lambda = 1.5$ µm. Sample cells, 16 µm thick, made with indium tin oxide (ITO)-coated glass slabs, were filled with the same initial chemical syrup. This was prepared by diluting the NLC 5CB (Merck, $\approx 30$ wt%) in the prepolymer system Norland Optical Adhesive NOA-61. The POLICRYPS grating was cured by a total UV intensity of 11 mW cm$^{-2}$, acting on the sample for $\tau \approx 1000$ s at high temperature (e.g. above the nematic–isotropic transition point of the 5CB liquid crystal), these being the optimal conditions for achieving a high diffraction efficiency and a morphology of good quality.[18] Almost the same UV intensity and curing time proved also to be adequate for the realization of the PDLC grating, but in this case the sample was cured at room temperature. In order to explore the performances of both gratings, we used a weak ($P \approx 1$ mW) He–Ne laser radiation ($\lambda_R = 633$ nm), with its angle of incidence adjusted for satisfying the Bragg condition for the first-order diffracted beam. With the aim of performing a comparison under the same experimental conditions, before starting the curing process of each sample, we measured the intensity $I_{in}$ of the impinging probe beam (before the sample) and the transmitted intensity $I_{tr}$. Then, once the curing process had been completed and the UV light switched off, we measured both the intensity $I_0$ of the zero-order (direct transmitted) probe beam and the intensity $I_1$ of the first-order diffracted probe beam. In this way we were able to calculate the zero-order transmittivity $T_0 = I_0/I_{in}$, the first-order transmittivity $T_1 = I_1/I_{in}$, the total transmittivity $T_{tot} = T_0 + T_1$ and the first-order diffraction efficiency, which is usually calculated as $\eta_1 = I_1/I_{tr}$. During all the experiments, the intensity of the probe beam was maintained at a fixed value (the value of the initial impinging intensity before the curing process started). We measured the first-order diffraction efficiency at room temperature both for POLICRYPS and HPDLC gratings, obtaining $\eta^1_{POLICRYPS} = 88\%$ and $\eta^1_{HPDLC} = 41.2\%$. We stress that the value of $\eta^1_{POLICRYPS}$ is not the highest that we can get since, by using other POLICRYPS gratings (not involved in comparisons with HPDLC ones), we have obtained $\eta^1_{POLICRYPS}$ values as high as 98%. The electro-optic response of the two gratings was investigated by exploiting a low frequency (500 Hz) square-wave voltage, and results are reported in Figure 11.4.

Figure 11.4(a) represents the switching curve of the POLICRYPS grating: the behavior of the first-order transmittivity $T_1$ (circles), zero-order transmittivity $T_0$ (squares) and total transmittivity $T_{tot}$ (triangles) is reported *versus* the root mean square applied electric field.

It is worth noting that $T_{tot}$ is only slightly lower than 1 and remains approximately the same for all values of the applied field; this indicates that the grating exhibits negligible scattering losses. The situation is quite different for the HPDLC grating (Figure 11.4(b)): the total transmittivity is well below 1 and increases as the applied field increases. We also note that the switching efficiency $h_{sw} \equiv \dfrac{T^1_{on} - T^1_{off}}{T^1_{on}}$, where $T^1_{on}$ and $T^1_{on}$ are the first-order

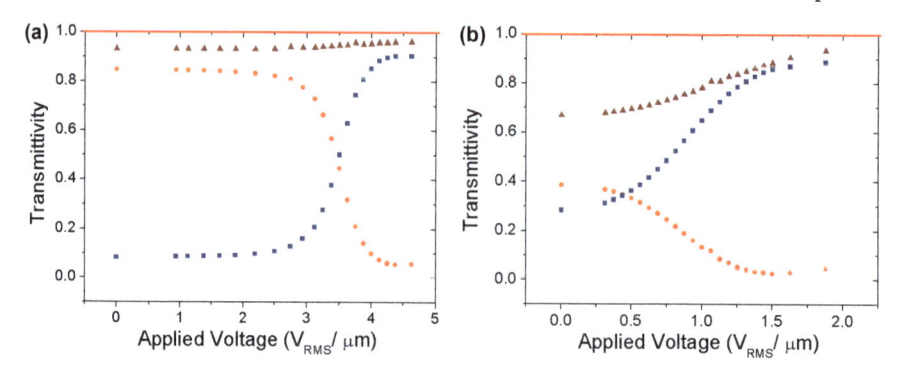

**Figure 11.4** Dependence on applied voltage of the zero-order transmittivity $T_0$ (squares), first-order transmittivity $T_1$ (circles) and total transmittivity $T_{tot}$ (triangles) for (a) a POLICRYPS grating and (b) an HPDLC grating at room temperature. Error bars are of the order of the dot size. The pictures in the inset show respectively a typical POLICRYPS and HPDLC grating morphology observed under a polarizing optical microscope.

**Table 11.1** Measured values of the switching times for a POLICRYPS and an HPDLC grating obtained from the same initial mixture.[13] Reproduced by permission of the Optical Society of America.

|  | $\tau_{fall}$ (ms) | $\tau_{rise}$ (ms) |
|---|---|---|
| POLICRYPS | $1.12 \pm 0.03$ | $0.88 \pm 0.03$ |
| HPDLC | $10.53 \pm 0.18$ | $1.36 \pm 0.04$ |

transmittivities in the switch-on and switch-off condition, respectively, is almost the same (93.3%) for both gratings. Where the switching fields are concerned, the first diffracted beam is almost completely switched off by a field of about 1.5 V μm$^{-1}$ applied to the HPDLC grating, while a value of about 4.3 V μm$^{-1}$ is needed to obtain the same effect in the POLICRYPS one. This particular difference can be due to the average size of NLC droplets in the HPDLC; evidently, this size is large enough to enable low switching fields. This is confirmed by the values of the switching times shown in Table 11.1: both the rise and fall times of the HPDLC grating are longer than those of the POLICRYPS; this suggests a very large average size of PDLC droplets. Here, we stress that the electro-optic behavior shown in Figure 11.3(a), and its noticeable difference with the one of Figure 11.3(b), represents the best evidence of the good performances of POLICRYPS gratings; indeed, people working with HPDLCs of nanosized droplets also find for these materials behaviors that are comparable to the one shown in Figure 11.3(a), but for values of the switching fields which are about four-fold higher.[19]

### 11.3.1 Tunable Phase Modulator

The preferential orientation and the good alignment assumed by the molecular director **n** of the LC material within a POLICRYPS structure recently suggested a possible use of these systems as switchable phase modulators.[20] Examples of such devices are already present in literature. A basic embodiment is obtained by enclosing a NLC with a positive dielectric anisotropy in a cell made of two ITO-coated glasses, treated to give a planar alignment to the NLC director. Since the liquid crystal is birefringent, light with wavelength $\lambda$, propagating through the structure, is separated into an ordinary and an extraordinary component. If $L$ is the thickness of the sample and $\Delta n_{LC}$ indicates its birefringence, the phase difference $\delta_{LC}$ between these two waves, measured at the exit of the sample, depends on the value of $\Delta n_{LC}$: $\delta_{LC} = 2\pi L \Delta n_{LC}/\lambda$. By applying an external electric field **E** with direction perpendicular to the glass slabs of the cell, **n** tends to reorient along the same direction as **E**, thus producing a change in the phase difference. However, this simple device presents some drawbacks. The orientation of **n** is, indeed, sensitive to temperature changes.[21] This can represent a serious limit for an eventual device when the power of the impinging radiation is high. Moreover, the switching times of such devices are usually quite long (2–8 ms), thus limiting the field of possible applications. In order to overcome the above-mentioned problems, the NLC layer is often stabilized by means of polymeric chains;[22] their presence improves the response times of the device but, unfortunately, drastically increases the operating voltages (due, probably, to the torque exerted by the polymer on the nematic director). Moreover, due to the irregularity of morphology induced by the presence of polymeric chains, visible light is strongly scattered. Therefore, these systems are suitable only for wavelengths in the infrared range. Several features of POLICRYPS structures make them an attractive alternative to the discussed system. First, they exhibit limited scattering losses when illuminated by visible light. Secondly, the polymer slices confine and stabilize the NLC molecules, thus also influencing their alignment, and thirdly, POLICRYPS structures can be driven by low voltages and exhibit short switching times. We expect that the better the alignment of the NLC director in the nematic layer of the POLICRYPS, the higher the value of $\Delta n_{LC}$; then, the phase retardation introduced by the grating will depend on the angle that the polarization vector of the impinging light forms with the nematic director within the LC layers, Figure 11.5.

It is important to underline that, because of the diffractive nature of a POLICRYPS structure, the light impinging on the device will not only experience a phase modulation but will also undergo a dichroic absorption as explained in ref. 23. This double behavior can be taken into account by considering the POLICRYPS, in terms of the Jones matrix formalism, as both a retardation plate and a dichroic absorber. This is done by multiplying the Jones matrix of the generic phase retarder by a new matrix $L$ given by:

$$L = \begin{pmatrix} H & 0 \\ 0 & V \end{pmatrix} \quad (11.1)$$

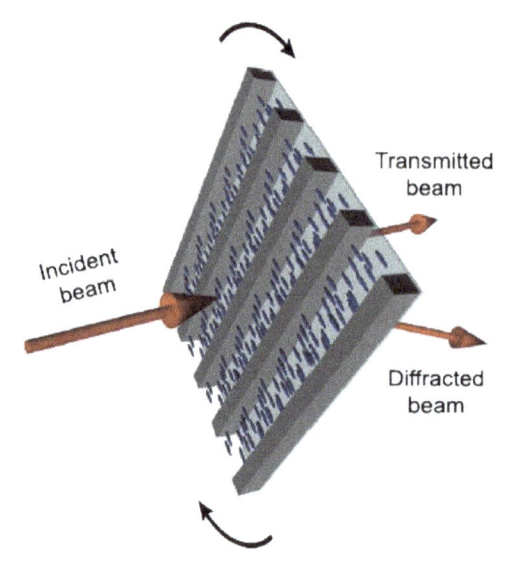

**Figure 11.5** Sketch of a POLICRYPS grating in transmission configuration used as a phase modulator.[11]
Reproduced by permission of IOP Publishing. All Rights Reserved.

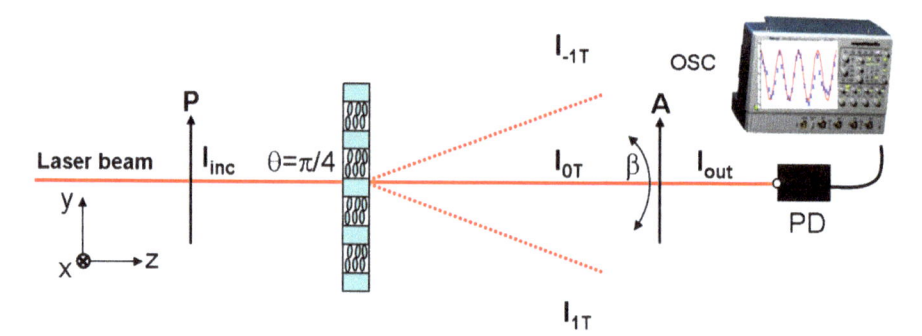

**Figure 11.6** Experimental geometry utilized for measuring the intensity transmitted by the system composed of a birefringent/dichroic sample put between two polarizers. P polarizer, A analyzer, $I_{inc}$ total incident intensity, $I_{out}$ output intensity, $I_{0T}$ and $I_{\pm 1T}$ zeroth- and first-order transmitted intensities, respectively. $\theta$ is the angle between the light polarization direction (y-axis) and the grating optical axis (lying in the xy-plane), PD Photodetector, OSC oscilloscope. The probe beam is from a He–Ne laser at the wavelength $\lambda = 632.8$ nm. S is the POLICRYPS sample.[23]
Reproduced by permission of the Optical Society of America.

where $H$ and $V$ parameter values depend on the considered material and can reflect a broad range of situations. The POLICRYPS used for experiments as a phase modulator has a thickness $L = 3.03$ μm and a fringe spacing $\Lambda = 1.22$ μm. In order to check the phase retardation properties of this structure, we used the experimental setup reported in Figure 11.6. The

POLICRYPS is put between a polarizer P and an analyzer A, with its optical axis oriented at an angle $\theta = \pi/4$ with respect to the first polarizer; in this position, the field components have the same amplitude ($E_\parallel = E_\perp$) and the sample introduces the maximum retardation.

During experiments, the position of the sample remains fixed while the analyzer is rotated (in steps of 10°) around the axis of propagation of the probe light (z-axis in Figure 11.6). We define $\beta$ as the angle between directions of analyzer and incident polarization (therefore $\beta = 0$ when the analyzer A is parallel to the polarizer P). If we indicate with $I_{inc}$ the intensity of the impinging beam, by means of eqn (11.2) and (11.3) (derived in ref. 23), it is possible to calculate the complex electric field $\tilde{E}_{out}(\beta)$ and hence the intensity $I_{out}(\beta)$ of light transmitted by the analyzer A in our experimental geometry.

$$\tilde{E}_{out}(\beta) = \frac{\sqrt{2}}{2}\sqrt{I_{inc}}\begin{pmatrix} -He^{i\frac{\delta}{2}}\sin^2\beta - Ve^{-i\frac{\delta}{2}}\sin\beta\cos\beta \\ He^{i\frac{\delta}{2}}\sin\beta\cos\beta + Ve^{i\frac{\delta}{2}}\cos^2\beta \end{pmatrix} \quad (11.2)$$

$$I_{out}(\beta) = \tilde{E}_{out}(\beta) \cdot \tilde{E}^*_{out}(\beta) = \frac{I_{inc}}{2}[H^2\sin^2\beta + V^2\cos^2\beta + HV\sin 2\beta \cos\delta] \quad (11.3)$$

Parameters $H$ and $V$ are given by:

$$H = \sqrt{\frac{2I_{out}(\beta = \pi/2)}{I_{inc}}} \quad (11.4)$$

$$V = \sqrt{\frac{2I_{out}(\beta = 0)}{I_{inc}}} \quad (11.5)$$

While the phase retardation $\delta$ introduced by the sample can be calculated as:

$$\cos\delta = \frac{1}{HV}\left(\frac{2I_{out}(\beta = \pi/4)}{I_{inc}} - \frac{H^2 + V^2}{2}\right) \quad (11.6)$$

By substituting the obtained data in eqn (11.4)–(11.6), we obtain: $H = 0.727$, $V = 0.406$ and $\delta = 1.26$ rad. In Figure 11.7, the experimental value of $I_{out}$ as a function of the angle $\beta$ (crosses) is compared with the theoretical behavior predicted by eqn (11.3) (solid line).

The different values of $H$ and $V$ show that, even at normal incidence, the diffraction efficiency of the POLICRYPS grating is significant. As for the birefringence of the structure, the obtained value of $\delta$ yields $\Delta n = 0.042$. By considering that the periodicity of the grating is much larger than the probe wavelength we can exclude that this considerably high value is due to the

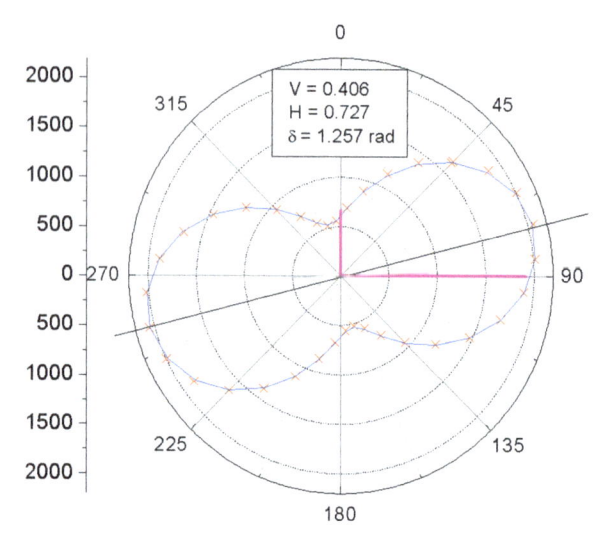

**Figure 11.7** Behavior of the intensity transmitted by the analyzer put after a POLICRYPS grating as a function of the angle $\beta$ between the electric field of the impinging wave and the axis of the analyzer itself. Two segments in the graph evidence output intensity values for the analyzer positions $\beta = 0$ and $\beta = \pi/2$, respectively.[23]
Reproduced by permission of the Optical Society of America.

birefringence and hence to the geometrical features of the grating. We are confident, instead, that this value indicates that the stabilizing and confining action exerted by polymer slices on the NLC molecules has a direct influence on their alignment.

## 11.3.2 Microsized Organic Lasing System

A new intriguing scenario of applications emerges if we explore the possibility of obtaining a lasing action in POLICRYPS structures. Indeed, in recent years, many efforts have been spent in research for the realization of lasing devices based on organic systems: good candidate materials for achieving this result are cholesteric liquid crystals (CLCs). It is well known that CLC materials possess a helical periodic superstructure that provides a 1D spatial modulation of the refractive index.[21] This system behaves as a photonic bandgap (PBG), *i.e.* it exhibits a window in the electromagnetic spectrum where wave propagation is forbidden. This is due to a mechanism known in the literature as distributed feedback (DFB), and has the consequence that the system behaves as a mirrorless optical resonator. If the CLC material is doped with fluorescent guest molecules, a gain enhancement of the radiation, propagating in the structure, is possible. Kogelnik and Shank[24] were the first to report laser action in mirrorless periodic Bragg DFB structures, while Goldberg and Schnur predicted laser action in chiral liquid crystals.[25] There are many advantages in using POLICRYPS as a host structure for dye doped CLC helices. The sharp and parallel channels of POLICRYPS can behave as an array of optical resonators, each of them working as a microlaser. The length of the single channel is not limited by the sample geometry; in principle the single cavity can be several centimeters long, thus containing thousands of periods of the CLC helices. At the same time, its volume can be reduced at will by changing the periodicity of the structure. Optical resonators with these two features present a high quality factor $Q$ and correspond to very efficient microcavity lasers. Such an array of microlasers has been experimentally realized in a POLICRYPS structure.[26] A slightly different chemical mixture was used: a small amount (0.7 wt%) of Irgacure 2100 and Darocur 1173 photoinitiators (1:1 wt%, Ciba Specialty Chemicals) was used to reinforce the polymeric network and a 0.09 wt% of pyrromethene dye (Exciton) was added, which represented the gain medium of our system. Other components were 29.9 wt% BL088 cholesteric liquid crystals (Merck), and 69.3 wt% of NOA-61 monomer (Norland). The mixture was introduced by capillarity between ITO-coated glass plates separated by 13.5 μm thick mylar spacers. The sample was then prepared by following the typical recipe for obtaining POLICRYPS. The only difference is that the curing temperature was sensitively higher in order to bring the CLC material in the isotropic phase during curing. At the end of the whole process, an almost complete phase separation was obtained, giving rise to helixed liquid-crystal channels periodically separated by polymer walls. A scanning electron microscopy analysis of the sample showed a periodicity of 5 μm with the microcavity width of about 1.5 μm. The system was optically pumped with the second harmonic ($\lambda = 532$ nm) of a Nd:YAG laser. The laser beam was focused onto the sample by means of a cylindrical lens ($f = 100$ mm) and linearly polarized perpendicularly to the microchannels. The long axis of the section was oriented perpendicularly to the orientation of the polymeric walls; therefore, the profile obtained (long axis of approximately 5 mm) ensured the simultaneous excitation of multiple microchannels. Above a certain pump power, stimulated emission was achieved, emerging from the microcavities in a direction parallel to the glass plates and along the microchannels. At their end, highly sensitive emission measurements were performed in a restricted cone angle of about 0.1 rad. The sketch in Figure 11.8 shows this lasing scenario of the microlaser array.

The stimulated emission emerging from the microchannels was circularly polarized, demonstrating that the distributed feedback mechanism due to the CLC helices is the cause of the observed phenomenon. The dependence of the emitted intensity and spectral linewidth (FWHM) on the input pump energy are reported in Figure 11.9.

At low excitation energies, both the emission intensity and the linewidth show a quasilinear dependence on the pump energy. Above a characteristic threshold (the pump energy per excited sample area was about 5 mJ cm$^{-2}$, which corresponds to about 25 nJ per pulse), the emitted intensity suddenly starts to increase nonlinearly. Also, above this threshold, the emission linewidth breaks off from the previous trend and begins to decrease

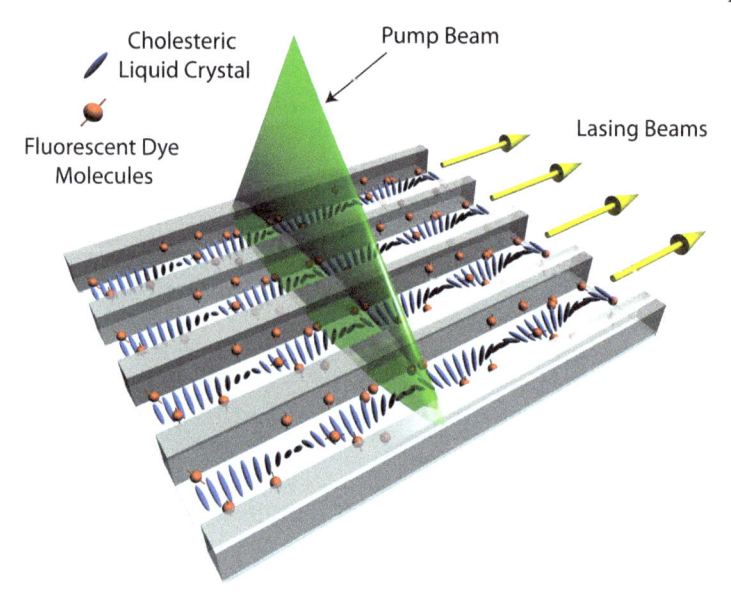

**Figure 11.8** Sketch of a multilaser array realized in a POLICRYPS structure.[11]
Reproduced by permission of IOP Publishing. All Rights Reserved.

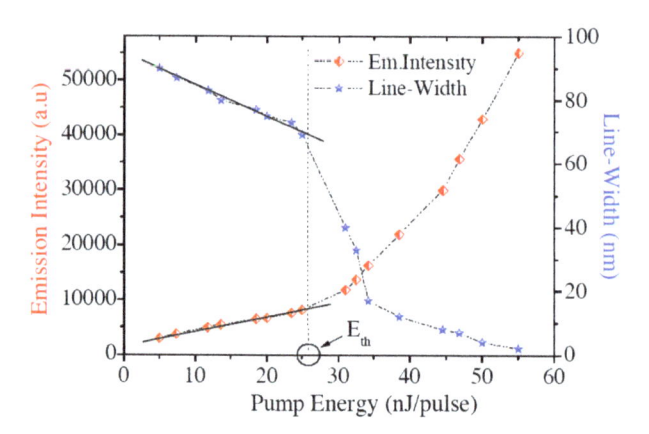

**Figure 11.9** Emitted intensity and linewidth dependence on input pump energy. Above a threshold of 25 nJ per pulse the reported curves change from initial regimes while lasing occurs.[26]
Reproduced by permission of the Optical Society of America.

significantly. The observed pump energy value at which the power-explosion and line-narrowing effects occur (25 nJ per pulse) is one order of magnitude lower than in the case of other conventional dye-doped systems in a similar environment and under the same pumping conditions.

A striking scenario is presented in Figure 11.10, showing the spatial distribution of the laser emission emerging from the microcavity laser array. A high sensitivity and resolution (1390×1024 12-bit PixelFlyQe, PCO)

POLICRYPS: A Multipurpose, Application-Oriented Platform 213

**Figure 11.10** Spatial distribution of the laser emission emerging from the mirrorless microcavity laser array. The periodicity of maximum intensities is 5 µm. This value is in agreement with the tailoring distance between the polymeric microchannels.[26]
Reproduced by permission of the Optical Society of America.

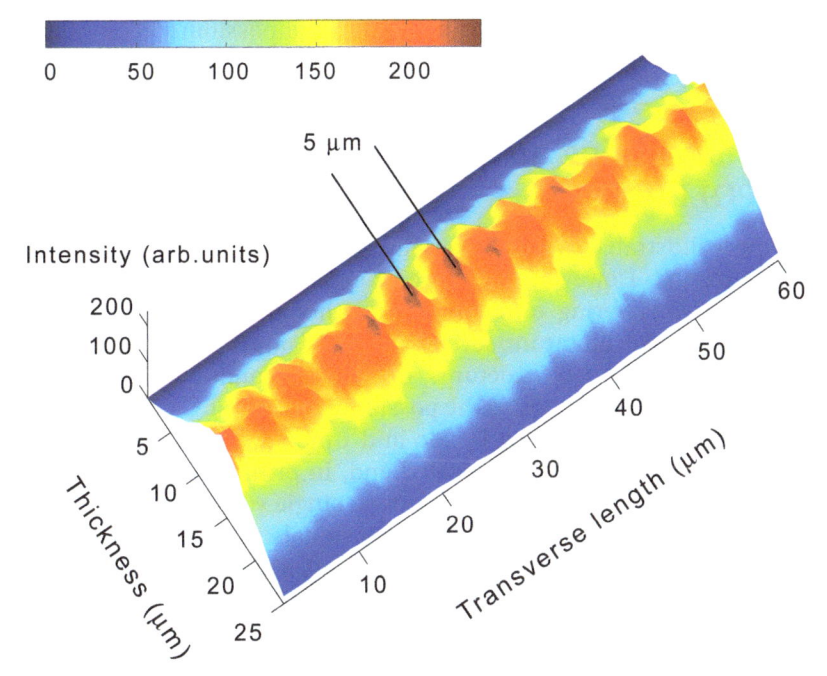

imaging CCD camera was employed in order to check the near-field modal profile of the stimulated emission. Images were acquired by scanning in the proximity of the output edge of our sample cell, in a direction perpendicular to the microchannels. The mapped intensity profile hereby obtained indicates that the maxima of lasing intensities have a spatial recurrence, with a periodicity that is found to be about 5 µm; this value is in perfect agreement with the initial tailoring configuration (*i.e.* the distance between the polymeric walls).

Therefore, we can definitely conclude that POLICRYPS microchannels act as miniaturized mirrorless cavity lasers, where the emitted laser light propagates along the liquid-crystal helical axis, which behaves as a Bragg resonator.

This level of integration can lead to new photonic chip architectures and devices, such as a zero-threshold microlaser, phased array, discrete cavity solitons, filters, and routers. Furthermore, tailoring a proper array of electrodes, which enables the application of a local electric field, would give rise to electrically programmable phase holograms with interesting light-polarization properties. Then, by including different dyes in neighbor channels, and by using proper microfibers connected at the exit, the result should be a multicolor microlaser array with the possibility to control the intensity of each channel separately.

## 11.4 Photocuring an Arbitrary Morphology POLICRYPS

As previously illustrated, the exploitation of POLICRYPS technique can bring noticeable results. However, the range of achievable morphologies is very limited when considering only interference holography as a photocuring tool. Indeed, in the literature several examples are reported of two- and three-dimensional HPDLC structures, obtained by multibeam interference holography[27,28] and recently also the possibility of realizing a 2D POLICRYPS structure has been explored by De Sio and Umeton.[29] However, these procedures are generally very sophisticated and, in principle, time wasting. As such, it would be very convenient to master a single-step procedure enabling the fabrication of more exotic, two-dimensional, periodic and aperiodic structures taking advantage of the typical POLICRYPS features. In order to achieve this result, we exploited the use of a reflective spatial light modulator (SLM). The main feature of this device is that it can modify the wavefront of an impinging laser beam in such a way that any desired two-dimensional light pattern can be obtained. Some of these patterns are impossible to realize by using standard interference holography; on the contrary, they can be obtained by just reconfiguring, in real time, the SLM display. However, the transition from interference holography to SLM in photocuring POLICRYPS materials even if apparently straightforward, in practice, it is not. The first encountered limitation is that the phase-only SLM (Pluto-VIS by Holoeye) cannot be exposed to UV light, being operative only in the 400–600 nm wavelength range, while the POLICRYPS curing technique has usually been studied and optimized by only using UV light as a curing source. Therefore, we chose a Nd:YAG laser doubled in frequency, emitting at $\lambda = 532$ nm, and started a series of attempts to individuate the proper materials to realize the POLICRYPS visible curing, which has never been tried before. The detailed illustration of these attempts is reported elsewhere.[30] Under an experimental viewpoint, the photocuring process takes place by using the 4f Fourier lenses setup sketched in Figure 11.11.

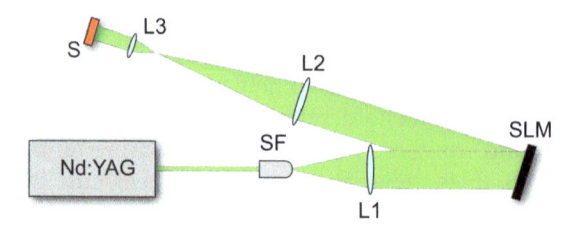

**Figure 11.11**   4f Fourier set-up for image reconstruction.[37]
Reproduced by permission of the Optical Society of America.

The beam, coming from a Nd:YAG laser doubled in frequency ($\lambda = 532$ nm), is spatially filtered, enlarged and collimated by the lens L1; then, it is used to illuminate the SLM, which modifies the wavefront. Finally, the light pattern coming from the SLM is magnified and relayed to the sample plane by the two lenses L2 and L3. The magnification ratio $M = f3/f2$ depends on the focal lengths $f2$ and $f3$ of the lenses L2 and L3, respectively. A photosensitive prepolymer syrup, made of the prepolymer system NOA61 (70–72 wt%, by Norland), the nematic liquid crystal E7 (28–30 wt% by Merck) and the photoinitiator Irgacure 784 (1–2 wt% by BASF Resins) is used to fill in, by capillarity, the sample cell, obtained by putting two glass substrates at a controlled distance. In these experiments, the exploitation of the POLICRYPS technology consisted of exposing the sample to the light pattern produced by the SLM at a temperature higher than the nematic–isotropic transition point. It is worth noting that the NLC concentration in the prepolymer mixture has a solubility threshold of about 30 wt%; when this value is overcome, NLC droplets appear during the polymerization process, thus affecting the typical POLICRYPS morphology. Moreover, in order to avoid polymeric branches being formed in correspondence of the bright areas of the curing pattern,[31] it is convenient to adjust, *a priori*, the ratio between bright and dark areas as the one corresponding to the maximum achievable phase separation (70 : 30); the possibility of doing this kind of choice represents an innovative advantage of using a SLM for fabricating POLICRYPS structures.

In order to provide an example of the new scenario of possibilities offered by the use of an SLM, we realized a fork grating (Figure 11.12(a)) that, to the best of our knowledge, is impossible to realize by using standard holographic techniques. A fork grating is a diffractive optical element whose diffraction pattern is composed by several orders, each of them being an optical vortex (Figure 11.12(c)). A vortex is a light beam characterized by a helical wavefront described by the phase function $\psi_1 = \exp(iq\theta)$, where $\theta$ is the azimuthal angle of a cylindrical coordinate system $(r, \theta, z)$ around the z-axis, which indicates the beam propagation direction. For points belonging to the helical axis ($r = 0$), the phase is not defined and the corresponding field amplitude is zero: an optical vortex projected on a screen appears therefore like a ring with a zero intensity region in the center (Figure 11.12(d)). The integer value $q$ that appears in the $\psi_1$ expression indicates the number of phase winding around the dark spot. Optical vortices are widely used for optical trapping of microscopic dielectric particles. Indeed, in conventional optical trapping, a Gaussian laser beam can trap those particles whose refractive index is greater than that of the surrounding medium. On the contrary, due to the gradient of intensity with the dark center, optical vortexes can trap particles whose refractive index value is either higher or lower than the one of the surrounding medium.[32,33] We reckon, therefore, that the realization of cheap devices endowed with POLICRYPS electro-optical characteristics and able to produce switchable optical vortexes represents quite an interesting application.[34] The pattern

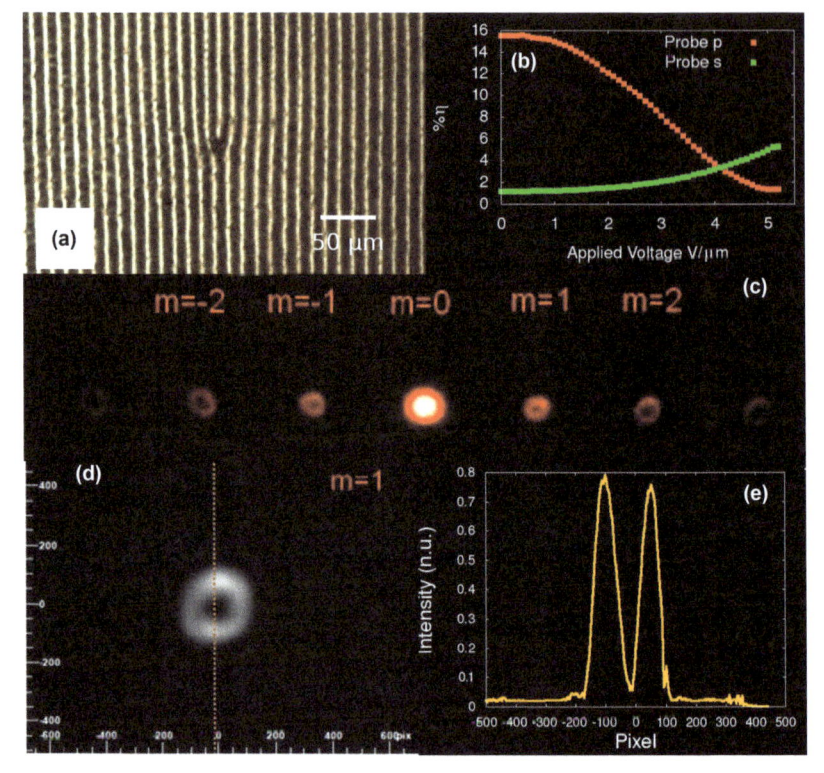

**Figure 11.12** (a) Optical microscopy image of a fork grating with a pitch of 13 μm. (b) Switching curves for s- (red squares) and p-polarized (green squares) incoming light. (c) Far-field diffraction pattern for a red probe. (d) Beam-profiler acquisition of the first diffracted order and (e) the related cut along a radius ($\theta = \pi/2$).[37]
Reproduced by permission of the Optical Society of America.

designing the fork grating has been calculated *via* computer-generated holography.[35,36] More precisely, it has been obtained by plotting the intensity distribution $I_{fork}$ calculated as the interference of two waves (eqn (11.7)): the object beam containing the vortex phase $\psi_1$ and a plane-wave reference beam with phase $\psi_2$.

$$I_{fork} = |\psi_1 + \psi_2|^2 = |\exp(iq\theta) + \exp(ikz)|^2 = 2[1 - \cos(kz - q\theta)] \quad (11.7)$$

The produced image (calculated for $q = 1$) has been implemented on the SLM *via* the direct imaging method.[37] The resulting structure is, in fact, a POLICRYPS fork grating able to produce optical vortexes that can be easily switched on and off by applying an external electric field of 5 V μm$^{-1}$. The grating periodicity in this case is 13 μm.

A complete characterization of the first-order diffracted beam is reported in the following: in Figure 11.12(d) a beam profiler acquisition of the first diffracted order is presented, its intensity profile is similar to the Gaussian profile but with a zero intensity region in the center (Figure 11.12e). The switching behavior of the first diffracted order is presented in Figure 11.12(b). The green and the red squares are related, respectively, to the diffraction efficiency for s- and p-waves and show a strongly selective response in polarization. A measurement of the characteristic response times of the system to an applied electric field has been performed. Switching times necessary to turn off and on the structure are $t_{off} = 800$ μs and $t_{on} = 4$ ms, respectively. The diffraction efficiency of the fork grating is 16%. By comparing these results with the one presented in ref. 34 it is evident that the fork-grating performances have been improved by the exploited POLICRYPS technique, resulting in higher diffraction efficiency and smaller switching voltages.

## 11.5 Conclusions

In the framework of photocurable materials, POLICRYPS represents a very reliable nano/microstructure with several possibilities of application. Indeed, a few main features of this system act as a common denominator for these applications: the sharpness of the structure and the uniformity of the LC films minimize light scattering losses, while the application of a suitable, relatively low, external voltage can determine, in a millisecond timescale, a reorientation of the LC director and hence the tunability of the device. Depending on the way a light beam propagates through the POLICRYPS, it can be used as a switchable diffraction phase grating, for light impinging at a given angle with the structure; a switchable optical phase modulator (with a light beam impinging almost perpendicularly to the structure); an array of mirrorless optical microresonators devoted to obtain a tuneable lasing effect (if the NLC is substituted with a mixture of dye-doped CLC and the system is optically pumped). Finally, a new scenario of possibilities appears when a spatial light modulator is utilized as a photocuring source. A typical example of these possibilities is the one-step fabrication of a fork grating exploiting the POLICRYPS technology. Performances exhibited in all the above applications are very interesting and stimulate further investigations in the different fields.

## Acknowledgements

Our sincere thanks go to Prof. Giuseppe Strangi and his group for the fruitful collaboration in the realization and characterization of microlaser arrays in POLICRYPS structures.

## References

1. J. D. Margerum, A. M. Lackner, E. Ramos, G. W. Smith, N. A. Vaz, J. L. Kohler and C. R. Allison, *US Pat.* 5096282, 1992.

2. R. L. Sutherland, L. V. Natarajan, V. P. Tondiglia and T. J. Bunning, *Chem. Mater.*, 1993, **5**, 1533.
3. A. Y.-G. Fuh, C.-Y. Huang, M.-S. Tsai, J.-M. Chen and L.-C. Chien, *Jpn. J. Appl. Phys.*, 1996, **35**, 630.
4. M. Date, Y. Takeuchi, K. Tanaka and K. Kato, *J. Soc. Inf. Disp.*, 1999, **7**, 17.
5. M. Jazbinsek, I. D. Olenik, M. Zgonik, A. K. Fontecchio and G. P. Crawford, *J. Appl. Phys.*, 2001, **90**, 3831.
6. M. J. Escuti, J. Qi and G. P. Crawford, *Appl. Phys. Lett.*, 2003, **83**, 1331.
7. X. Sun, X. Tao, T. Ye, P. Xue and Y.-S. Szeto, *Appl. Phys. B: Lasers Opt.*, 2007, **87**, 267.
8. R. L. Sutherland, V. P. Tondiglia, L. V. Natarajan, P. F. Lloyd and T. J. Bunning, *Proc. SPIE 7050, Liquid Crystals XII*, 2008, 705003.
9. D. E. Lucchetta, R. Karapinar, A. Manni and F. Simoni, *J. Appl. Phys.*, 2002, **91**, 6060.
10. L. De Sio, A. Veltri, R. Caputo, A. De Luca, G. Strangi, R. Bartolino and C. P. Umeton, *Liq. Cryst. Rev.*, 2013, **1**(1), 2.
11. R. Caputo, A. De Luca, L. De Sio1, L. Pezzi, G. Strangi, C. Umeton, A. Veltri, R. Asquini, A. d'Alessandro, D. Donisi, R. Beccherelli, A. V. Sukhov and N. V. Tabiryan, *J. Opt. A: Pure Appl. Opt.*, 2009, **11**, 024017.
12. R. Caputo, A. V. Sukhov, C. P. Umeton and R. F. Ushakov, *J. Exp. Theor. Phys.*, 2000, **91**, 1190.
13. R. Caputo, L. De Sio, A. V. Sukhov, A. Veltri and C. P. Umeton, *Opt. Lett.*, 2004, **29**, 1261.
14. R. Caputo, C. P. Umeton, A. Veltri, A. V. Sukhov and N. Tabiryan, *Eur. Pat.* 1649318; *US Pat.* 2007/0019152A1, 2007.
15. L. De Sio, R. Caputo, A. De Luca, A. Veltri, A. V. Sukhov and C. P. Umeton, *Appl. Opt.*, 2006, **45**, 3721.
16. L. De Sio, A. Veltri, A. Tedesco, R. Caputo, A. V. Sukhov and C. P. Umeton, *Appl. Opt.*, 2008, **47**, 1363.
17. R. L. Sutherland, V. P. Tondiglia, L. V. Natarajan, T. J. Bunning and W. W. Adams, *Appl. Phys. Lett.*, 1994, **64**, 1074.
18. R. Caputo, A. V. Sukhov, N. V. Tabyrian, C. P. Umeton and R. F. Ushakov, *Chem. Phys.*, 2001, **271**, 323.
19. D. E. Lucchetta, L. Criante and F. Simoni, *J. Appl. Phys.*, 2003, **93**, 9669.
20. L. De Sio, N. Tabiryan, R. Caputo, A. Veltri and C. P. Umeton, *Opt. Exp.*, 2008, **16**, 7619.
21. P. G. de Gennes and J. Prost, *The Physics of Liquid Crystals*, Oxford University Press, 2nd edn, 1995.
22. Y. H. Wu, Y. H. Lin, Y. Q. Lu, H. Ren, Y. H. Fan, J. Wu and S. T. Wu, *Opt. Exp.*, 2004, **12**, 6382.
23. R. Caputo, I. Trebisacce, L. De Sio and C. P. Umeton, *Opt. Exp.*, 2010, **18**, 5776.
24. H. Kogelnik and C. V. Shank, *Appl. Phys. Lett.*, 1971, **18**, 152.
25. L. S. Goldberg, J. M. Schnur, *US Pat.* 3 771 065, 1973.

26. V. Barna, R. Caputo, A. De Luca, N. Scaramuzza, G. Strangi, C. Versace, C. Umeton, R. Bartolino and G. N. Price, *Opt. Exp.*, 2006, **14**, 2695.
27. R. L. Sutherland, V. P. Tondiglia, L. V. Natarajan, S. Chandra, D. Tomlin and T. J. Bunning, *Opt. Exp.*, 2002, **10**, 1074.
28. V. P. Tondiglia, L. V. Natarajan, R. L. Sutherland, D. Tomlin and T. J. Bunning, *Adv. Mater.*, 2002, **14**, 187.
29. L. De Sio and C. Umeton, *Opt. Lett.*, 2010, **35**, 2759.
30. M. Infusino, A. Ferraro, A. De Luca, R. Caputo and C. Umeton, *J. Opt. Soc. Am. B*, 2012, **29**(11), 3170.
31. A. Veltri, R. Caputo, C. Umeton and A. V. Sukhov, *Appl. Phys. Lett.*, 2004, **84**, 3492.
32. K. T. Gahagan and G. A. Swartzlander Jr., *Opt. Lett.*, 1996, **21**, 827.
33. K. T. Gahagan and G. A. Swartzlander Jr., *J. Opt. Soc. Am. B*, 1998, **15**, 524.
34. Y. J. Liu, X. W. Sun, Q. Wang and D. Luo, *Opt. Exp.*, 2007, **15**, 16645.
35. A. Kumar, P. Vaity, Y. Krishna and R. P. Singh, *Opt. Lasers Eng.*, 2010, **48**, 276.
36. A. V. Carpentier, H. Michinel, J. R. Salgueiro and D. Olivieri, *Am. J. Phys.*, 2008, **76**, 916.
37. M. Infusino, A. De Luca, V. Barna, R. Caputo and C. Umeton, *Opt. Exp.*, 2012, **20**(21), 23138.

CHAPTER 12

# SU-8 for Microsystem Fabrication

YI CHIU*[a] AND YU-TING CHENG*[b]

[a] Electrical and Computer Engineering Department, National Chiao Tung University, Taiwan; [b] Microsystems Integration Laboratory, Electronics Engineering Department, National Chiao Tung University, Taiwan
*Email: yichiu@mail.nctu.edu.tw; ytcheng@faculty.nctu.edu.tw

## 12.1 Introduction

SU-8 is a negative-tone photoresist that can be used to fabricate thick, high aspect ratio structures. The thickness of SU-8 structures ranges from several micrometers to several hundred micrometers or up to millimeters by direct spin coating or stacking of multiple layers of dry films. It is often used as a mold for metallic parts in a LIGA-like process. Moreover, due to its excellent chemical resistance and mechanical strength, SU-8 is also used as a permanent structural material for free-moving parts, fluidic channels, or packaging substrates with thickness in the range of submillimeters. Since SU-8 has a low Young's modulus compared with those of silicon, glass, or metal, it is used in actuators such as microgrippers or RF switches to lower the driving voltage. It is also used to fabricate soft springs to lower the resonance frequency in various sensors and actuators. Being a negative resist, SU-8 can be used to fabricate complex three-dimensional structures such as sealed microchannels or tilted optical surfaces by multiple exposures. Another feature of SU-8 is that its properties such as index of refraction, glass-transition temperature, and internal stress can be controlled and modified during processes by the exposure doses, baking

---

RSC Smart Materials No. 13
Photocured Materials
Edited by Atul Tiwari and Alexander Polykarpov
© The Royal Society of Chemistry 2015
Published by the Royal Society of Chemistry, www.rsc.org

temperature, additives, *etc.* This provides possibilities for novel device design and fabrication without complex fabrication processes.

SU-8 has relatively little absorption in the visible spectrum range. Therefore, it has been used in a variety of optical applications. Refractive and diffractive components such as lenses and gratings have been fabricated with SU-8. Curved optical surfaces for refractive components can be fabricated by gray-scale lithography or electrostatic pulling of the resist while in the liquid form. Special exposure techniques and equipments were also developed to fabricate tilted surface for reflectors, prisms, and beam splitters. SU-8 is also used for optical waveguides that are often integrated with optical sensors or interferometers for bioMEMS sensing applications. Complex diffractive structures such as photonic crystals have been fabricated in SU-8 by holographic photolithography or direct laser beam writing. Such structures can be used in various optical or RF applications.

SU-8 has been integrated in RF components such as switches and inductors due to its low loss in the RF range. It is also used as substrates for RF systems in a package. Depending on the design, SU-8 can also be used as flexible substrates for flexible or wearable electronics.

## 12.2 Material Properties and Fabrication Processes

### 12.2.1 Material Properties

The SU-8 comprised of three key ingredients, Epon SU-8 epoxy, gamma-butyrolactone (GBL) solvent, and a photoacid generator taken from the family of the triarylium-sulfonium salts is an epoxy-based negative photoresist. Irradiation will induce the formation of a low concentration of a strong acid acting as a catalyst for following crosslinking process. The photoacid generator decomposes to form hexafluoroantimonic acid that protonates the epoxides on the oligomer. After baking, the protonated oxonium ions will react with neutral epoxies in a series of crosslinking reactions resulting in high mechanical and thermal stability of the lithographic structures. Meanwhile, the SU-8 resist can be added with different additives or diluted with different solvents to have a variety of property changes. For instance, the commercial SU-8 2000 formulated with cyclopentanone as solvent exhibits better adhesion characteristic with substrates and faster processing times. Table 12.1 summarizes the physical properties of SU-8 in general.[1-3]

### 12.2.2 Microfabrication Processes

Two kinds of SU-8 processes, which are spin coating and dry-film lamination, respectively, have been developed for making microstructures. The spin-coating process is similar to the conventional IC photolithography process. The thickness of SU-8 film coated on the substrate depends on the rotation speed of the wafer holder, which is around 1000–3000 rpm, and the viscosity of SU-8. The thickness ranges from several μm up to 1 mm. Multiple

**Table 12.1** Physical properties of SU-8.[1–3]

| | |
|---|---|
| Young's modulus, $E$ (postbaked at 95 °C) | 4.02 GPa |
| Young's modulus, $E$ (hardbaked at 200 °C) | 4.95 ± 0.42 GPa |
| Biaxial modulus of elasticity, $E/(1 - \nu)$ | 5.18 ± 0.89 GPa |
| Poisson ratio ($\nu$) | 0.22 |
| Maximum stress (hardbaked at 200 °C) | 34 MPa |
| Glass temperature, $T_g$ (fully crosslinked) | >200 °C |
| Degradation temperature (fully crosslinked) | ~380 °C |
| Thermal expansion coefficient (postbaked at 95 °C) | 52 ± 5.1 ppm K$^{-1}$ |
| Dielectric constant | 4.5 @ 10 MHz, 4.2 @ 10 GHz |
| Loss (postbaked at 100 °C, (+8 h at 65 °C in KOH)) | 0.08 @ 100 GHz |
| Refraction index | 1.67 |
| Breakdown voltage | >4×10$^7$ V m$^{-1}$ |

spin coating is required for realizing thick SU-8 coating. Meanwhile, the wettability of SU-8 on the substrate is critical for obtaining uniform and reliable coatings since SU-8 has hydrophobic characteristics.[1] Therefore, a commercial adhesive promoter, such as HMDS, OmniCoat from Microchem Inc., *etc.*, will be coated first onto the inorganic substrate before spin coating SU-8. Previous reports have also investigated the related influences of surface pretreatment and postprocessing on the physical characteristics of SU-8 microstructures. Morikaku *et al.* found the adhesion of SU-8 onto a substrate could be strengthened by hard baking at 200 °C for 60 min. With the coating of the adhesive promoter, which is 100 wt% hexamethyldisilazane (Tokyo Ohka Kogyo Co., Ltd; OAP) and enables the reduction of the SU-8 polymer chains breakage, SU-8 would exhibit higher yield strength and higher tensile strength in primer-coated tensile specimens.[4] Wouters and Puers found that the SU-8 epoxy will swell on being immersed in a solution because of the absorption or desorption of water to result in the volume change, or alternatively a change in built-in stress.[5] In addition, experimental results showed the built-in stress of SU-8 reached its maximum value when submersed in propylene glycol methyl ether acetate and in isopropyl alcohol and the stress will be strongly correlated with the formation of cracks and delamination in SU-8.

Two main kinds of approaches have been developed for making a freestanding microstructure based on spin-coated SU-8. One is double exposure and the other is sacrificial layer release. For making the SU-8 bridge-type structure developed by Yoon *et al.*, a large optical dose is first applied through the column mask onto a SU-8 coated substrate and postexposure baked at 95 °C on a hot plate.[6,7] Then, a small optical dose is applied through the lateral bridge mask, and the structure is postexposure baked at 95 °C in an oven. After the two-step exposure and bake, the SU-8 has the desired top bridge portion crosslinked at the top of the crosslinked columns. Regarding the approach of sacrificial layer release, it is more like a conventional surface micromachining process.[8,9] A sacrificial layer, such as photoresist AZ4620, polydimethylglutarimide (PMGI), *etc.*, is deposited and patterned on a substrate first. Then, SU-8 is coated on the top of the sacrificial layer. After patterning the SU-8 by optical exposure and baking at 95 °C for 30 min, the sacrificial layer can be removed by developer and free-standing SU-8 microstructures can be formed. In addition to these two common approaches to fabricate free-standing SU-8 microstructures, stereolithography[10,11] to form 3D structures layer by layer and two-photon-absorption technique[12,13] to trigger polymerization in 3D forms have also been developed for microstructure fabrication. However, these techniques are both time consuming with serial process characteristics.

The general fabrication processes employing dry film lamination is first to fabricate open structures using standard multilayered photolithography on silicon or glass substrates. Metal can be deposited and patterned between these layers for electrical contact. A dry film resist is then laminated on top to close the structure. Further photolithography can be performed on the laminated film and these steps can be repeated multiple times as necessary. Compared with liquid resist, dry-film resist has advantages such as large thickness, good uniformity, good flatness, no edge beads, and lower processing cost. Lamination of multiple dry-film resist allows the fabrication of complex closed structures such as wells or channels in microfluidics that are not possible in conventional micromachining using sacrificial release. Another feature of the dry-film lamination process is that photolithography is performed after lamination, thus allowing accurate level-to-level alignment without special alignment equipments, as are needed in wafer-bonding processes.

In addition to the microstructure fabrication, SU-8 can be also utilized as a packaging layer or substrate for microsystem integration. For example, Xiao et al.[14] utilized ultrathin silicon wafers combined with SU-8 bonding and deep reactive ion etching (DRIE) technology to fabricate folded MEMS varactor devices. Instead of SOI wafers, two silicon substrates bonded using the SU-8 can enable a more flexible design including low parasitic substrate capacitance and low manufacturing cost. Chao et al.[15] presented a low-cost heterogeneous integration technology to assemble a CMOS chip with an organic substrate (SU-8/PDMS) using low temperature Au–Au thermocompressive bonds (<200 °C) for flexible wireless microsystem fabrication. −15 dB return loss and −0.25 dB insertion loss @ 40 GHz of a microstrip-to-coplanar waveguide interconnect transition between CMOS chip and SU-8 substrate was also successfully developed for the application. In the following sections, various kinds of SU-8 microstructures based on the fundamental processes and their related applications in biomedical, RF and optical microsystems will be reviewed.

## 12.3 SU-8 Structures

### 12.3.1 Parts and Actuators

Ultrathick SU-8 resist has been used as molds or structural material for free-moving microparts such as gears and rotors. Rotary parts with thickness of several hundred micrometers and up to 1 mm were demonstrated by using UV photolithography[16,17] and X-ray photolithography,[18] respectively. Whereas postprocessing assembly might be needed to assemble a number of parts together[17,18] or cap the fabricated parts on the substrate,[16] a self-sacrificial multilayer SU-8 process was developed by Foulds and Parameswaran to fabricate self-capped free-moving parts on the substrate without post-processing assembly.[19] In this process, a modified SU-8 layer that absorbed UV heavily was coated on top of a regular SU-8 layer to form a bilayer. Each bilayer was coated and exposed lithographically. The UV-absorbing SU-8 on the top layers prevented the exposure of the SU-8 on the bottom layers. After all the layers were coated and exposed, a single development step released the structures by using the unexposed SU-8 as the sacrificial layer. Stapled hinges and rotary gears were demonstrated, as shown in Figure 12.1.

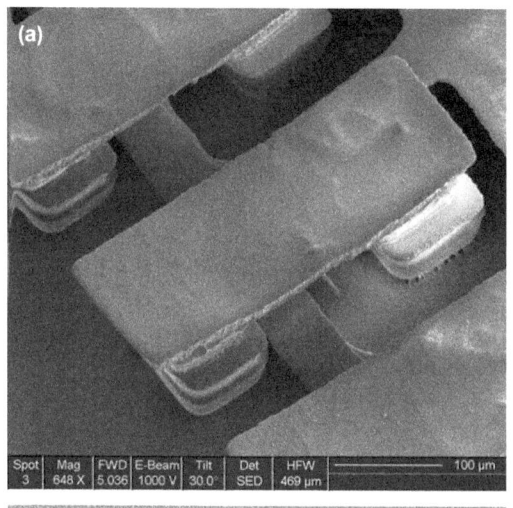

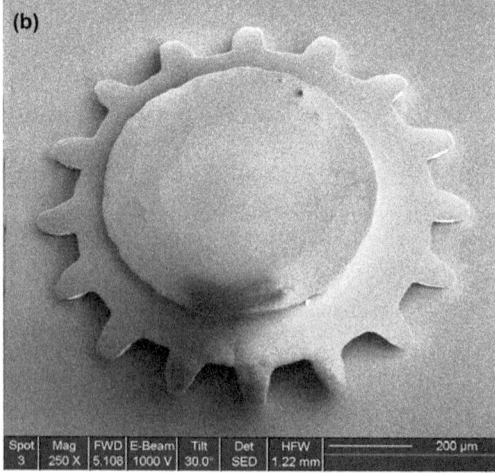

**Figure 12.1** (a) Stapled hinge and (b) gear fabricated in SU-8. (Reprinted from Ref. 19 with permission from IOP Publishing.)

SU-8 can be integrated in silicon-based processes to provide additional choice of materials and thus more design flexibility. Since SU-8 and silicon have different material properties such as density, rigidity, and fragility, devices fabricated with both materials can be optimized more independently by proper combination of the two materials. For example, SU-8 was integrated in a SOI process by Conradie and Moore where SU-8 microposts were fabricated on silicon cantilever beams as added mass to adjust the resonance frequency.[20] Because of its low Young's modulus, SU-8 was used to build soft torsional springs by Fujita et al. for large deflection angles in a dual-axis microscanning mirror, while the mirror plate and the gimbal frame structures were fabricated in the silicon substrate for its mechanical strength, rigidness, and flatness.[21] The maximum scanning angles at resonance for the inner and outer frames were $\pm 40°$ and $\pm 30°$, respectively. A single-crystalline MEMS scanner supported by soft SU-8 springs was fabricated by Bachmann et al. by using a multiwafer bonding technique.[22] The scanner was driven by vertical comb-drive actuators fabricated in two stacked silicon substrates using deep RIE. A maximum scan angle of 25° was obtained at 30 V ac input voltage.

Compliant SU-8 structures were used in 3D out-of-plane micro assembled devices as locking mechanisms. As demonstrated by Tsang et al.,[23] various SU-8 planar components attached to long serpentine springs were fabricated and anchored to the substrate. The out-of-plane assembly of these components was accomplished by pushing the planar structures laterally until they rotated to the desired 90° position. The distorted serpentine springs caused a frictional force at the contact between the substrate and the upright structures to hold the structures at the desired angular position. SU-8 locking mechanisms proposed by Chiu et al. were used to fix the angular positions of single-crystalline micromirrors fabricated in SOI substrates,[24] as shown in Figure 12.2. The locking mechanisms were implemented on the mirror plate edges as well as the rotational hinges. The mirrors were assembled by simple vertical push on the assembly pad through the through-wafer holes etched from the back of the substrate. Vertical mirrors could be assembled in about 30 s with an angular accuracy of $89.8 \pm 0.3°$. The integration of SU-8 as a second structural layer on SOI substrates provides a low-temperature alternative compared with standard Si/Poly-Si surface micromachining processes.

With active control imposed on SU-8 structures, various kinds of microactuators have been demonstrated. Since SU-8 has a low Young's modulus and large thermal expansion coefficient compared with those of silicon, glass, or metal, SU-8 actuators can be operated at lower control voltage or power. For example, SU-8 thermal actuators based on the "hot-arm–cold-arm" structure can be integrated with microgrippers to manipulate microobjects such as living cells. Since SU-8 is not conductive, a resistive heating film is usually deposited on top (or bottom) of the SU-8 layer to heat the structure. The low operation voltage/temperature, low Young's modulus and thus more compliant structure, and biocompatibility of SU-8 make SU-8 microthermal actuators and grippers ideal for biomedical applications. Such

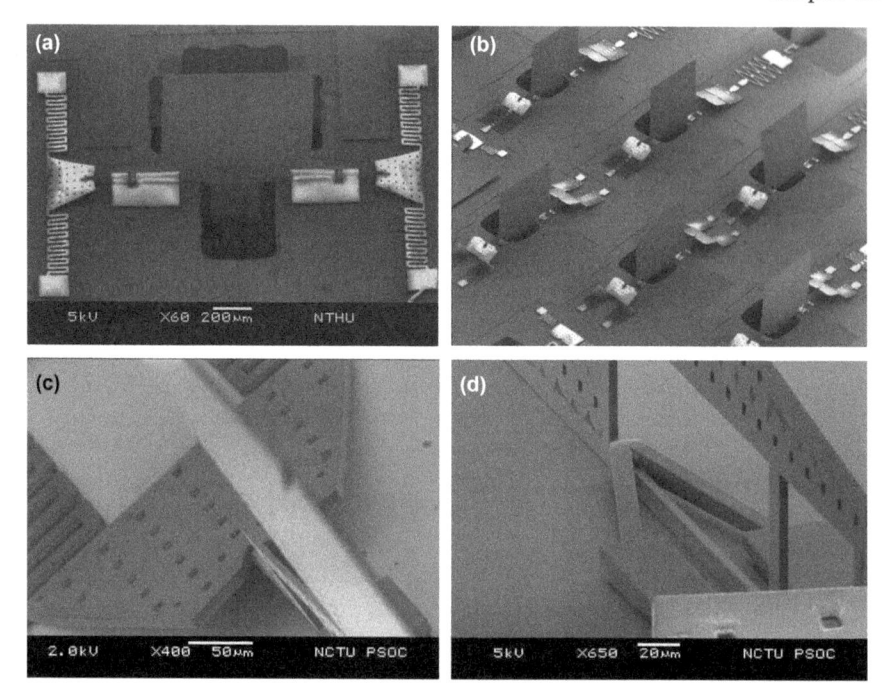

**Figure 12.2** (a) Silicon micro mirror without release etching holes latched by SU-8 mechanisms, (b) micro mirror arrays, (c) SU-8 side latch, (d) SU-8 V-shaped compliant staple on silicon hinge.
(Reprinted from Ref. 24 with permission from IEEE.)

an electrothermally actuated SU-8 microgripper was demonstrated by Chronis and Lee to manipulate living cells and other biological species in aqueous solution,[25] as shown in Figure 12.3. The 20 μm thick SU-8 gripper was heated by a Cr/Au film on the bottom. With an applied voltage of 2 V, a temperature rise of 30 °C and gripper opening of 10 μm were obtained. Duc et al. used SU-8 to fill the gaps of a meander silicon skeleton heater.[26] The large thermal conductivity of silicon ensured uniform and rapid heating of the SU-8 blocks to increase its response speed. With the uniform heating in the vertical direction, unintended displacement in the out-of-plane direction could be reduced. Piezoresistive sensors were implemented at the ends of the actuators to enable automatic control without the need for a monitoring camera during the gripper operation. The minimum detectable gripper displacement and force were 1 nm and 770 nN, respectively.

Electrostatic actuators are attractive due to their low power consumption and less demanding of special materials. A gap-closing electrostatic SU-8 optical scanner was demonstrated by Moon et al.[27] SU-8 was first coated on a wet-etched silicon mold wafer with a thickness of 130 μm and then patterned for the scanning mirror and springs. The top SU-8 structure was released from the mold, metallized, and then bonded with a second silicon substrate with etched recesses as the bottom electrodes to form the optical scanner. Dai et al. developed a two-mask process to fabricate a monolithic

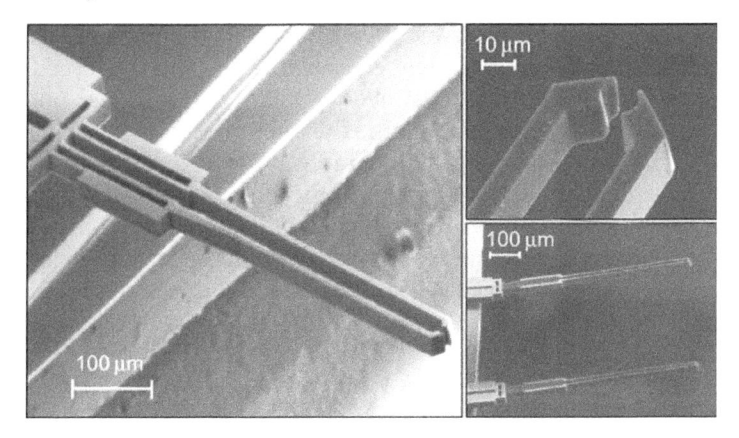

**Figure 12.3** Electrothermal SU-8 microgripper.
(Reprinted from Ref. 25 with permission from IEEE/ASME.)

electrostatic comb-drive actuator using thick SU-8 as the structural material and electroplated copper as the sacrificial material.[28,29] Experiments showed agreement between measured and simulated actuation displacement.

### 12.3.2 Microfluidic Structures and Systems

SU-8 is biocompatible and can be processed with low cost to fabricate deep microchannels with smooth sidewalls. Therefore, it is a very attractive material for microfluidic applications. The smoothness is especially important for optical waveguides to reduce scattering losses and for microfluidic components such as valves to prevent leakage. Many microfluidic components and systems for on-chip liquid handling, mixing, and detection have been demonstrated in SU-8. Seidemann et al. demonstrated a check valve fabricated in a 360 μm deep and 200 μm wide SU-8 flow channel using copper as the sacrificial material.[30] Ribeiro et al. demonstrated a microfluidic mixer with a channel depth of 600 μm and width of 500 μm in a single SU-8 layer spin-coated on a glass substrate.[31] The inlets and outlets for liquid were drilled in another top glass substrate that was glued on the SU-8 microchannels. Carlier et al. fabricated multilayered SU-8 microfluidic channels for electrospray-ionization mass spectroscopy.[32] A 150 μm thick layer of SU-8 was first coated and patterned for the capillary tube housing. On top of the thick SU-8 layer, a 55 μm thick layer of SU-8 was spin coated and patterned for the microchannels with a cross section of 30 μm×55 μm. Finally, a 3 μm thick SU-8 was coated on a Pyrex cover that was used to seal the channels.

Complex microfluidic structures can be fabricated by more elaborated processes. An all-SU-8 microfluidic chip with integrated in-channel micromeshes was demonstrated by Sato et al.[33] The chip was composed of two SU-8 parts. First, the inlets and outlets were fabricated in the 50 μm thick top cover layer that was released from the substrate as a standalone part. The SU-8 layers for fabricating the microchannels were coated on a second substrate. The top cover layer was then bonded to the thick SU-8 channel layers during the soft bake. The bonded structure was then exposed from the front side to define the microchannels by using conventional photolithography and from the back side to define the micromeshes in the channels by using multiangle inclined photolithography. After exposure and development, the standalone all-SU-8 microfluidic chip was released from the substrate by using a lift-off process. The embedded channels had width and height of 800 μm and 120 μm, respectively. The diameter and height of the micropillars forming the meshes were 10 μm and 120 μm, respectively. The rhomboidal mesh pores had a dimension of 10 μm×20 μm, as shown in Figure 12.4.

SU-8 was also used to fabricate microflow cytometers for analyzing microparticles such as cells and bacteria.[34–37] In the cytometer presented by Lin and Lee,[35] flow channels and tapered fiber alignment slots were fabricated in a 55 μm thick SU-8 layer on an ITO glass substrate. Another ITO substrate coated with a 1 μm thick photoconductive amorphous silicon layer was bonded to the SU-8 structures to seal the channels. Microparticles

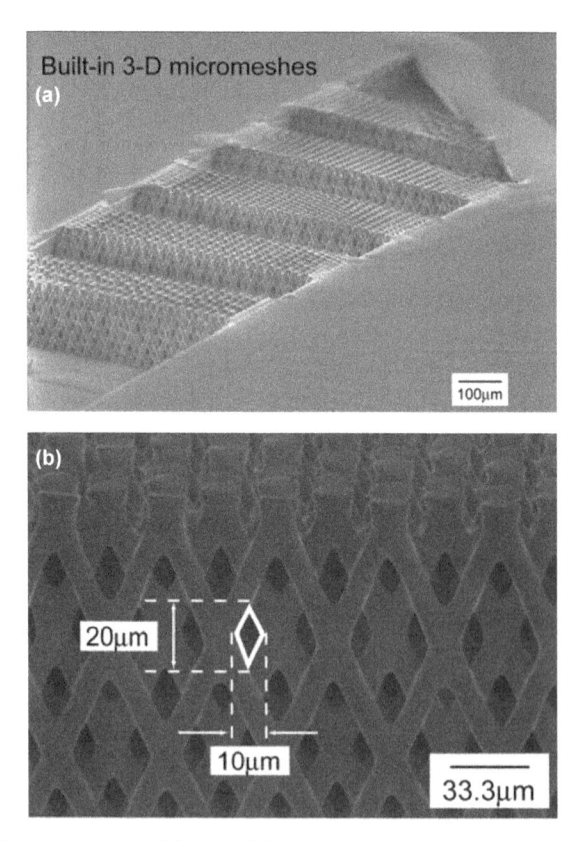

**Figure 12.4** (a) Micromesh fabricated in SU-8 microchannel, (b) close-up of micromesh structures.
(Reprinted from Ref. 33 with permission from IOP Publishing.)

in the channels were driven by gravity and focused by optically induced dielectrophoretic (ODEP) forces generated by projected optical virtual electrodes. Opposing optical fibers were placed perpendicularly to the channels to launch and collecting light to detect passing particles. In the downstream, ODEP switches were further used to sort the two kinds of particles according to the detection results. In similar devices presented by Barat et al.[36] and Watts et al.,[37] the light sources were coupled into waveguides and then collimated by microlenses fabricated in the SU-8 layer. Optimal optical design could reduce the optical spot size and divergence angle in the channel to improve detection sensitivity.

Active optical components were further integrated with SU-8 waveguides and microchannels for an integrated lab-on-a-chip system presented by Balslev et al.,[38] as shown in Figure 12.5. This device included an optically pumped liquid dye laser that emitted light at 576 nm, waveguides that delivered light to various locations on the cuvette, microfluidic channels with passive diffusive mixers, and photodiodes fabricated on the silicon substrate. Successful distinction of output photodiode currents for solutions containing different dye concentrations was demonstrated. This device demonstrated the feasibility of monolithic integration of optical, fluidic, and electronic components for optofluidic applications by using SU-8.

Complex three-dimensional sealed microchannels can be fabricated by using laminated dry films for good uniformity and almost unlimited layers. In the work presented by Abgrall et al.,[39] SU-8 was first coated on a PET release liner on a rigid substrate. The backed SU-8/PET stack was peeled from the substrate and used as a dry-film resist. Oxygen plasma treatment was used to improve adhesion between the SU-8 layers during the lamination. Microfluidic channels with width of 10 to 500 μm and height of 20 to

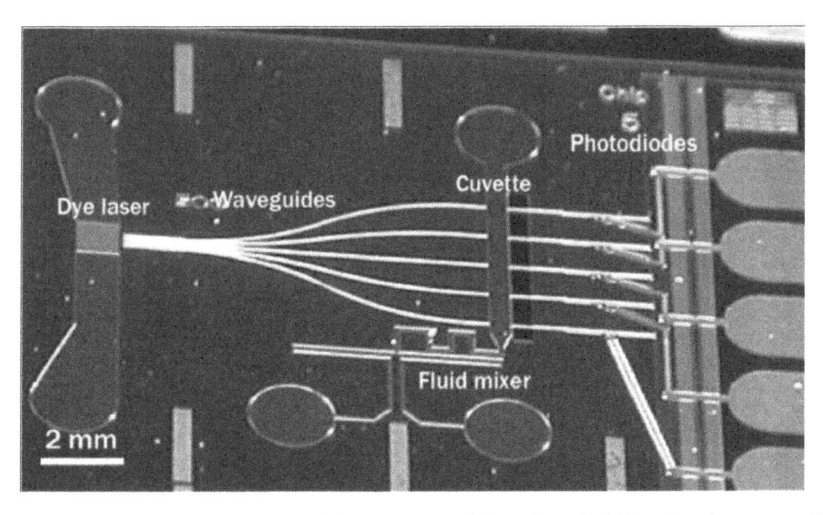

**Figure 12.5** Integrated lab-on-chip device with microfluidic dye laser, optical waveguides, microfluidic channels network and photodiodes. (Reproduced from Ref. 38.)

100 µm were demonstrated. In hydraulic tests, no variation of channel dimensions was detected for interlevel thickness larger than 35 µm. In the three-dimensional microfluidic system presented by Peng et al.,[40] the SU-8 dry film was fabricated by spin coating SU-8 on a polyimide film. After a prebake, the SU-8 film was bonded to another SU-8 layer with fabricated open SU-8 structures on a substrate. For a test channel with dimensions of 8 µm (W)×8 µm (H)×4000 µm (L), observation of continuous flow of DI water for over 24 h showed no evidence of liquid leakage, demonstrating the robustness of the fabricated structures.

## 12.4 SU-8 Applications

### 12.4.1 SU-8 Sensors

#### 12.4.1.1 Biosensors

Since the detection of many biochemical species or molecules is based on fluorescence or optical scattering, the feasibility of integrating high-quality SU-8 waveguides with microfluidic channels enables the implementation of microfluidic detection chips based on optical detection.

Optical signals of SU-8-based sensors can be detected by using integrated optical devices. In the biochemical sensor presented by Shew et al.,[41] two parallel SU-8 waveguides were connected at both ends by Y-junctions to form a Mach–Zehnder interferometer (MZI). A difference of 0.004 in the refractive indices of cured and uncured SU-8 was measured and exploited to build the claddings and cores of the waveguides, respectively. Sample solutions with various target molecule concentration caused changes in the effective index of the functionalized sensing branch of the MZI and thus its interferometric output. Experiments showed a maximum sensitivity of $10^{-9}$ g ml$^{-1}$ in antibody detection. A similar device was presented by Sepúlveda et al.[42] for DNA detection. Experiments showed a 300 pM limit of detection of direct (labelless) DNA hybridization by using this integrated MZI. With further optimization and integration, the detection limit was expected to be reduced to the 10 pM range.

Bio/chemical species can also be detected with simple and robust cantilever-based sensors *via* optical or other readout mechanisms. The cantilever surfaces are first coated and functionalized with a sensing film. When the target molecules to be detected bind with the sensing film, the increased mass causes a reduction of the resonance frequency of the cantilever. It is also possible to operate the sensor in a static mode by detecting the cantilever bending caused by the surface stress in the sensing film bound with the molecules. Calleja et al. used cantilever sensors to measure the adsorption of single-stranded DNA.[43] The thickness of the sensors was about 1.5 um. The bending of the cantilevers was measured from displacement of a laser spot reflected from the back of the cantilever tip. In similar sensors developed by Nordström et al., the signals were obtained from the coupling intensity modulation of bending waveguides or the on-chip metal

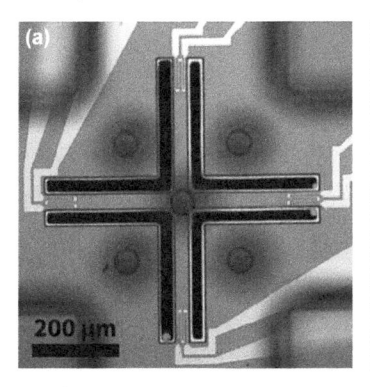

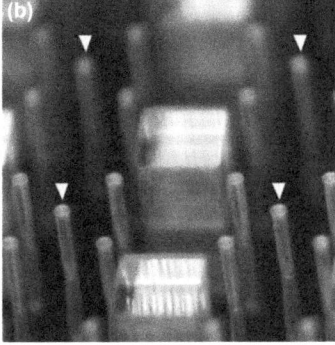

**Figure 12.6** (a) Top view of a sensing pillar, (b) array of 2×2 sensing elements surrounded by passive pillars and posts.
(Reproduced from Ref. 47.)

piezoresistors.[44,45] Schmid et al. presented a biosensor based on SU-8 resonant beams.[46] The 1.45 µm thick single- and double-clamped SU-8 microbeams were driven into resonance electrostatically with a typical resonance frequency of about 300 kHz. For streptavidin detection, the minimum detectable concentration was 0.025 µg ml$^{-1}$, corresponding to a frequency shift of 1.9%. For DNA detection, the minimum detectable concentration was 0.13 nmol ml$^{-1}$, corresponding to a frequency shift of 0.9%.

Another SU-8-based sensor with electronic readout was presented by Doll et al. for biological force measurement at microscales,[47] as shown in Figure 12.6. SU-8 pillars with a height of 350 µm and diameters from 45 to 70 µm were fabricated on the center of two crossing SU-8 bridges with a thickness of 5 µm. When a force was applied to the tip of the sensing pillar, the pillar tilted and caused the SU-8 bridges to distort and bend. The bending was measure by the Cr/Au strain gauges embedded at the base of the beams. The forces generated by free-moving *C. elegans* worms were measured by using this device and a peak contact force of 2.5 ± 2.5 µN was found. The force resolution in the 1 Hz to 1 kHz range was 260 nN.

A SU-8 cap mounted on the tip of an optical fiber was developed for *in vivo* pressure monitoring in the work by Hill et al.,[48] as shown in Figure 12.7. The SU-8 cap was composed of three SU-8 layers: a 2 µm thick pressure sensing diaphragm, a 50 µm thick annulus with a 100 µm diameter hole that defined the Fabry–Perot cavity, and a 100 µm thick SU-8 sheath to adapt the optical fiber that was fixed with a biocompatibile glue. The sensor showed a linear range from 0 to 125 mmHg with a resolution of 1–2 mmHg.

### 12.4.1.2 Other Sensors

In the MZI-based pressure sensor presented by Pelletier et al.,[49] the SU-8 sensing waveguide was built on top of a wet-etched silicon membrane, while the reference waveguide was on the rigid substrate. The pressure difference between the two sides of the membrane deforms the sensing waveguide and

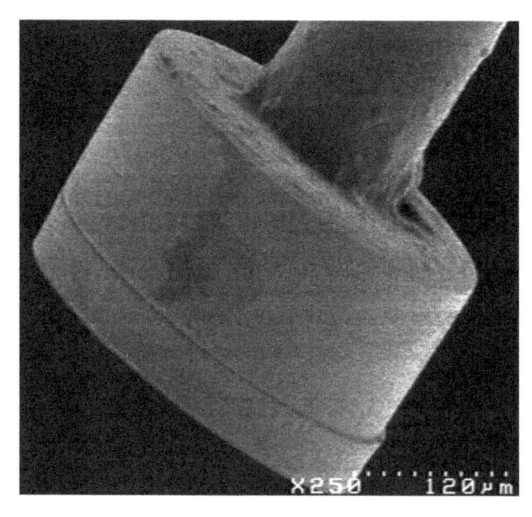

**Figure 12.7** SU-8 pressure cap mounted on fiber tip.
(Reprinted from Ref. 48 with permission from Elsevier.)

thus induced a phase change of the light therein. The experiments showed that an applied pressure of $2\times10^5$ Pa could induce a $\pi$ rad phase shift.

In the shear force sensor presented by Tseng and Lin,[50] SU-8 thick resist was used to fabricate optical fiber grooves and a vertical metal-coated floating reflective mirror on a silicone membrane. When the sensor was placed in a flow field, the fluid exerted a shear stress force on the membrane and induced a displacement of the floating vertical mirror. The Fabry–Perot interferometers formed by the fiber end faces and the vertical mirror were used to detect the mirror displacement and thus the shear stress on the fluid/membrane interface. A resolution of 0.065 Pa was demonstrated.

Noninterferometric sensors based on SU-8 waveguides have also been demonstrated. Chang-Yen and Gale[51] presented an integrated oxygen sensor by immobilizing a dye layer on top of a SU-8 waveguide. The evanescent optical field that extended from the waveguide core excited the dye in the top layer. The fluorescent light emitted by the dye correlated linearly to the dissolved oxygen concentration when the sensor was placed in water. With an excitation at 450 nm, a resolution of 0.6 mg l$^{-1}$ dissolved oxygen concentration was obtained over the range of 0.8 to 24.8 mg l$^{-1}$.

Llobera et al. presented an all-SU-8 optical accelerometer where the input/output waveguides were anchored to the substrate and the sensing waveguide was embedded in the SU-8 moving mass,[52] as shown in Figure 12.8. Upon the action of the external acceleration, the sensing waveguide was displaced and the misalignment of the three waveguides reduced the coupling efficiency and thus the optical power at the output. Experiments showed a reproducible sensitivity of at least 13.1 dB g$^{-1}$.

In the work presented by Seena et al.,[53] SU-8 was first mixed with carbon black (CB) to form a piezoresisitve composite material. SU-8 microcantilevers with embedded SU-8/CB piezoresistors were functionalized with

*SU-8 for Microsystem Fabrication*  233

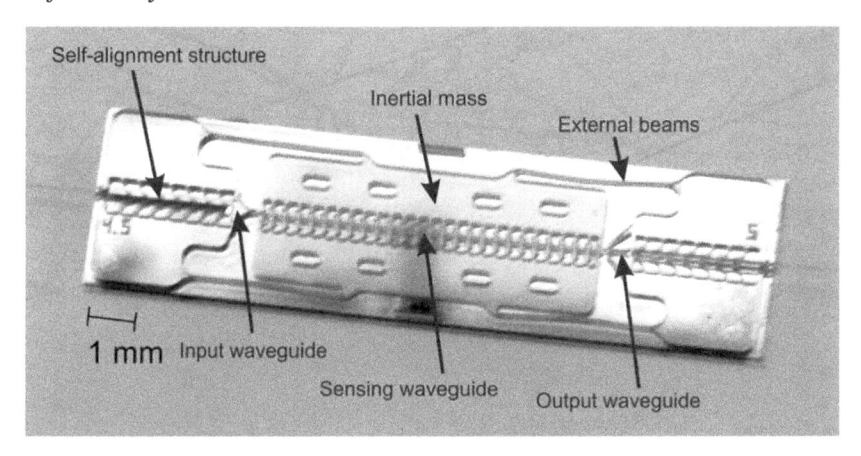

**Figure 12.8** SU-8 accelerometer with optical waveguides for readout. (Reprinted from Ref. 52 with permission from IEEE/ASME.)

a receptor layer to detect explosive molecules. The adsorbed molecules caused a surface stress and resulted in a change in resistance in the sensing SU-8/CB composite. The signal was read by a DC bridge circuit with a detection limit of 6 ppb.

### 12.4.2 SU-8 Micro-Optical Components

#### 12.4.2.1 Deformable Mirror

SU-8 has been used in deformable micromirrors for adaptive-optical systems. The choice of SU-8 as the deformable membrane reduces the driving voltage due to the low Young's modulus of SU-8. Most of the deformable mirrors are driven electrostatically since the mirror membranes can be used as one of the electrodes in the electrostatic piston actuators. Conédéra et al.[54] and Liotard et al.[55] have demonstrated a continuous metallized SU-8 membrane as the mirror plane fabricated on top of an array of SU-8 piston actuators. In this all-SU-8 device, the first 10 μm thick layer was used for the actuator plate and the springs and the second 10 μm thick layer was used for the mirror membrane. A 2 μm piston motion can be obtained with a 30 V applied voltage. With well-characterized actuation curves, a 21 nm standard deviation between the desired and actual displacement of the actuator was achieved.

Friese and Zappe developed a SU-8 deformable mirror using a two-wafer fabrication process.[56] Segmented control electrodes and a SU-8 spacer layer were fabricated on the bottom wafer. The top wafer was first metalized and then a thin layer of SU-8 for the mirror membrane was coated. The two wafers with the SU-8 layers facing each other were bonded and the SU-8 deformable membrane was released by etching the top wafer from its back side, as shown in Figure 12.9. The fabricated mirror was tested in a feedback adaptive optical system; the wave front error could be reduced by 75% within 30 control steps. Lukes and Dickensheets fabricated

**Figure 12.9** SU-8 deformable mirror for adaptive optical applications. (Reprinted from Ref. 56 with permission from IEEE/ASME.)

a surface-micromachined circular SU-8 deformable membrane using the oxide or silicon substrate as the sacrificial material.[57] After the membrane was released, concentric electrodes were patterned on the membrane as the top electrode while the substrate was used as the bottom electrode. The deformable mirror was used in an optical microscope for focus control and aberration compensation.

### 12.4.2.2  Microlens

Microlenses or microlens arrays are an important component in miniaturized optical systems in telecommunication, data storage, display, imaging, energy, and biodetection applications. The minimum size of lenses manufactured by traditional lens-making technology such as grinding/polishing or molding is about 1 mm. With the advance of semiconductor and MEMS technology, it becomes feasible to fabricate submillimeter or smaller lenses or lens arrays by MEMS technology. The motivation and advantage of this approach is the accurate dimension control and potential for batch fabrication. However, fabrication of curved or slanted surfaces still remains a challenge due to the quasitwo-dimensional nature of MEMS technology. SU-8 has good optical transmission in the visible spectrum range[58–60] and can be fabricated with enough thickness for refractive components. Therefore, various SU-8 micro-optical components have been reported.

Microcylindrical lenses focus light in one dimension. It can be used in laser-beam shaping and optical detection in microfluidic systems. Hsieh *et al.* fabricated microcylindrical lenses and self-alignment mechanism for optical fiber alignment using SU-8.[60] Both vertically focusing and horizontally focusing lenses were fabricated in a one-mask process, and additional assembly steps were needed to rotate the vertically focusing lenses by 90°. The transmission of 488 nm light in SU-8 with a thickness up to 250 μm was measured to be more than 92%. The fiber holder and microlens were further integrated with a SU-8 microfluidic channel in a laser-induced fluorescence system. A minimum detection limit of 5 μM dye concentration was achieved.

Whereas cylindrical lenses with curvature in only one dimension can be fabricated simply by conventional photolithography, curved surfaces with curvatures in two dimensions are more difficult to fabricate in the semiconductor or MEMS processes. One possible solution is by using grayscale photolithography. Hung et al. fabricated grayscale photomasks by reflowing AZ4620 photoresist on a quartz or glass plate.[61] The reflowed structures had varying thickness and could absorb different amounts of UV light at different location during exposure. However, the AZ4620 structures also focused the UV light and complicated the exposure processes. Therefore, glycerol as an index-matching fluid was used to fill the gap between the photomask and the SU-8 resist during exposure to reduce the refraction at the surfaces of the AZ4620 structures.

Curved surfaces can also be fabricated by direct writing with electron beams (e-beam)[62,63] or proton beams.[58,64] Shields et al. fabricated Ø250 μm f/9 and Ø60 μm f/2 microlenses in 8 μm thick SU-8 with a 30 kV electron beam.[63] A rms surface variation of 1/10 λ could be achieved. Blazed gratings and Fresnel lenses were both fabricated in 10 μm thick SU-8 on silicon substrates by Bettiol et al.[58] Refractive microlenses were also fabricated by proton-beam writing. In the work presented by Sow et al., an array of micro-SU-8 cylinders was first fabricated by 2.0 MeV proton-beam lithography on a glass substrate.[64] After chemical development, the cylinder array was heated to allow SU-8 to reflow and form an array of microscale planoconvex lenses. The fabricated lenses had a diameter of 180 μm and thickness of 24 μm. The micrometer lens array was used in an optical tweezer setup to generate multiple optical spots to trap microbeads simultaneously.

Nonphotolithographic techniques have also been developed to fabricate microlenses. In the work demonstrated by Kuo et al., a PMMA film was first coated on a SU-8 array to form a soft concave mold.[65] A 110 μm thick SU-8 was then spin coated on a glass substrate that was then placed on top of the mold. The SU-8 was soft baked and exposed through the glass substrate for polymerization. After the exposure and postexposure bake, the SU-8 microlens array could be stripped from the PMMA mold. Microlenses with diameters from 50 μm to 200 μm and numerical apertures from 0.3 to 0.75 could be achieved. Measurement showed that the surface roughness was only 10.2 nm. Kuo and Lin exploited the difference in glass-transition temperatures of polymerized and unpolymerized SU-8 to fabricate microlenses in a novel stamping process.[66] A 500 μm thick SU-8 layer was first applied on the surface of a soda-lime glass substrate. The top SU-8 surface was partially exposed with an estimated depth of polymerization of 300–400 μm by using a mask with circular openings. A second full exposure was performed to define the edges of the reservoir of unpolymerized SU-8. To fabricate microlens arrays by using this reservoir as a stamp, a clean substrate was brought in contact with the top surface with circular openings and heated to a temperature between the glass-transition temperature between polymerized SU-8 (225 °C) and unpolymerized (55 °C) SU-8. At this temperature, the polymerized enclosure of the reservoir stayed rigid, while the unpolymerized SU-8 in the reservoir started to flow. Under pressure, the melted SU-8 was squeezed out of the reservoir and adhered to the glass substrate. The stamp was then removed the stamped substrate was heated to obtain uniform and smooth microlenses, as shown in Figure 12.10. The surface roughness of a Ø200 μm lens was measured to be 3.84 nm. Number apertures of the fabricated microlenses ranged from 0.22 to 0.58.

Curved surfaces of microlenses can also be controlled by electrostatic forces, as demonstrated by O'Neill et al.[67] When a liquid droplet of UV-curable resin is deposited on a flat surface, the final shape and curvature of the droplet surface are determined by an energy-minimization process. The application of an external electrostatic field can alter the contact angle of the liquid/air/substrate interface and thus change the shape of the droplet. The droplet is then cured in the presence of the applied field to form a microlens. In most of the setup for this fabrication process, the electric field is applied by using a pair of parallel-plate electrodes. In the work presented by Zhan et al., a series of voltage was applied to the droplet and the shape changed from the initial spherical to parabolic to cone shapes with increasing field strength.[68] The fabricated lens with a diameter of 2.6 mm had a focal length of 2.005 mm, corresponding to a f/# of 1.67. In the MTF measurement, the lens showed a cutoff frequency of 800 lp mm$^{-1}$ and a Strehl ratio of 0.742. Hung et al. patterned conductive concentric annular electrodes on the bottom electrode so that the distribution of the applied electrostatic field could be further controlled.[69] The annular electrodes also allowed the change of the base diameter of the droplet through electrowetting, thus providing another degree of freedom in the lens design and manufacturing. In addition, the transparent ITO electrodes were designed as a Fresnel zone plate to add focusing power to the microlens. Microlenses with Strehl ratios up to 0.84 and NA up to 0.66 were achieved by this fabrication process.

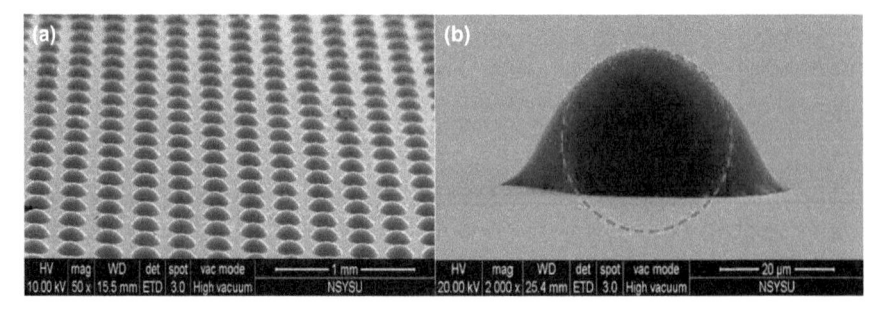

**Figure 12.10** (a) Array of microlenses, (b) close-up view of a Ø50 μm microlens. (Reprinted from Ref. 66 with permission from IOP Publishing.)

## 12.4.2.3 Microprisms and Mirrors

Whereas lenses change the wavefront curvature, prisms and mirrors change the propagation direction of light. In many optical systems, prisms and mirrors are used to fold the optical path to reduce the total dimension of the system. Prisms and mirrors for integrated micro-optical systems pose a different challenge for fabrication as compared with microlenses. The most important specifications of a microprism are the prism angles and surface flatness and roughness. Due to the limit of the quasitwo-dimensional MEMS fabrication technology, it is very difficult to fabricate three-dimensional prisms with good surface quality, precise angles and shapes, and dimensions in the mm or sub-mm range. Since the prism or mirror surfaces in micro-optical systems are usually tilted at particular angels from the substrates, nonconventional fabrication processes or out-of-plane assembly are needed to integrate the system on planar substrates.

In the optical switch presented by Liu *et al.*, the vertical sidewall of a photolithographically defined SU-8 thin plate was used as a reflector for optical switching.[70] The 150 μm high vertical SU-8 mirror was fabricated on a prestressed electrostatic $Si_3N_4$/Poly-Si bimorph actuator to control the optical path. A Cr/Au layer was sputtered on the mirror surface to improve the reflectivity. The surface roughness of the vertical mirror measured by AFM was 53.5 Å; the verticality measured by SEM was 89.1°. The insert losses of the assembled switch in the transmission and reflection states were 2.9 dB and 4.1 dB, respectively. Similar mirrors were also employed by Pan[71] and Guerre *et al.*[72] for optical switching applications.

In conventional photolithography, the developed sidewalls are nominally vertical because of the vertical incidence of UV light. Even some special processes or fabrication parameters[73,74] can be used to change the profile of developed structures, the achievable range of angle control is small. For large mirror tilt angles, such as 45° for folding mirrors or prisms in many micro-optical systems, new fabrication technologies are needed. The most-often employed fabrication technology for these micromirrors or prisms is inclined photolithography. The UV light in inclined photolithography is incident upon the photoresist-coated substrate at an oblique angle by either tilting the substrate or by bending the UV light with a prism. Discrete or integrated micromirrors or prisms with dimensions up to 1 mm can be fabricated by this method. In the setup proposed by Hung *et al.*,[75] tilted substrates were exposed in a tank filled with glycerol to reduce the refraction on the mask surface to achieve a wider angular range of exposed structures, as shown in Figure 12.11. The surface roughness of developed prisms was measured to be less than 10 nm. The surfaces of two 45° prisms were coated with metal layers with different thicknesses to function as a mirror and a beam splitter. Finally the space between the two prisms was refilled with SU-8 resist to obtain a solid and robust micro-optical system with embedded reflecting surfaces for optical-storage applications. Similar concepts were employed in Ling and Lian in which water, instead of glycerol, was used due

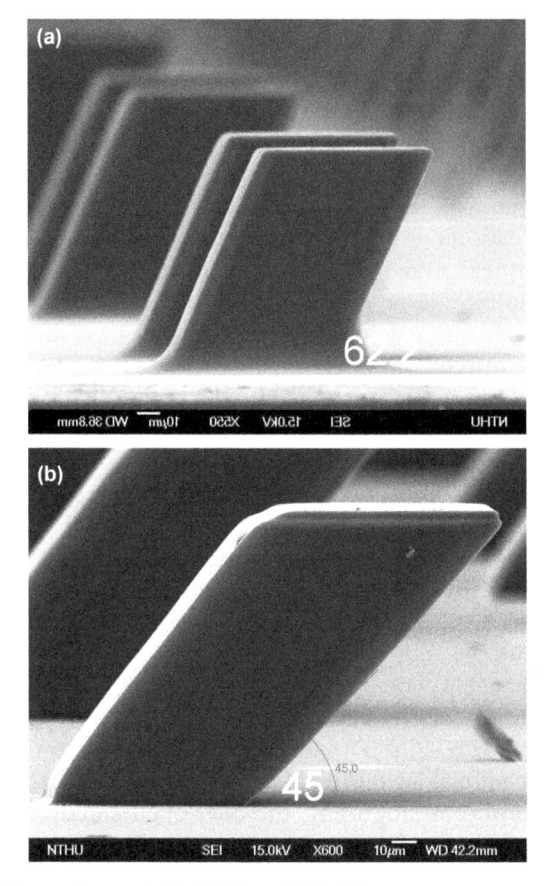

**Figure 12.11** Microprisms at (a) 62.2° and (b) 45°.
(Reprinted from Ref. 75 with permission from IOP Publishing.)

to its more stable index of refraction.[76] Huang and Tseng utilized inclined photolithography to fabricate prisms with angles from 25° to 30° to guide light into the quartz substrate of a total internal reflection based fluorescence sensing chip.[77] Dou et al. fabricated SU-8 premold for optical waveguides with 45° mirrors at the ends for optical-interconnection applications.[78] A repeatable angular error control of 0.5° could be achieved.

### 12.4.2.4 Photonic Crystals

Photonic crystals are periodic two-dimensional or three-dimensional structures with subwavelength features. They have strong diffractive characteristics and exhibit optical dispersion similar to the electronic band structures of semiconductor crystals. Propagation and cut-off of optical modes in the dispersion relationship can be tailored by suitable design of the periodicity. Therefore, much interest for optical, communication and scientific applications has been found in this artificial material.

Layer-by-layer lamination processes can be used to fabricate photonic crystals in SU-8. Leong et al. developed a SU-8 dry film on water-solvable backing substrate such as PVA.[79] After the SU-8/PVA dry film was bonded onto the previously fabricated SU-8 structure, the PVA backing substrate was removed by immersing the bonded stack in DI water. Eight layers of 100 μm pitch gratings stacked in a woodpile A-B-C-D structure could be obtained consistently by using this technique. FTIR studies showed a broadening of the transmission dip and increased absorption with increasing number of stacked layers, implying that the 3D SU-8 multilayered structures were potentially suitable for broadband THz absorbers.

Photonic crystals have been fabricated in SU-8 by using direct laser writing (DLW) or holographic lithography. In DLW, a laser beam is focused and scanned across the soft-baked resist to expose it in small volumes (voxels) sequentially. To obtain subwavelength resolutions, two-photon absorption (TPA) is often employed. The photon energy of the exposing light source in TPA is smaller than the polymerization threshold of the resist. Therefore, it takes two (or more) photons to expose and polymerize the resist. In such a case the voxel where the light intensity is above the TPA threshold is smaller than the diffraction-limited focused optical spot and subwavelength voxels can be obtained.

Deubel et al. used a 120 fs Ti-sapphire laser at 800 nm to write three-dimensional photonic crystal structures in SU-8 for telecommunication applications.[80] Large-scale FCC layer-by-layer structures with stop bands ranging from 1.3 to 1.7 μm were demonstrated. Typical rod diameters in these woodpile structures were in the 180–250 nm range; the rod spacing $a$ was from 0.8 to 1.0 μm and the lattice constant $c$ was $(c/a)^2 = 2$. The versatility of LDW was further exploited in Seet et al. beyond simple layer-by-layer woodpile structures.[81] A cubic structure composed of spiral units was fabricated by a femtosecond Ti-sapphire laser at 800 nm. The voxel had an ellipsoidal shape with major and minor diameters of 1.17 μm and 0.475 μm, respectively. A typical spiral had an arm length of 2.7 μm, a vertical pitch of 3.04 μm, and a two-dimensional lattice period of 1.8 μm. Adjacent layers with 180° structural phase shift and L-shaped defect waveguides implemented with missing spirals were demonstrated, as shown in Figure 12.12, showing the capability of LDW for defect engineering that was not possible with holographic lithography.

Another technology for photonic-crystal fabrication is holographic lithography (HL), which is also called interference lithography (IL).[82] Multiple coherent laser beams with different propagation directions, intensities, phases, or polarizations are overlapped spatially and temporally to produce an interference pattern in the photoresist. After exposure and development of the resist, the microstructure is a direct replica of the three-dimensional interference pattern. Compared with direct writing by focused beams, HL/IL is a more efficient way to fabricate complex periodic microstructures. However, the periodic nature of optical interference prohibits fabrication of arbitrary geometries by this technology.

**Figure 12.12** (a) L-shaped defect waveguide and (b) adjacent layers with 180° phase shift fabricated in SU-8 by using laser direct writing.
(Reprinted from Ref. 81 with permission from John Wiley and Sons.)

The multiple exposing beams are usually derived from splitting a single laser beam by mirrors, prisms, or gratings. To generate three-dimensional periodic patterns, at least four beams are needed. In the work presented by Miklyaev et al.,[83] a frequency-tripled Q-switched single-mode Nd:YAG laser at 355 nm with about 6 ns pulses was split into four beams with the central beam circularly polarized and the other beams linearly polarized. A custom-made prism was placed on top of the photoresist to compensate for the refraction effect on the surface to obtain an FCC structure with nominal lattice constant $a = 550$ nm, as shown in Figure 12.13. The measured transmission spectrum along the (111) direction ($\Gamma\Lambda$) showed a dip around 680 nm, consistent with the band structure calculation after parameter fitting. Kondo et al. fabricated large-area (about 1 mm in diameter) two- and three-dimensional microstructures in SU-8 by using a diffraction grating to split the beam from a femtosecond 800 nm Ti-sapphire laser.[84] The diffracted beams were then collimated and focused on the SU-8 resist. Two-dimensional arrays of rods with a diameter of 0.43 μm and a height of 1.3 μm

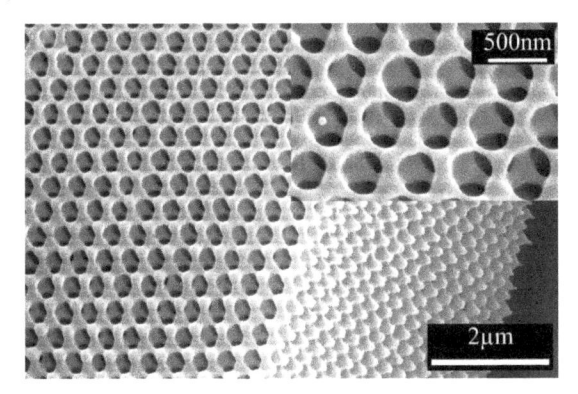

**Figure 12.13** FCC photonic crystal fabricated by using holographic lithography. (Reprinted from Ref. 83 with permission from AIP Publishing LLC.)

were fabricated by a 1 min exposure. Three-dimensional arrays of microrods were also demonstrated in 25 µm thick SU-8. Pang et al. demonstrated periodic spiral structures obtained by HL/IL.[85] A 325 nm He–Cd laser beam was split by a grating into one central circularly polarized and six circumpolar linearly polarized side beams. The amplitude, phase and polarization of the beams were adjusted by wave plates and neutral-density filters placed in the optical path. A prism was placed on top of the resist with matching fluid to reduce interference effects. For a 15 µm resist thickness, the optimum exposure time was 3 s. The developed structure showed a large shrinkage (about 70%) in the vertical direction possibly caused by the surface tension of liquid during the development process. Except for this large shrinkage, the fabricated spiral structures resembled calculation results quite well.

### 12.4.3 SU-8 for RF Applications

Si-based system-on-a-package (SOP) schemes have been proposed for microsystem fabrication in this decade. Discrete RF components are designed and fabricated on separate chips, and then fully integrated onto a silicon substrate to form a microsystem with the design flexibility without having any material and process limits for better performance. In these studies,[86,87] a critical integration concept is disclosed in the development of wireless transceivers. The silicon instead of ceramic or printed circuit board (PCB) material as the packaging substrate can enhance the transceiver performance in terms of small form factor, small parasitic effect, low noise and low power consumption in the future. Not only RF CMOS circuit chips including RF transceiver and baseband chips can be integrated but also high-performance RF passive components such as inductors, capacitors and filters can be directly fabricated on the silicon substrate using conventional IC or MEMS processes. Based on the silicon SOP architecture, SU-8 with the characteristics of high aspect ratio and photodefinability has been introduced to realize the passive component fabrication on the silicon substrate.

Meanwhile, recent technology development in personal portable devices mainly focuses on the design and fabrication of wearable wireless sensing and communicating systems where flexible electronics are the essence in the systems. In the fabrication of flexible electronics, the fabrication temperature is restricted, i.e. $<\sim 200$ °C, due to the low glass transition point of polymer substrate. Thus the existing semiconductor processing techniques cannot be directly adapted for the device fabrication. Based on the SOP scheme, the low-temperature SU-8 process addressed in the prior sections can also enable the realization of cost-effective MEMS components for flexible RF application.

### 12.4.3.1 Inductors

In RF circuitry, the inductor is a key element pervasively applied in the design of bandpass filters, oscillators, impedance matching networks, and power splitters. The quality factor, $Q$, of a conventional on-chip spiral inductor can be depicted as the following equation:[88]

$$Q = \frac{\omega L_s}{R_s} \cdot \frac{R_p}{R_p + [(\omega L_s / R_s)^2 + 1] R_s} \cdot \left[ 1 - \frac{R_s^2 (C_s + C_p)}{L_s} - \omega^2 L_s (C_s + C_p) \right]$$

(12.1)

where $L_s$, $R_s$, $C_s$, $C_p$, $R_p$, and $\omega$ are the inductance, series resistance, and series feedforward capacitance of the inductor, parasitic shunt capacitance, substrate resistance, and signal frequency, respectively. There are two main factors, which are the induced eddy current loss in the substrate, i.e. $R_p$, and the ohmic loss in the inductor, i.e. $R_s$, respectively, to affect the $Q$ value. Because the employment of inductors with high $Q$ value in the aforementioned RF circuits can facilitate the realization of low-power and low-noise performance, several process techniques have been developed for high-$Q$ inductor fabrication including removing the silicon substrate underneath the inductors, changing the inductor design from spiral to solenoid, thickening the conducting wire of the inductor, etc.

Yoon et al.[7] demonstrated a three-dimensional, high aspect ratio, high-$Q$ solenoid-type RF inductor using a SU-8 epoxy core and subsequent metal electrodeposition. In the inductor as shown in Figure 12.14, the conducting wire on the side wall of the inductor was formed by the serial connected high aspect ratio (up to 10 : 1) SU-8 columns electroplated with Cr/Au. The serial-interconnect to these columns was a SU-8 bridge fabricated by a double-exposure and single-development scheme. Experimental measurement showed that a single-turn inductor, 900 μm in height and 600 μm in width in the air-core area, exhibited a maximum $Q$-factor of 84 and an inductance of 1.17 nH at 2.6 GHz. The approach to implement SU-8 core allows for relatively simple formation of extremely high aspect ratio columns in the side wall of 3D solenoid inductors. Ghannam et al.[89] presented a low-cost multilayer 3D

SU-8 for Microsystem Fabrication 243

**Figure 12.14** SU-8 solenoid inductor coated with 10–15 μm electroplated Cu metal.
(Reprinted from Ref. 7 with permission from IEEE/ASME.)

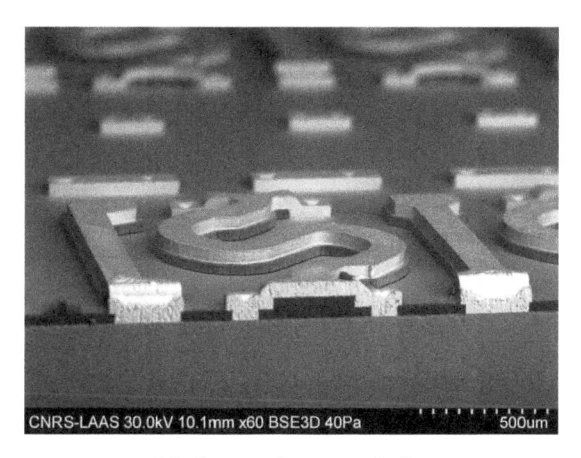

**Figure 12.15** Interconnected inductors integrated above a 50 W LDMOS transistor using multiplayer SU-8 process technology.
(Reprinted from Ref. 89 with permission from IEEE.)

copper interconnect process for high $Q$ inductor fabrication. In the process, two kinds of photoresists, BPN and SU-8, were used to form multilevel 3D interconnects in a single metallization step. The SU-8 was patterned to form a thick dielectric layer underneath specific locations and coated with a 3D seed layer to insure 3D electroplating current flow for the fabrication of desired 3D interconnect shapes. The BPN was applied as a mold structure with an aspect ratio of 16:1 for the electrodeposition of copper windings. A high-$Q$ (55 @ 5 GHz) power inductor, as shown in Figure 12.15 has been developed and integrated above a 50 W RF power LDMOS device using this process.

In addition to high-$Q$ inductors, for short-range wireless system applications like radio-frequency identifier (RFID) antenna, wireless power transmission units, and DC/DC converters, on-chip inductors with large inductance are needed for the system miniaturization. Wang et al.[90] developed a CMOS-compatible process to fabricate on-chip inductors with low DC resistance and high power density. They employed a silicon mold for electroplating thick copper windings and used SU-8 as another mold for electroplating the NiFe magnetic core as well as an isolation material between windings and cores. Because the silicon deep reactive ion etching (DRIE) barely attacked SU-8, the SU-8 mold would not be damaged and could be used as the mold for following through-wafer copper/permalloy electroplating. The SU-8 provided the functions as an isolator between the core and winding and as a buffer to absorb thermal stresses, simultaneously. *Via* the SU-8 process, a high power density inductor could be realized with significant low DC resistance. The research team has demonstrated a pot-core inductor with a low-frequency inductance of 134 nH and a dc resistance of 9.1 mΩ as shown in Figure 12.16. A research group in Cornell applied two kinds of photoresists, SU-8 and SPR-220, to fabricate stacked on-chip spiral

**Figure 12.16** (a) Top view, (b) cross-sectional view of a high-power density inductor fabricated using Si and SU-8 molds for electroplating the Cu windings and NiFe core.
(Reprinted from Ref. 90 with permission from IEEE.)

inductors. By taking advantage of the strong chemical resistance of SU-8, Reissman *et al.* used SU-8 as a permanent photodefined structural layer to build the coil layers, as well as the dielectric material between the layers, and SPR-220 was used as a photodefined mask for patterning evaporated copper layer to form the inductor using wet-etching.[91] Such single-level lithography instead of the traditional two-level approach can accomplish low-cost stacked on-chip spiral inductors, as shown in Figure 12.17, with the inductance over tens of µH.

## 12.4.3.2 RF-filters

Regarding RF filters, Jiang *et al.*[92] have presented a coaxial dual mode Ka band (Kay-Ay band, 26.5–40 GHz) filter comprised of five 700 µm conductive layers where an air-filled coaxial transmission line was designed and built with low-dispersion and low-loss characteristics due to its compact form. In the fabrication process, UV exposed SU-8 microstructures exhibited an aspect ratio of 40 : 1 and a sidewall of $90 \pm 0.1°$. Although there was no measurement data related to this SU-8 filter, the authors pointed out that the SU-8 microfabrication technology could facilitate volume production of the filter in such high frequencies. The Ka band filter is typically fabricated using conventional laser or electric discharge machining (EDM) ablation on a metal sheet, which has an inevitable weakness, *i.e.* less structural tolerance accompanied with uneven edges and oxide residuals on the structure edges, resulting in a poor filter performance in high-frequency regimes. These requirements can, however, be met by the SU-8 process. In fact, another research team from Georgia Institute of Technology has reported a metal-transfer-micromolding (MTM) technique for the fabrication of RF passive components comprised of metallized molded device structures with very high aspect ratios.[93] The fabrication process was proposed to realize a system-on-a-package integration scheme where RF components were air lifted on an organic RF circuit board. There were a total of three major process steps developed in the MTM technique, including master fabrication, mold preparation and metallization, and metal transfer molding, respectively. In the master fabrication process, SU-8 was photolithographically patterned to make high aspect ratio 3D structures layer-by-layer, which was utilized as the master. The first layer of SU-8 was patterned to define the base structure of the planar RF components such as transmission lines. Then, several layers of SU-8 were patterned one by one to define the 3D structures with differing heights. After the master fabrication, a negative mold was formed by casting and curing a suitable molding material such as polydimethylsiloxane (PDMS), around the master. Meanwhile, the layer-stacked SU-8 structures could then be either coated with a conductive metal or used as molds for producing copper structures using an electroplating process. A coplanar waveguide (CPW) based bandstop filters was realized, as shown in Figure 12.18, by the technique and it showed an insertion loss of less than 2.5 dB in passband and 20 dB in stopband over the entire passband centred at 37 GHz. Moreover, since the MTM technique can be utilized to fabricate 3D structures with differing height, a monopole antenna array at various resonant

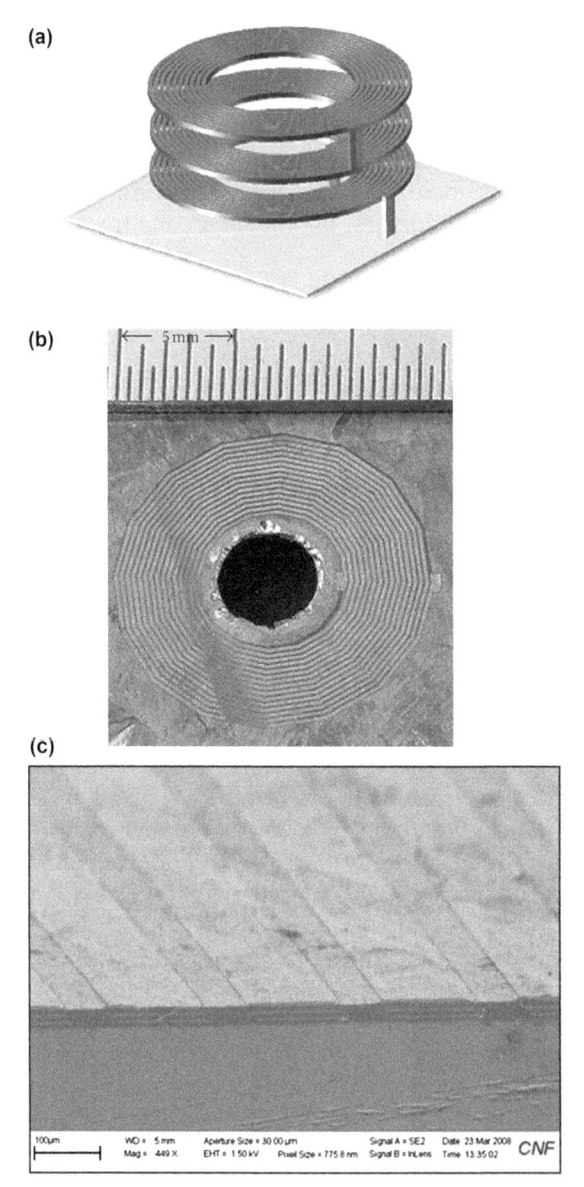

**Figure 12.17** Multilayered, stacked inductor (MLSI). (a) A schematic diagram, (b) top view of as-fabricated triple-layer inductor with air core, (c) cross-sectional view of as-fabricated triple-layer inductor. (Reprinted from Ref. 91 with permission from Hindawi.)

frequencies, as shown in Figure 12.18, has been demonstrated by the same research team. The monopole antennas showed good radiation performance and the 10 dB bandwidths at the resonant radiation frequencies were as large as 21.5% for the tested structures.

SU-8 for Microsystem Fabrication 247

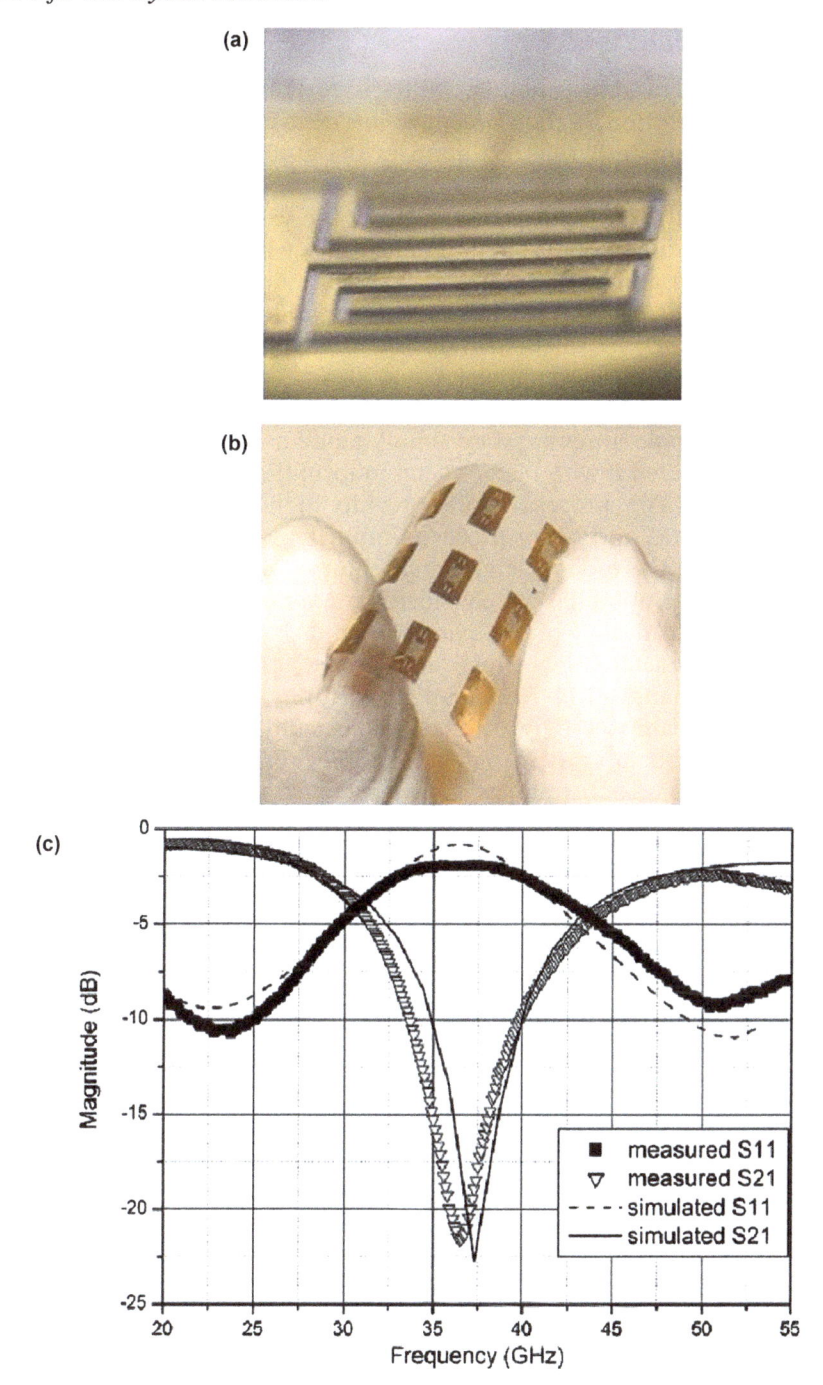

**Figure 12.18** Coplanar waveguide (CPW) based bandstop filters fabricated using the MTM technique on a molded epoxy substrate (a) inclined view, (b) all filters on a substrate, and (c) RF characteristics with 35 μm air gap. (Reprinted from Ref. 93 with permission from IEEE.)

## 12.4.3.3 Antennas

In wireless systems, the antenna, which is responsible for signal transmission and reception, is the element to interface the transceiver with the outside environment. A rule of thumb in the antenna design is that the antenna size should be comparable with the signal wavelength for better performance. In the RF/microwave regime, the wavelength has been reduced to the centimeter or even millimeter scale, which makes the concept of using on-chip antennas more practical while developing miniaturized wireless transceivers. Previously, we have introduced the polymer-core process for high-Q 3D solenoid inductor fabrication developed by Yoon *et al.*[6,7] The same process approach could also be utilized for the antenna fabrication. Instead of using the SU-8 mold and electroforming process to form the monopole antenna, they simply made a high aspect ratio SU-8 column and covered it with plated metal to form the antenna, as shown in Figure 12.19.[94] The process could provide higher aspect ratio, lower metallization time, and better process simplicity.

In addition to the monopole antenna requiring a high aspect ratio process, planar patch antennas are considered more practical than other types of antennas. The design of the patch antenna is fairly simple and can provide the characteristics of dual and circular polarizations, frequency agility, broad bandwidth, feedthrough flexibility, and omnidirectional patterning by beam steering with good process uniformity. Lin *et al.* have demonstrated the microstrip patch antenna fabricated on the flexible polymer SU-8/PDMS substrate, as shown in Figure 12.20,[95] using a flexible heterogeneous chip-integration scheme to form a miniature wireless microsystem. It featured smaller form factor, lower manufacture cost, better RF characteristics such as low dielectric constant and loss tangent, and ease to be integrated with CMOS chips. The measured bandwidth and maximum gain were 3% from 6.2 to 6.4 GHz and 2.17 dBi, respectively. Meanwhile, the related bending effects while the antenna was used for flexible tag applications have been investigated. It was shown that the frequency shift and the variation on the radiation pattern in the *xz*-plane

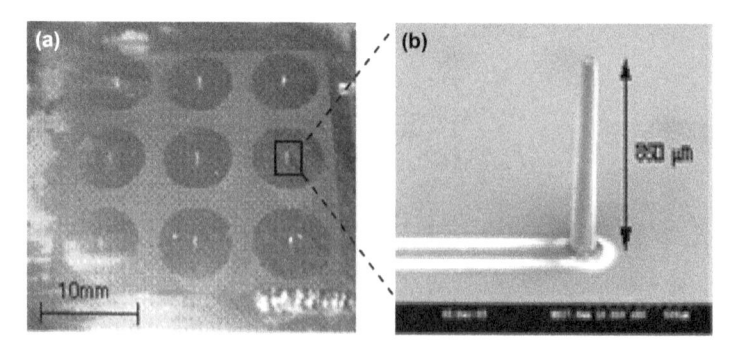

**Figure 12.19** (a) 3×3 monopole array, (b) single SU-8 monopole antenna fed with a CPW.[94]

*SU-8 for Microsystem Fabrication* 249

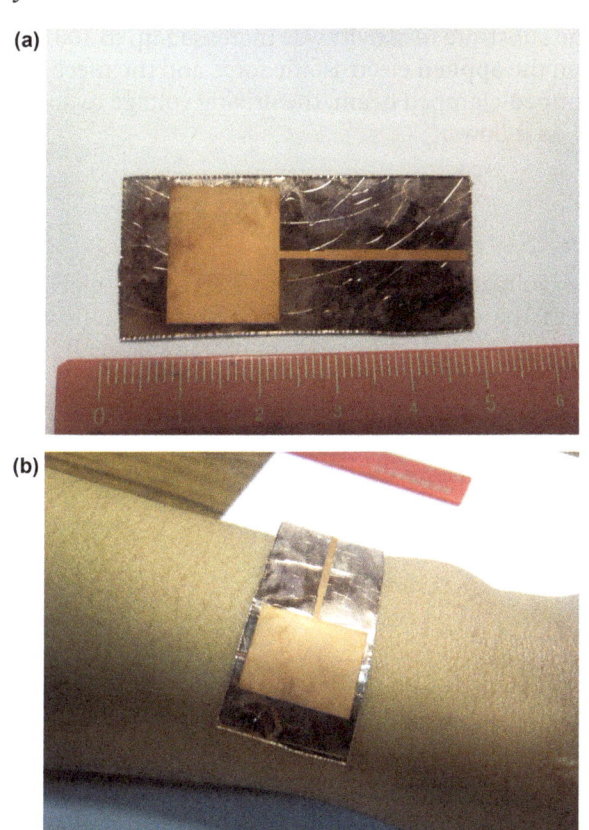

**Figure 12.20** Photographs of (a) a patch antenna on a SU-8/PDMS flexible substrate and (b) an antenna circled around the wrist to show the antenna flexibility for wearable device applications.
(Reprinted from Ref. 95 with permission from IEEE.)

were due to the reduction of capacitance and the uneven fringing E-fields on the edge of the patch, respectively.

### 12.4.3.4 Other RF MEMS Components

Instead of being a molding or packaging structure for wireless system applications, SU-8, in fact, can be utilized as the micromechanical structure for the fabrication of RF MEMS devices such as MEMS switch, MEMS varactor, MEMS power divider, *etc*. This can be attributed to its low Young's modulus and good mechanical and chemical stability. For instance, Chao *et al*.[96] demonstrated a SU-8 micromachining process for MEMS series switch fabrication. The switch designed with a clamped-clamped SU-8 (5 µm)/Cu (2 µm)/SU-8 (3 µm) beam structure was driven by electrostatic force and performed better than 0.75 dB insertion loss and −28.2 dB isolation up to

12 GHz once the substrate resistivity was increased up to 100 Ω cm. From the relation between the applied electrostatic force and the mechanical restoring force of the clamped-clamped beam, the driving voltage to actuate the switch can be derived as follows:[97]

$$V = \sqrt{\frac{2kg^2(g_0 - g)}{\varepsilon_0 A}} \quad (12.2)$$

where $V$ is the voltage applied between the driving electrodes, $A$ is the actuation area of the electrodes, $k$ is the spring constant of the clamped-clamped beam, $\varepsilon_0$ is the dielectric constant of free space, and $g_0$ and $g$ are the spacing between two driving electrodes with zero bias and biased at $V$, respectively. While the switch is in the "on" state, the spacing, $g$, is already smaller than $1/3g_0$, which makes the switch beam operate at an unstable point. This means that the switch would collapse to form the contact between the switch and the CPW once the driving voltage is slightly larger than the following $V_p$,

$$V_p = V(2g_0/3) = \sqrt{\frac{8kg_0^3}{27\varepsilon_0 A}} \quad (12.3)$$

Thus, the actuation voltage can be estimated as the same as the $V_p$ and the driving voltage for the switch is proportional to the square root of the product of the spring constant $k$ and the cube of $g_0$ for microwave and RF applications. In this work, ~63 V instead of several hundred volts driving voltage has revealed that the SU-8 switch is practical for wireless CMOS system applications.

Li et al.[98] presented a variable-ratio radio-frequency (RF) power divider with a large deflection MEMS actuator directly integrated in a waveguide. They utilized SU-8 springs to support a tilting microplate driven by a vertical comb drive actuator. Owing to the low Young's modulus characteristics of the SU-8 spring, the resonance frequency of the torsional mode of the mirror was only 370 Hz. The mirror were tilted into the waveguide to achieve the function of a variable-ratio RF MEMS power divider designed to work at a frequency of 80 GHz with low dissipative loss, maintaining good characteristics in terms of impedance match and port isolation. For a dc actuation voltage of 30 V, the fabricated microplate devices could achieve a deflection angle of 5.9°, making variable power split ranging from equal division up to a ratio of more than 2.

## 12.5 Conclusion

The SU-8, a photodefinable polymer, has been widely utilized for microsystem fabrication due to the characteristics of high aspect ratio, low dielectric constant and loss, better chemical resistance, *etc*. With increasing types of SU-8 series epoxy-based photoresists developed for various process applications, it is our belief SU-8 can facilitate the advancement of microtechnology for improving device functionality and performance to satisfy the needs in various applications.

# References

1. A. del Campo and C. Greiner, *J. Micromech. Microeng.*, 2007, **17**, R81.
2. http://memscyclopedia.org/su8.html.
3. J. M. Shaw, J. D. Gelorme, N. C. LaBianca, W. E. Conley and S. J. Holmes, *IBM J. Res. Dev.*, 1997, **41**, 81.
4. T. Morikaku, Y. Kaibara, M. Inoue, T. Miura, T. Suzuki, F. Oohira, S. Inoue and T. Namazu, *J. Micromech. Microeng.*, 2013, **23**, 105016.
5. K. Wouters and R. Puers, *J. Micromech. Microeng.*, 2010, **20**, 095013.
6. Y.-K. Yoon, J.-W. Park and M. G. Allen, *Solid-State Sensor, Actuator, and Microsystems Workshop*, 2002, Hilton Head, U.S.A., 374.
7. Y.-K. Yoon, J.-W. Park and M. G. Allen, *IEEE/ASME J. Micromech. Syst.*, 2005, **14**, 886.
8. J. M. Bustillo, R. T. Howe and R. S. Muller, *Proc. IEEE*, 1998, **86**, 1552.
9. K. H. Lau, A. Giridhar, S. Harikrishnan, N. Satyanarayana and S. K. Sinha, *Microsyst. Technol.*, 2013, **19**, 1863.
10. S. Corbel, O. Dufaud and T. Roques-Carmes, *Stereolithography*, Springer e-book, 2011, pp. 141–159.
11. J. S. Choi, H. W. Kang, I. H. Lee, T. J. Ko and D. W. Cho, *Int. J. Adv. Manuf. Technol.*, 2009, **41**, 281.
12. W. H. Teh, U. Dürig, G. Salis, R. Harbers, U. Drechsler, R. F. Mahrt, C. G. Smith and H.-J. Güntherodt, *Appl. Phys. Lett.*, 2004, **84**, 4095.
13. S. Juodkazis, V. Mizeikis, K. K. Seet, M. Miwa and H. Misawa, *Nanotechnology*, 2005, **16**, 846.
14. Z. Xiao, W. Peng, R. F. Wolffenbuttel and K. R. Farmer, *Sens. Actuators A: Phys.*, 2003, **104**, 299.
15. T. Y. Chao, C. W. Liang, Y. T. Cheng and C.-N. Kuo, *IEEE Trans. Electron Devices*, 2011, **58**, 906.
16. V. Seidemann, J. Rabe, M. Feldmann and S. Büttgenbach, *Microsyst. Technol.*, 2002, **8**, 348.
17. C.-H. Ho, K.-P. Chin, C.-R. Yang, H.-M. Wu and S.-L. Chen, *Sens. Actuators A: Phys.*, 2002, **102**, 130.
18. J. Kouba, R. Engelke, M. Bednarzik, G. Ahrens, H.-U. Scheunemann, G. Gruetzner, G. B. Loechel, H. Miller and D. Haase, *Microsyst. Technol.*, 2007, **13**, 311.
19. I. G. Foulds and M. Parameswaran, *J. Micromech. Microeng.*, 2006, **16**, 2009.
20. E. H. Conradie and D. F. Moore, *J. Micromech. Microeng.*, 2002, **12**, 368.
21. T. Fujita, K. Maenaka and Y. Takayama, *Sens. Actuators A: Phys.*, 2005, **121**, 16.
22. D. Bachmann, S. Kühne and C. Hierold, *IEEE Int. Conf. Micro Electro Mech. Syst.*, 2007, Kobe, Japan, 21.

23. S. H. Tsang, D. Sameoto, I. G. Foulds, R. W. Johnstone and M. Parameswaran, *J. Micromech. Microeng.*, 2007, **17**, 1314.
24. Y. Chiu, C.-S. Wu, W.-Z. Huang and J.-W. Wu, *IEEE J. Sel. Top. Quantum Electron.*, 2009, **15**, 1338.
25. N. Chronis and L. P. Lee, *IEEE/ASME J. Microelectromech. Syst.*, 2005, **14**, 857.
26. T. C. Duc, G.-K. Lau and P. M. Sarro, *IEEE/ASME J. Microelectromech. Syst.*, 2008, **17**, 823.
27. S.-H. Moon, M.-H. Jun, J.-Y. An and J.-H. Lee, *Microsyst. Tech.*, 2011, **17**, 1439.
28. W. Dai, K. Lian and W. Wang, *Microsyst. Tech.*, 2007, **13**, 271.
29. W. Dai, *Design and fabrication of micro transducers using cured SU-8 polymer as main structural material*, PhD Thesis, Louisiana State University, LA, USA, 2006.
30. V. Seidemann, S. Bütefisch and S. Büttgenbach, *Sens. Actuators A: Phys.*, 2002, **97**, 457.
31. J. C. Ribeiro, G. Minas, P. Turmezei, R. F. Wolffenbuttel and J. H. Correia, *Sens. Actuators A: Phys.*, 2005, **123**, 77.
32. J. Carlier, S. Arscott, V. Thomy, J. C. Fourrier, F. Caron, J. C. Camart, C. Druon and P. Tabourier, *J. Micromech. Microeng.*, 2004, **14**, 619.
33. H. Sato, H. Matsumura, S. Keino and S. Shoji, *J. Micromech. Microeng.*, 2006, **16**, 2318.
34. C.-H. Lin and G.-B. Lee, *J. Micromech. Microeng.*, 2003, **13**, 447.
35. Y.-H. Lin and G.-B. Lee, *Biosens. Bioelectron.*, 2008, **24**, 572.
36. D. Barat, G. Benazzi, M. C. Mowlem, J. M. Ruano and H. Morgan, *Opt. Commun.*, 2010, **283**, 1987.
37. B. R. Watts, T. Kowpak, Z. Zhang, C.-Q. Xu, S. Zhu, X. Cao and M. Lin, *Micromachine*, 2012, **3**, 62.
38. S. Balslev, A. M. Jorgensen, B. Bilenberg, K. B. Mogensen, D. Snakenborg, O. Geschke, J. P. Kutter and A. Kristensen, *Lab Chip*, 2006, **6**, 213.
39. P. Abgrall, C. Lattes, V. Conédéra, X. Dollat, S. Colin and A. M. Gué, *J. Micromech. Microeng.*, 2006, **16**, 113.
40. Z.-C. Peng, Z.-G. Ling, M. Tondra, C.-G. Liu, M. Zhang, K. Lian, J. Goettert and J. Hormes, *IEEE/ASME J. Microelectromech. Syst.*, 2006, **15**, 708.
41. B. Y. Shew, Y. C. Cheng and Y. H. Tsai, *Sens. Actuators A: Phys.*, 2008, **141**, 299.
42. B. Sepúlveda, J. S. del Rio, M. Moreno, F. J. Blanco, K. Mayora, C. Domínguez and L. M. Lechuga, *J. Opt. A: Pure Appl. Opt.*, 2006, **8**, S561.
43. M. Calleja, M. Nordström, M. Álvarez, J. Tamayo, L. M. Lechuga and A. Boisen, *Ultramicroscopy*, 2005, **105**, 215.
44. M. Nordström, D. A. Zauner, M. Calleja, J. Hübner and A. Boisen, *Appl. Phys. Lett.*, 2007, **91**, 103512.
45. M. Nordström, S. Keller, M. Lillemose, A. Johansson, S. Dohn, D. Haefliger, G. Blagoi, M. Havsteen-Jakobsen and A. Boisen, *Sensors*, 2008, **8**, 1595.

46. S. Schmid, P. Wägli and C. Hierold, *IEEE Int. Conf. Micro Electro Mech. Syst.*, 2009, Sorrento, Italy, 300.
47. J. C. Doll, N. Harjee, N. Klejwa, R. Kwon, S. M. Coulthard, B. Petzold, M. B. Goodman and B. L. Pruitt, *Lab Chip*, 2009, **9**, 1449.
48. G. C. Hill, R. Melamud, F. E. Declercq, A. A. Davenport, I. H. Chan, P. G. Hartwell and B. L. Pruitt, *Sens. Actuators A: Phys.*, 2007, **138**, 52.
49. N. Pelletier, B. Bêche, N. Tahani, J. Zyss, L. Camberlein and E. Gaviot, *Sens. Actuators A: Phys.*, 2007, **135**, 179.
50. F.-G. Tseng and C.-J. Lin, *IEEE Sens. J.*, 2003, **3**, 812.
51. D. A. Chang-Yen and B. K. Gale, *Lab Chip*, 2003, **3**, 297.
52. A. Llobera, V. Seidemann, J. A. Plaza, V. J. Cadarso and S. Büttgenbach, *IEEE/ASME J. Microelectromech. Syst.*, 2007, **16**, 111.
53. V. Seena, A. Fernandes, P. Pant, S. Mukherji and V. R. Rao, *Nanotechnology*, 2011, **22**, 295501.
54. V. Conédéra, L. Salvagnac, N. Fabre, F. Zamkotsian and H. Camon, *J. Micromech. Microeng.*, 2007, **17**, N52.
55. A. Liotard, F. Zamkotsian, V. Conédéra, N. Fabre, P. Lanzoni, H. Camon and F. Chazallet, *Proc. SPIE*, 2006, **6113**, 61130R.
56. C. Friese and H. Zappe, *IEEE/ASME J. Microelectromech. Syst.*, 2008, **17**, 11.
57. S. J. Lukes and D. L. Dickensheets, *IEEE/ASME J. Microelectromech. Syst.*, 2013, **22**, 94.
58. A. A. Bettiol, T. C. Sum, J. A. van Kan and F. Watt, *Nucl. Instrum. Methods Phys. Res. B*, 2003, **210**, 250.
59. R. Yang and W. Wang, *Sens. Actuators A: Phys.*, 2004, **113**, 71.
60. J. Hsieh, C.-J. Weng, H.-L. Yin, H.-H. Lin and H.-Y. Chou, *Microsyst. Tech.*, 2005, **11**, 429.
61. K.-Y. Hung, F.-G. Tseng and H.-P. Chou, *Microsyst. Tech.*, 2005, **11**, 365.
62. V. Kudryashov, X.-C. Yuan, W.-C. Cheong and K. Radhakrishnan, *Microelectron. Eng.*, 2003, **67–68**, 306.
63. E. A. Shields, F. Williamson and J. R. Leger, *J. Vac. Sci. Technol. B*, 2003, **21**, 1453.
64. C. H. Sow, A. A. Bettiol, Y. Y. G. Lee, F. C. Cheong, C. T. Lim and F. Watt, *Appl. Phys. B*, 2004, **78**, 705.
65. J.-N. Kuo, C.-C. Hsieh, S.-Y. Yang and G.-B. Lee, *J. Micromech. Microeng.*, 2007, **17**, 693.
66. S.-M. Kuo and C.-H. Lin, *J. Micromech. Microeng.*, 2008, **18**, 125012.
67. F. T. O'Neill, G. Owen and J. T. Sheridan, *Optik*, 2005, **116**, 158.
68. Z. Zhan, K. Wang, H. Yao and Z. Cao, *Appl. Opt.*, 2009, **48**, 4375.
69. K.-Y. Hung, F.-G. Tseng and T.-H. Liao, *IEEE/ASME J. Microelectromech. Syst.*, 2008, **17**, 370.
70. S. Liu, M. Bu, X. Ye, Z. Zhou, D. Zhang, T. Li, Y. Hao and Z. Tan, *Proc. SPIE*, 2001, **4601**, 354.
71. C. T. Pan, *J. Micromech. Microeng.*, 2004, **14**, 129.
72. R. Guerre, C. Hibert, Y. Burri, P. h. Flückiger and P. h. Renaud, *Sens. Actuators A: Phys.*, 2005, **123–124**, 570.

73. J. Zhang, M. B. Chan-Park and S. R. Conner, *Lab Chip*, 2004, **4**, 646.
74. W.-J. Kang, E. Rabe, S. Kopetz and A. Neyer, *J. Micromech. Microeng.*, 2008, **16**, 821.
75. K.-Y. Hung, H.-T. Hu and F.-G. Tseng, *J. Micromech. Microeng.*, 2004, **14**, 975.
76. Z. Ling and K. Lian, *Microsyst. Technol.*, 2007, **13**, 245.
77. S.-H. Huang and F.-G. Tseng, *J. Micromech. Microeng.*, 2005, **15**, 2235.
78. X. Dou, X. Wang, H. Huang, X. Lin, D. Ding, D. Z. Pan and R. T. Chen, *Opt. Exp.*, 2010, **18**, 378.
79. E. S. P. Leong, S. Y. Yew, T. S. Kustandi, Y. J. Liu, H. Tanoto, Q. Y. Wu, W. W. Loh, S. L. Teo and J. Teng, *Appl. Mater. Interfaces*, 2013, **5**, 5898.
80. M. Deubel, G. von Freymann, M. Wegener, S. Pereira, K. Busch and C. M. Soukoulis, *Nat. Mater.*, 2004, **3**, 444.
81. K. K. Seet, V. Mizeikis, S. Matsuo, S. Juodkazis and H. Misawa, *Adv. Mater.*, 2005, **17**, 541.
82. J.-H. Jang, C. K. Ullal, M. Maldovan, T. Gorishnyy, S. Kooi, C. Y. Koh and E. L. Thomas, *Adv. Funct. Mater.*, 2007, **17**, 3027.
83. Y. V. Miklyaev, D. C. Meisel, A. Blanco, G. von Freymann, K. Busch, W. Koch, C. Enrich, M. Deubel and M. Wegener, *Appl. Phys. Lett.*, 2003, **82**, 1284.
84. T. Kondo, S. Juodkazis, V. Mizeikis and H. Misawa, *Opt. Exp.*, 2006, **14**, 7943.
85. Y. K. Pang, J. C. W. Lee, H. F. Lee, W. Y. Tam, C. T. Chan and P. Sheng, *Opt. Exp.*, 2005, **13**, 7615.
86. J. Knickerbocker, P. S. Andry, L. P. Buchwalter, E. G. Colgan, J. Cotte, H. Han, R. R. Horton, S. M. Sri-Jayantha, J. H. Magerlein, D. Manzer, G. McVicker, C. S. Patel, R. J. Polastre, E. S. Sprogis, C. K. Tsang, B. C. Webb and S. L. Wright, *IEEE Electron. Compon. Technol. Conf.*, 2006, San Diego, U.S.A., 418.
87. T.-Y. Chao, C.-H. Li, Y. C. Chen, H.-Y. Chen, Y. T. Cheng and C.-N. Kuo, *IEEE Trans. Electron Devices*, 2010, **57**, 928.
88. C. Y. Yue and S. S. Wong, *IEEE J. Solid-State Circuits*, 1998, **33**, 743.
89. A. Ghannam, D. Bourrier, L. Ourak, C. Viallon and T. Parra, *IEEE Trans. Compon. Packag. Manuf. Technol.*, 2013, **6**, 935.
90. M. Wang, K. D. Ngo and H. Xie, *IEEE IECON 2008*, 2008, Orlando, U.S.A., 2672.
91. T. Reissman, J.-S. Park and E. Garcia, *Act. Pass. Electron. Compon.*, 2012, **2012**, 871620.
92. K. Jiang, M. J. Lancaster, I. Llamas-Garro and P. Jin, *J. Micromech. Microeng.*, 2005, **15**, 1522.
93. Y. Zhao, Y.-K. Yoon and M. G. Allen, *IEEE Electron. Compon. Technol. Conf.*, 2007, John Ascuaga's Nugget Sparks, U.S.A., 1877.
94. B. Pan, Y. Yoon, P. Kirby, J. Papapolymerou, M. M. Tentzeris and M. Allen, *IEEE MTT-S Int. Microwave Symp.*, 2004, Fort Worth, U.S.A., **3**, 1935.

95. C.-P. Lin, C.-H. Chang, Y. T. Cheng and C. F. Jou, *IEEE Ant. Wire. Pro. Lett.*, 2011, **10**, 1108.
96. T.-Y. Chao, M. C. Hsu, C.-D. Lin and Y. T. Cheng, *J. Micromech. Microeng.*, 2011, **21**, 025010.
97. G. M. Rebeiz, *RF MEMS Theory, Design and Technology*, John Wiley & Sons, New Jersey, 2003.
98. Y. Li, S. Kühne, D. Psychogiou, J. Hesselbarth and C. Hierold, *J. Micromech. Microeng.*, 2011, **21**, 074013.

CHAPTER 13

# UV-Based Dual Mechanism for Crosslinking and Stabilization of PAN-Based Carbon-Fiber Precursors

MARLON S. MORALES AND AMOD A. OGALE*

Department of Chemical Engineering, and Center for Advanced Engineering Fibers and Films, Earle Hall, Clemson University, Clemson, SC 29634-0910, USA
*Email: ogale@clemson.edu

## 13.1 Introduction

### 13.1.1 Polyacrylonitrile Precursors for Carbon Fibers

Polyacrylonitrile (PAN)-based carbon fibers possess outstanding mechanical strength, and have found numerous applications in ultrahigh performance structural composites for aerospace applications.[1] These fibers have a low density of about 1.8 g cm$^{-3}$ and can provide energy efficiency for various applications, including automotive, civilian aeronautical, energy generation, and sporting goods.[2] However, the higher cost of carbon fibers compared with other reinforcing materials (namely, glass fibers) limits their use for the high-end applications where strength-to-density ratio is critical. The demand for carbon fibers is expected to grow to about 150 000 metric tons over the next decade from the current demand of 40 000 metric tons. However, a large fraction of the increase is anticipated in the industrial sector that is cost-sensitive.[3] For cost-sensitive products, such as automotive applications, there is a need for the development of novel processes and precursors to reduce the carbon-fiber production cost and expand the use of carbon fibers.

A rate-limiting step in the conversion of PAN-based precursors into carbon fibers is the stabilization step. Thermo-oxidative stabilization typically involves the heating of the polymer precursor fibers at 200–300 °C in air.[4-6] This heat treatment involves the conversion of the PAN polymer chains into a condensed ring structure with a carbon–nitrogen double bond, often know as a "ladder" structure.[7] This is considered the most critical step during the production of carbon fibers because it determines the final structure and mechanical properties of the resulting carbon fibers.[8,9]

The thermo-oxidative reactions in PAN are highly exothermic.[1,8] Further, the thermal conductivity of PAN precursor (like other polymers) is relatively small ($\sim$0.26 W m$^{-1}$ K$^{-1}$).[10] Thus, the rate of heat dissipation during the stabilization is a rate-limiting step. If the heat of reaction is not adequately removed during stabilization, the core of the fibers can degrade and result in poor final mechanical properties of the carbon fibers.[11-13]

Reducing the rate of reaction and the temperature of initiation of the exothermic reactions can be used to control the heat flux of the fiber during the thermal oxidation step. This is one of the reasons for the use of copolymers of PAN containing a comonomer such as methyl acrylate (MA), acrylic acid (AA), or itaconic acid (IA). These comonomers disrupt the order in PAN leading to a reduction of the temperature of initiation of the cyclization reactions producing carbon fibers with higher mechanical properties than carbon fibers produced from PAN homopolymers.[14]

The exact reactions and mechanisms involved with thermo-oxidative stabilization and carbonization steps are very complex, and still not fully understood. This has prompted numerous investigations, as documented in various literature studies,[10,11,15-18] which indicate that stresses must be applied during stabilization and carbonization steps in an attempt to regain the molecular orientation. This strategy results in carbon fibers with high strength, but the limited molecular orientation prevents the fibers from developing an ultrahigh modulus.[15,19-21]

## 13.1.2 Stabilization Routes

As noted above, the stabilization step is of utmost importance during the production of carbon fibers because creation of a crosslinked polymer network within the precursor fiber is critical to its survival during the subsequent harsh carbonization temperatures. All the precursors discussed above utilize thermo-oxidative reactions as the route for crosslinking. However, in the polymer industry, reactions involving the use of photo-sensitive species to induce polymerization and crosslinking (curing) reactions have become an accepted industrial practice.[22-25] For this purpose, various types of radiation, such as ultraviolet (UV), X-ray, gamma, and e-beam, can be used. Of these, UV-based methods have become more popular than other types because it is least hazardous.[26] UV radiation is all radiation between 100 and 400 nm. UV-induced polymerization and curing can proceed through two major ways: free radical or cationic.[27] Of the two, free-radical photoinitiation is more popular and widely used for photo-induced polymerization and curing.

Figure 13.1 shows the three major ways of generating free-radical intermediates from a photoexcited species.[27,28] Homolytic cleavage involves the bond scission of a single exited species to generate two radicals. This mechanism has been investigated in numerous literature studies.[22,27–29] Hydrogen abstraction consists of the interaction of one excited species with another hydrogen-donor molecule to generate two radicals. Electron transfer involves the interaction of a positively charged excited species with a negatively charged molecule to generate two radicals, or the interaction of an excited species with a hydrogen donor molecule to generate two radicals. Even though the latter two mechanisms require the interaction of two species, which is not convenient for solid-state processes, some molecules can be synthesized with both functional groups, *e.g.*, Michler's ketone for hydrogen abstraction.[27,28] Note that the location of the UV-absorption bands, solubility, atomic composition, compatibility of the photoinitiator with the system, *etc.* are some of the criterions needed for the selection of a suitable photoinitiator.

Photoinduced crosslinking of melt-spun PAN-based precursors and successful conversion into carbon fibers has been reported in previous Clemson and related studies.[30,31] These PAN-based terpolymers contained a photosensitive comonomer, such as acryloyl benzophenone, incorporated into the polymer chain. However, a modification of the previous method, where external photoinitiator is added to the conventional PAN-based precursor solution prior to wet spinning (*i.e.* without synthesizing the photoinitiator into the polymer chain itself) has not been studied.

It is known that polymerization and stabilization reactions in PAN proceed faster and at lower temperatures when initiated by UV radiation.[26,31,32] Reduced cyclization temperature can also help to maintain molecular orientation during thermal oxidation, which in turn is important to impart superior modulus and strength properties to the resulting carbon fibers.[15,20,21] Alternatively, the incorporation of this UV-based prestabilization step may lead to a reduction of the time required for the thermal

**Free radical photoinitiation**

| Homolytic cleavage | Hydrogen abstraction | Electron transfer |
|---|---|---|
| A-B* $\xrightarrow{h\nu}$ A• + B• | A* + RH $\xrightarrow{h\nu}$ AH• + R• | (A$^+$)* + X$^-$ $\xrightarrow{h\nu}$ A• + X• |
| | | or |
| | | A* + DH $\xrightarrow{h\nu}$ (AH$^-$ DH$^+$) $\xrightarrow{h\nu}$ AH• + D• |

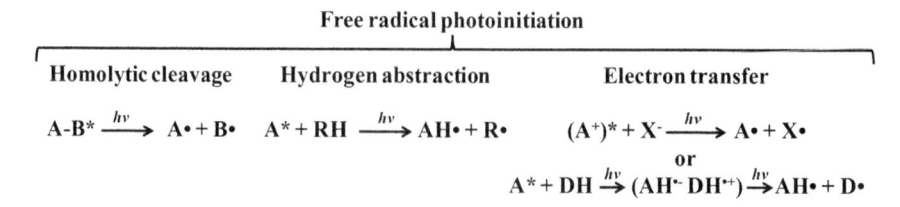

**Figure 13.1** Types of free-radical photoinitiation.

stabilization step, while maintaining the properties of resulting fibers. However, the role of external photoinitiator added to the polymer solution on the UV-induced cyclization and crosslinking reactions has not been systematically addressed in previous literature studies.

Therefore, the capability of these fibers to withstand further thermal stabilization and carbonization, including accelerated heating rates, and the properties of such fibers after the thermal-treatment steps was recently reported in our systematic study.[33–35] The overall goal was to investigate UV-induced stabilization reactions for PAN precursors and the effect of the addition of photoinitiator and further UV exposure on the properties of the resulting PAN-based prestabilized and final carbonized fibers. The fundamental study reported on the crosslinking and cyclization reactions in PAN-based samples induced by the addition of UV-sensitive photoinitiators that generate free radicals. Two different mechanisms, homolytic cleavage and hydrogen abstraction, were studied. The addition of photoinitiator and further UV treatment was investigated as an alternative to reduce the thermal oxidation processing time while maintaining the properties of the resulting fibers.

## 13.2 Experimental

### 13.2.1 Materials

During the first stage of this study involving UV-PAN-photoinitiator interaction, Poly(acrylonitrle-co-methylacrylate) with a nominal AN/MA ratio of 94 : 6, molecular weight ($M_w$) of 100 000, and glass-transition temperature ($T_g$) of ~82.5 °C was used. This PAN copolymer was obtained from Sigma-Aldrich (Aldrich Chemical Company, Inc. Milwaukee, WI). During the second stage of this study involving fiber spinning, carbon-fiber-grade polyacrylonitrle homopolymer with a molecular weight ($M_w$) of 233 000 and a glass-transition temperature ($T_g$) of 125 °C was used. This PAN homopolymer was obtained from Scientific Polymer (Ontario NY). Two different photoinititaors, 1-hydroxycyclohexyl phenyl ketone (denoted as "HPK") and 4,4′-bis(diethylamino)benzophenone (denoted as "BDP"), were used during this study, as shown in Figure 13.2. The solvent used in this work was dimethyl sulfoxide (DMSO). Photoinitiators and solvent were obtained from Sigma-Aldrich (Aldrich Chemical Company, Inc. Milwaukee, WI). Also, a standard silicon powder reference material was used for line position and line shape during X-ray diffraction studies (NIST reference material® 640d). All materials were used as received.

Figure 13.3 shows the mechanism by which these photoinitiator generate free radicals when exposed to UV radiation. HPK generates free radicals by homolytic cleavage and has a UV absorbance peak at 244 nm in the UVC region (250–260 nm). In contrast, BDP does so by hydrogen abstraction and has UV absorbance peak at 378 nm in the UVA region (320–390 nm).[27–29,36,37]

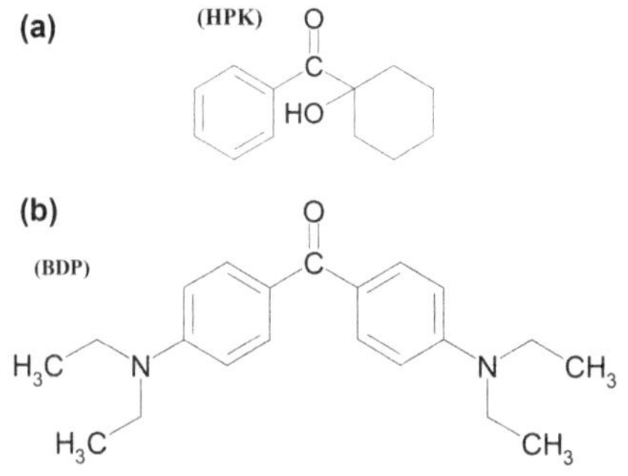

**Figure 13.2** 1-Hydroxycyclohexyl phenyl ketone (denoted as "HPK") and 4,4'-bis-(diethylamino)benzophenone (denoted as "BDP") photoinitiators used throughout the study.
(Reproduced from Morales and Ogale.[33])

### 13.2.2 UV Treatment

The PAN-based precursor and photoinitiator (where applicable) were dissolved in DMSO at 70 °C. The amount of solids in solution was kept at ~15 wt%. Films were cast from solution and dried at 70 °C in a conventional oven for ~24 h in air. The final thickness of the films was approximately 20 μm. To generate control samples, no photoinitiator was added; the remaining procedure was identical to the one used with the photoinitiator. A custom-built air-cooling system was placed inside the UV chamber to remove the excess of heat generated by the source during the UV-treatment. Thus, temperature was controlled during each experiment to study the reactions in two different regimes, 65 and 100 °C, that are respectively below and above the $T_g$ of the precursor. The films were placed on metal frames and UV exposed for 100, 300, and 600 s.

PAN-based precursor and photoinitiator (where applicable) were dissolved in a 99:1 mass ratio in DMSO at 70 °C. The amount of solids in solution was kept at ~16 wt%. Fibers were spun from solution using a custom-built wet-spinning unit fitted with a spinnerette that had 100 holes nominally 68 μm in diameter. A volumetric flow of 0.6 mL min$^{-1}$ was maintained during the spinning of the fibers. The coagulation bath consisted of 70 wt% DMSO/30 wt% distilled-deionized water. After the coagulation bath, the solidified fibers were passed through a distilled-deionized water washing bath; both baths were maintained at ~20 °C. The effective length of the coagulation and washing baths were 40 and 80 cm, respectively. A draw down ratio of ~1.2 was used during the wet spinning of the PAN-based fibers. The as-spun fibers were placed in an oven and dried at 70 °C for about 24 h. Next, the fibers were poststretched in a distilled-deionized water bath maintained at ~80 °C. The effective length of the poststretching bath was 80 cm. A draw down ratio of ~3.0 was used during the poststretching step. Pure PAN fibers as well as fibers containing 1 wt% of photoinitiator, with an

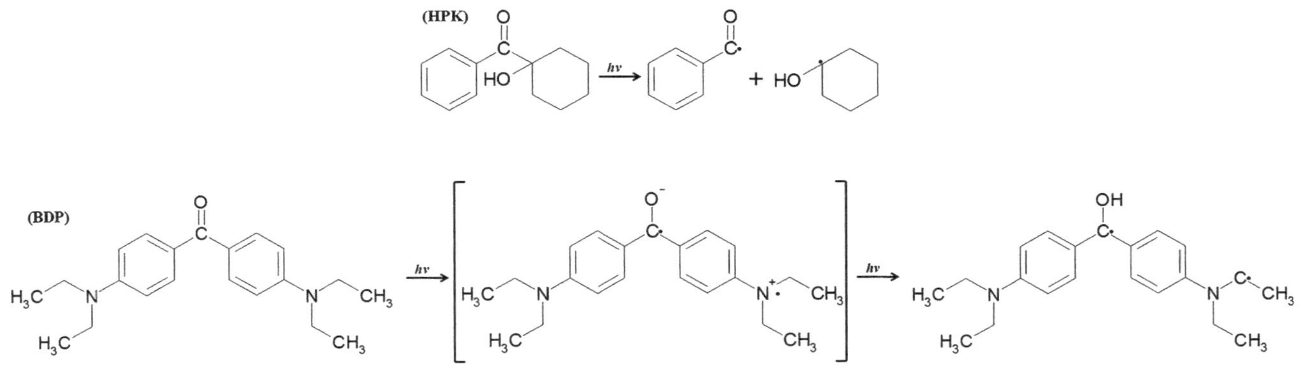

**Figure 13.3** Free-radical generation mechanism of HPK (homolytic cleavage) and BDP (hydrogen abstraction) when exposed to UV radiation.

average effective diameter of ~11.6 μm, were thus produced and used as the initial fibers throughout this study.

The samples were irradiated with a modified Nordson 4.5 kW UV-curing lamp (Model 111465A). A mercury arc bulb (model PM1163) was the UV source used throughout this work. Such UV sources are available commercially and widely used to provide a broadband output distribution in the four UV regions.[23,26,33] The chamber was modified to provide temperature control and the fibers were all UV-treated at approximately 150 °C. The intensity of this bulb was measured using a high-energy UV radiometer (model PP2000, Electronic Instrumentation and Technology, Inc.). Intensity values of approximately 0.228, 0.196, 0.032, and 0.095 W cm$^{-2}$ were measured for the UVA, UVB, UVC, and UVV ranges, respectively.

Fiber tows approximately 10 inches long were irradiated for 300 s. The bundle of fibers was held under approximately 0.1 g denier$^{-1}$ of tension during the UV treatment. The distance between the samples and the UV source was kept constant at approximately 20 cm.

All UV-treated films and fibers were compared against two types of controls. The first control consisted of pure as-produced PAN samples. The second control consisted of samples that were covered by a metal sheet (to avoid UV exposure) but ones that were treated in the same UV chamber to provide similar thermal exposure to that received by UV-treated samples (that also experienced the slightly elevated temperatures).

### 13.2.3 Thermal Treatment

PAN-based films were only thermo-oxidative stabilized (as in conventional processes) to produce control samples whose extent of cyclization could be compared with those of the UV-treated samples. The films were placed on metal frames and heated at 2.5 °C min$^{-1}$ to 320 °C and held there for 30 min. During this thermal oxidation, samples were removed from the oven at 150, 175, 200, 212.5, 225, 240, 255, 270, 280, 300, 320, and 320 °C (after 30 min) to analyze and quantify the partial extent of the cyclization reaction. A similar process was applied to PAN fibers. The fibers were stabilized under tension (0.1 g denier$^{-1}$ below 200 °C and 0.05 g denier$^{-1}$ above 200 °C), heated at 2.5 °C min$^{-1}$ to 300 °C, and held there for 30 min. During this thermal oxidation, samples were removed from the oven at 150, 175, 200, 225, 250, 275, 300, and 300 °C (after 30 min) to analyze and quantify the partial extent of the cyclization reaction.

In order to measure the properties of the produced carbon fibers by this novel process, fiber bundles were thermally stabilized in air atmosphere using a heating rate of 2.5 °C min$^{-1}$ from 25 °C to 300 °C, and held there for 30 min. Tension was applied to the samples during thermal oxidation by dead-weight loading. A tension of 0.1 g denier$^{-1}$ was applied at temperatures below 200 °C, and 0.05 g denier$^{-1}$ above 200 °C. As indicated in earlier studies, the external addition of small amounts (~1 wt%) of photoinitiator followed by a short (5 min) UV treatment increased the rate of the cyclization reaction.[33,34] Based on these results, UV-treated fibers with photoinitiator were placed in a preheated oven at 225 °C and rapidly heated. Then, using a heating rate of 2.5 °C min$^{-1}$, the fibers were heated to 300 °C, and held there for 30 min. A tension of 0.05 g denier$^{-1}$ was applied on these samples. Note that the total duration of this fast-thermal stabilization step was only 60 min, which was less than half of the 140 min used for the conventional thermal stabilization step.

Thermally stabilized fibers were then carbonized under a tension of 0.07 g denier$^{-1}$ in an inert environment (helium). This was accomplished by dead-weight loading using a custom-designed graphite fixture, which could be mounted inside of a graphite furnace ASTRO HP50-7010. A heating rate of 10 °C min$^{-1}$ was used to a maximum temperature of 1200 °C, and the fibers held at 1200 °C for 1 h. All UV-treated and thermally stabilized and carbonized samples were compared against three types of control fibers. The first two controls consisted of conventionally thermal stabilized pure PAN fibers and non-UV-treated fibers but containing photoinitiator. The third control consisted of fast-thermal-stabilized pure PAN fibers without UV treatment. These controls were produced to confirm that the improvement in the mechanical properties and the reduction in the thermal stabilization time are due to the combined positive effects of the presence of photoinitiator and UV treatment of the samples.

### 13.2.4 Characterization

Fourier transform infrared spectroscopy (Nexus 870 FT-IR ESP, Nicolet) was used to determine the chemical changes induced during the UV treatment and to estimate the extent of the cyclization reaction. Peaks located in the double-bond region between 1450 and 1700 wave numbers (cm$^{-1}$) are associated with the ladder structure.[8,38-40] The scans were conducted from 400 to 4000 cm$^{-1}$ and the transmittance intensities were normalized against the intensity of the 2940 cm$^{-1}$ peak (CH$_2$ asymmetric stretching) for comparison. The extent of the cyclization reaction was estimated from the following conversion index:[11,41,42]

$$\text{Conversion index}\,(\%) = \frac{I_{1600}}{I_{2240} + I_{1600}} \times 100\% \qquad (13.1)$$

where $I_{1600}$ is the measured intensity of the peak located at approximately 1600 cm$^{-1}$, which corresponds to carbon–nitrogen double bonds associated with the heterocyclic ring structure or ladder structure developed during the thermal oxidation of the PAN-based precursors. $I_{2240}$ corresponds to the measured peak intensity of the peak located at approximately 2240 cm$^{-1}$ associated to the carbon–nitrogen triple bond present in the original polyacrylonitrile structure. The peak intensities values used to calculate these conversion indices were raw intensity values given by the FTIR spectrometer without any normalization.

Differential scanning calorimetry (Pyris 1 DSC, Perkin Elmer Instruments) was conducted to observe the thermal behavior of the samples after being UV treated and to quantify the residual heat that a UV-treated sample would liberate upon further conventional thermal oxidation. Approximately 4 mg of film sample was heated to 350 °C at a rate of 5 °C min$^{-1}$ in an air environment. For each of the characterization techniques used during this study, at least three different measurements were conducted on the different samples.

Morphological analysis of fibers was conducted by scanning electron microscopy with a Mitsubishi 4800 SEM unit. To retain the original cross-sectional shape and other physical features, the fibers were cryofractured in liquid nitrogen. To determine the effective diameter of the fibers, Image-Pro Plus 7.0 (Media Cybernetics, Inc.) analysis software was used to measure the cross-sectional area of each type of sample. The effective diameter calculated from these area measurements is reported only as a reference. At least 40 cross-sectional areas were measured for each type of fiber.

A PHOENIX single-filament tensile testing unit (Measurements Technology Inc.) was used to measure the mechanical properties of the different fiber types produced during this work. The single fibers were mounted on 25 mm paper tabs. At least 20 samples per type of fiber were prepared and tested.

Wide-angle X-ray diffraction (WAXD) analysis was conducted on bundles of fibers using a Rigaku-MSC (Houston, TX) X-ray diffraction unit. The X-ray diffraction pattern was captured through an image plate and the WAXD generated images were analyzed using Polar v2.6.7 (Stonybrook Technology and Applied Research, STAR). The curve fitting of the spectra was done using OriginPro 7 (v7.0383, OriginLab Corporation). The unit generates an X-ray beam with a wavelength of 1.5406 Å (Cu target). The diameter of the X-ray beam hitting the sample was ~0.5 mm. The distances between the X-ray source and sample and the sample and the image plate detector were approximately 70 and 12 cm, respectively. The operational conditions of the X-ray source were 45 kV and 0.65 mA. The gas used for most of the beam path was helium. Each bundle of fibers was mounted on a 10 mm paper tab and sprinkled with silicon standard powder. The exposure time per sample was approximately 2 h. The plane spacing and crystal size were calculated using the equations below:

$$\lambda = 2d \sin \theta \tag{13.2}$$

$$\frac{1}{d^2} = \frac{4}{3}\left(\frac{h^2 + hk + k^2}{a^2}\right) + \frac{l^2}{c^2} \tag{13.3}$$

$$l_a = \frac{0.89\lambda}{B \cos \theta} \tag{13.4}$$

$$B^2 = B_M^2 - B_S^2 \tag{13.5}$$

for these equations, $\lambda$ corresponds to the wavelength of the X-ray beam, $d$ is the distance between adjacent planes in the set $(hkl)$, $\theta$ is the angle of incidence of the X-ray beam, $a$ and $c$ are the lattice parameters of the hexagonal structure, $l_a$ is the crystal size, $B$ is the corrected full width at half-maximum (FWHM) intensity of the spectrum peak (in terms of $2\theta$ scale), $B_M$ is the measured FWHM, and $B_S$ is the system broadening determined, in this case, by the silicon standard. Eqn (13.2) and (13.3) were employed to calculate the interplanar spacing of the crystals. Eqn (13.2) corresponds to the Bragg's law and eqn (13.3) corresponds to the lattice geometry equation for hexagonal crystal structure observed in PAN. Eqn (13.4) and (13.5) were employed to calculate crystal size. Eqn (13.4) corresponds to the Scherrer formula. Warren's broadening correction (eqn (13.5)) was used to correct the system broadening on each WAXD scan.[13,43–49] Also, conversion or aromatization index was determined using the following equation:[41,46,50]

$$\text{Conversion index (\%)} = \frac{A_A}{A_A + A_P} \times 100\% \quad (13.6)$$

where $A_A$ is the area of the peak associated with the aromatic structure around $2\theta = 25°$ and $A_P$ correspond to the area of the (1 0 0) crystal peak of PAN at $2\theta = 17°$. In addition, the Rigaku-MSC diffractometer is capable of measuring the orientation of the crystal in the fibers. These azimuthal intensity scans were measured for the most prominent peak of the initial PAN precursor fibers, which is the (1 0 0) peak, located at approximately 17°.

## 13.3 Results and Discussion

### 13.3.1 Influence of UV Radiation and Photoinitiator

Figure 13.4(a) displays the FTIR spectra of three control specimens: as-produced PAN and two non-UV irradiated but thermally treated samples at 100 °C for 300 and 600 s. These correspond to the longest treatment times for each set of samples discussed in the following sections. In Figure 13.4(b), the section between ~1450 and ~1700 cm$^{-1}$ displays the double-bond region associated with the carbon–oxygen, carbon–nitrogen, and carbon–carbon double bonds present in the ladder structure. This region includes the most significant differences among all the FTIR spectra. The peaks located at 2940, 2865, 2242, 1730, 1454, and 1055 cm$^{-1}$ have been assigned to CH$_2$ asymmetric stretching, CH$_2$ symmetric stretching, C≡N stretching, C=O stretching (due to the presence of methyl acrylate), CH$_2$ scissor vibration, and –C–C–C– backbone bending, respectively.[8,25,31,38,39] These results confirm that simply raising the temperature in the UV chamber has no discernible effect on the chemical modification of the nonirradiated samples.

Figure 13.5 displays FTIR spectra obtained from samples UV treated at 100 °C for 300 s, with and without photoinitiators. It is evident that the intensities of this double-bond region were higher for UV-treated samples

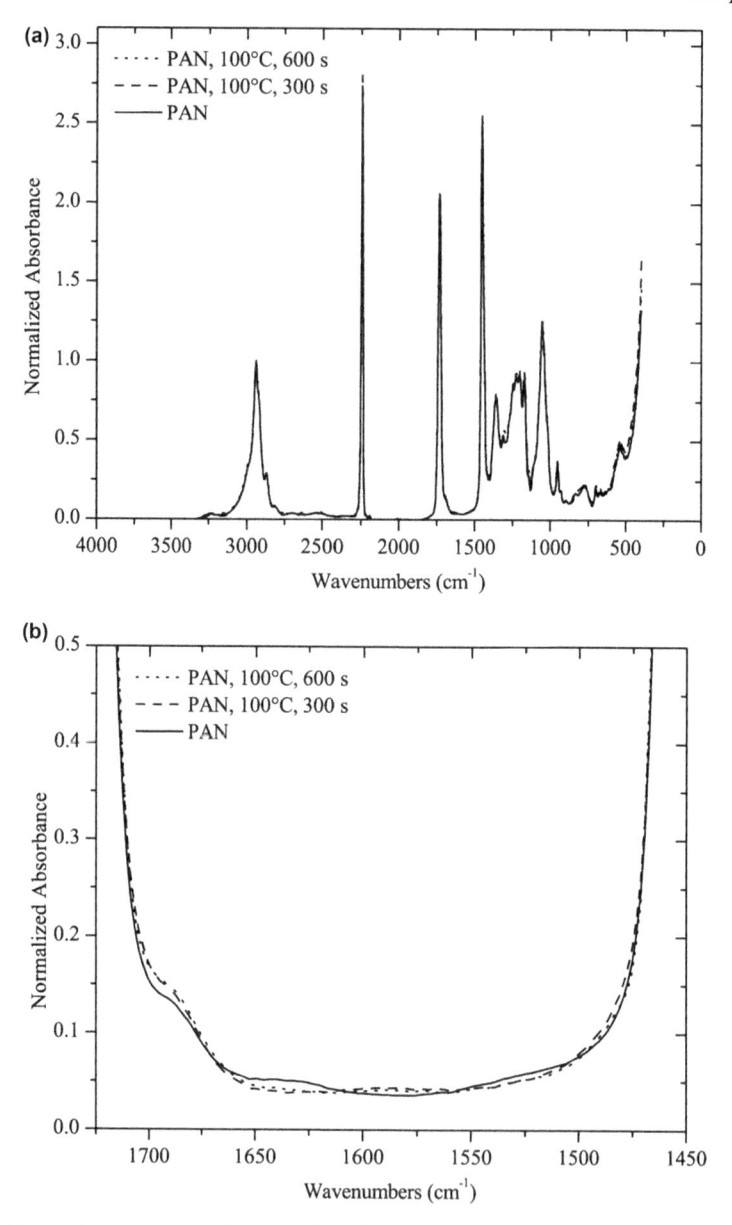

**Figure 13.4** FTIR spectra of as-produced PAN and two non-UV-exposed but thermally treated control samples at 100 °C for 300 and 600 s: (a) over extended wave numbers, and (b) zoomed-in spectra over the double-bond region.
(Reproduced from Morales and Ogale.[33])

containing 1 wt% BDP (hydrogen abstraction photoinitiator) as compared with samples containing 1 wt% HPK (homolytic cleavage) or the controls (pure, UV treated). The conversion index was calculated for each set

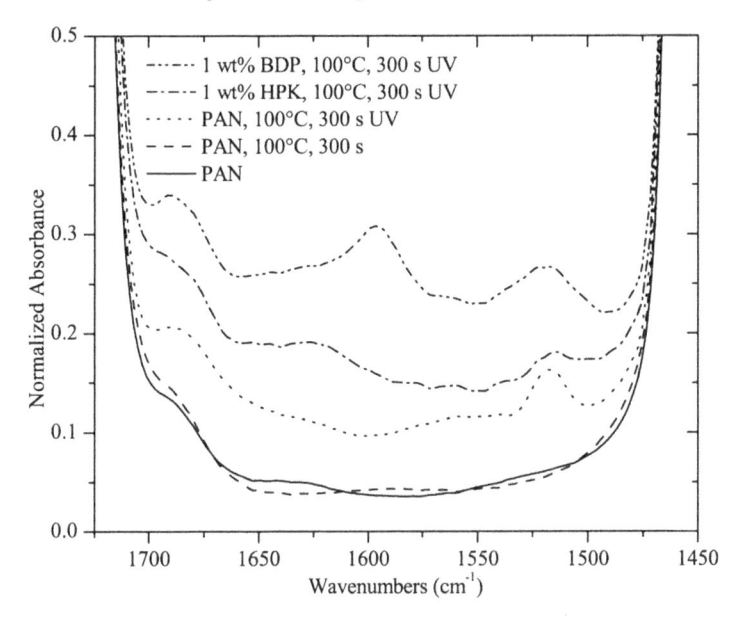

*UV-Based Dual Mechanism for Crosslinking and Stabilization* 267

**Figure 13.5** FTIR results of PAN-based copolymer containing 1 wt% BDP, 1 wt% HPK and without photoinitiator; UV treated at 100 °C for 300 s; compared against as-produced and thermally treated control samples. (Reproduced from Morales and Ogale.[33])

of samples at 9.0 ± 1.5, 4.6 ± 1.5, 2.2 ± 1.1, 0.1 ± 0.2, and 0 ± 0.2% for samples containing 1 wt% BDP, 1 wt% HPK, UV-treated pure, and the two control samples, respectively. At 95% confidence, the conversion index was higher for UV-treated samples containing 1 wt% BDP (hydrogen abstraction) compared with other samples containing 1 wt% HPK (homolytic cleavage) and UV-treated pure samples. Thermally treated samples without UV exposure show no ladder formation. Hence, the presence and type of photoinitiator influences the extent of cyclization of the samples. These results also confirm that UV-sensitive components can be physically added into the precursor, and not have to be present in the polymer backbone.

The greater double-bond formation during UV treatment afforded by the hydrogen-abstraction mechanism (over that by homolytic cleavage) can be attributed to the competition between the PAN-based precursor and the photoinitiator. It is known that mercury lamps provide a broadband output distribution in the UVA, UVB, UVC and UVV regions.[23] PAN has a UV absorption peak in the UVC region (~274 nm), which falls in the same UV absorption region as do most of photoinitiators dissociating by homolytic cleavage.[27,29] Specifically, HPK has a UV absorption peak at ~244 nm. Thus, the PAN-based precursor and photoinitiator compete for the same region of the UV spectrum, and energy absorption is inefficient. On the other hand, hydrogen-abstraction-type photoinitiators have UV-absorption bands at higher wavelengths in the UVA region, specifically ∼378 nm for BDP.[27] This reduces the competition between the PAN-based precursor and photoinitiator for the same region of the UV spectrum. This establishes that hydrogen-abstraction photoinitiators offer a more efficient mechanism, as compared to homolytic cleavage type for UV-induced reactions in PAN-based precursors.

Figure 13.6 displays the DSC results of the residual heat of reaction for a PAN-based precursor, with and without photoinitiator, UV-treated at 100 °C for 300 s. The exothermic heat of reaction observed is a measure of further thermal oxidation that the sample can undergo. It is evident from the results that UV-treated samples display a reduction in the residual heat of cyclization, which may possibly be attributed to three phenomena occurring during the UV-treatment of the precursor: cyclization, crosslinking, and scission of the polymer chains.[15,30,31,51,52] This reduction of the residual heat of reaction of UV-treated specimens as compared with samples without UV treatment means that some reaction has already taken place during UV irradiation, albeit at a lower temperature. Interestingly, specimens containing 1 wt% BDP (hydrogen-abstraction) show higher heat of reaction compared with that displayed by samples containing 1 wt% HPK (homolytic cleavage) and pure UV-treated samples. Thus, samples containing 1 wt% BDP will undergo additional thermal cyclization during

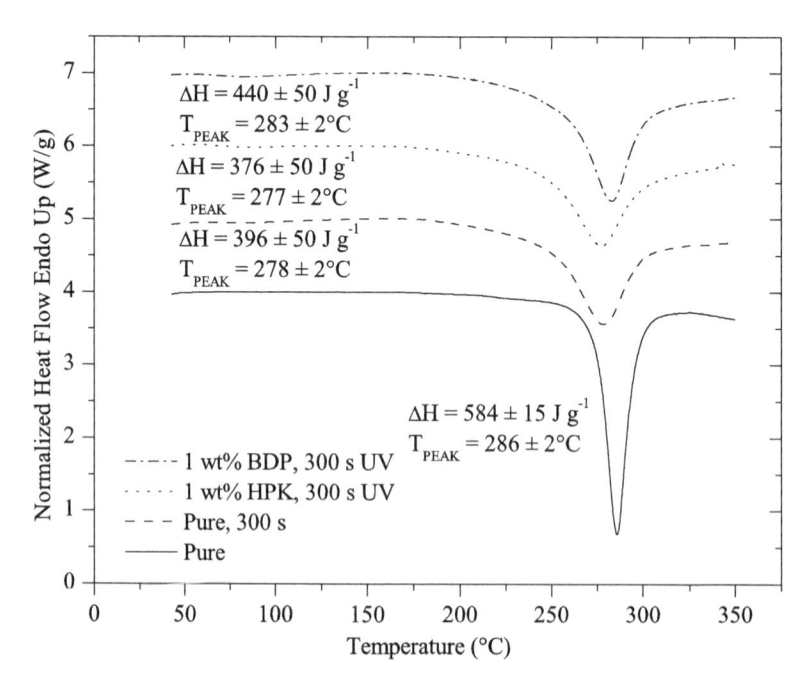

**Figure 13.6** DSC scans of PAN copolymer with and without photoinitiator UV treated at 100 °C for 300 s. $\Delta H$ represents the residual heat of cyclization. (Reproduced from Morales and Ogale.[33])

the further thermal oxidation step, which is desired, even though the same specimens had also displayed higher FTIR double-bond conversion. Thus, BDP is able to absorb more UV energy delivered by the source to react with the PAN-based precursor increasing double-bond formation. At the same time, reduction of chain scission and crosslinking that hinder ladder formation during the further thermal oxidation step is observed. By the selection of a suitable photoinitiator, the undesired side reactions (namely chain scission and crosslinking) can be reduced to favor ladder formation. Since BDP offered a better performance, it was further investigated with regards to different exposure temperatures and durations.

## 13.3.2 UV-Radiation Time and Temperature

Figure 13.7(a) and (b) display FTIR spectra of PAN-based copolymer without any photoinitiator (Figure 13.7(a)) and with 1 wt% BDP (Figure 13.7(b)). Both set of samples were UV treated at 100 °C for 100, 300 and 600 s and then compared with as-produced and thermally treated control samples. The conversion indices were measured to be $6.2 \pm 1.5$, $9.0 \pm 1.5$, and $15.2 \pm 1.5\%$ for samples containing 1 wt% BDP UV treated for 100, 300, and 600 s, respectively. In contrast, the conversion indices of pure samples were $0.6 \pm 1.1$, $2.2 \pm 1.1$, and $7.3 \pm 1.5\%$ for the three durations. Increased UV exposure time led to higher double-bond formation and higher conversion index in the double-bond region. These results are consistent with prior studies that show that PAN precursors can be cyclized by high-intensity UV radiation.[15,30] The cyclized polymer can subsequently be thermally stabilized for successful conversion into carbon fibers.[31] Although there is some evidence of cyclization reaction occurring in UV-treated samples without photoinitiator, the double-bond region intensity is significantly higher for samples containing photoinitiator.

In addition to the different intensities observed for the different samples, two other differences were observed. Samples without photoinitiator show three noticeable peaks located at 1680, 1630, and the clearest at 1520 cm$^{-1}$. In contrast, samples containing 1 wt% BDP have an extra peak located at 1600 cm$^{-1}$. This peak is very clear for samples containing photoinitiator after UV treatment for only 100 s, and it is not significant for samples without photoinitiator even after 600 s of UV treatment. It is known that radicals are more likely to add to unsaturated groups, such as the nitrile group present in the PAN structure.[31,51,52] Thus; it is believed that the generated photoinitiator radicals react with the carbon–nitrogen triple bonds to start the radical cyclization reaction. The peak located at 1600 cm$^{-1}$, clearly observed for samples containing photoinitiator, has been assigned to the carbon–nitrogen double bond. The presence of the 1520, 1600, 1630, and 1680 cm$^{-1}$ peaks is a clear indicator of the formation of the ladder structure after the UV treatment.[8,38–40]

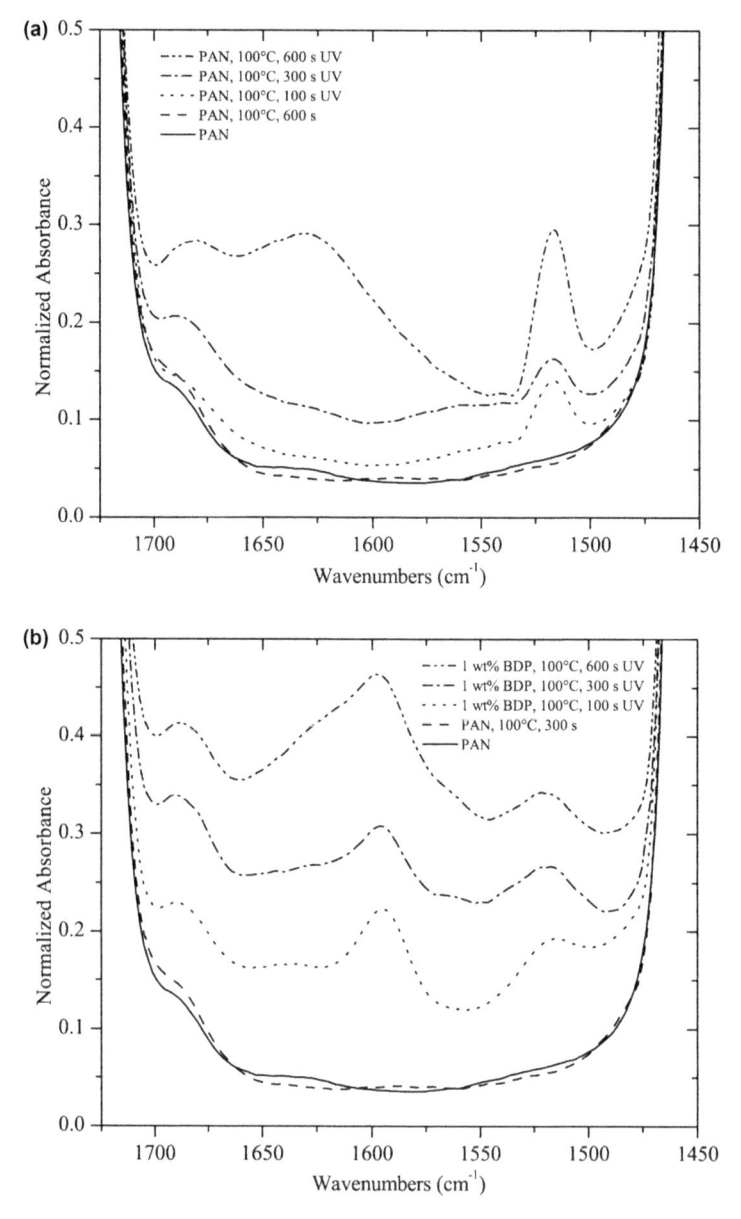

**Figure 13.7** FTIR analysis of PAN-based copolymer with and without photoinitiator UV treated at 100 °C for 0, 100, 300, and 600 s: (a) without photoinitiator, and (b) with 1wt% BDP; compared against as-produced and thermally treated control samples.
(Reproduced from Morales and Ogale.[33])

In contrast, due to the absence of photoinitiator in pure samples, a dehydrogenation reaction is believed to be the primary pathway during UV treatment. Hydrogen radicals produced during the UV treatment combine to form hydrogen, with carbon–carbon double-bond formation in the backbone of the precursor or even crosslinking and scission of the polymer chains, presumably in the amorphous region.[15,51,52] Based on these arguments, the two peaks located at 1630 cm$^{-1}$ and 1520 cm$^{-1}$ (the clearest and most intense for UV-treated samples without photoinitiator) are believed to be associated with the conjugated carbon–carbon double-bond system. The peak located at 1680 cm$^{-1}$ has been assigned to the carbon–oxygen double-bond formation. Although these are the most noticeable peaks, these overlap with other weaker peaks associated with the ladder structure. Conjugated carbon–oxygen and carbon–nitrogen double bonds, carbon–carbon and carbon–nitrogen single-bond stretching, and NH inplane bending have been assigned in the region between 1500 and 1700 cm$^{-1}$.[8,38] The presence of photoinitiator affects the extents and how favorable a specific reaction can become over other parallel reactions taking place during the UV treatment of the PAN-based precursors.

To study the effect of polymer mobility on the photoinduced cyclization of PAN-based specimens with and without photoinitiator, samples were UV treated below and above the $T_g$ of the PAN-based precursor. Figure 13.8(a) and (b) summarize these results. Figure 13.8(a) shows PAN-based specimens without photoinitiator UV treated at 65 or 100 °C for 600 s. These two temperatures are below and above the glass-transition temperature of the PAN-based copolymer ($T_g \sim 82.5$ °C), respectively. Figure 13.8(b), displays another set of samples containing 1 wt% BDP UV treated at the same temperatures as the pure precursor. The conversion indices of the samples without photoinitiator UV treated at 65 and 100 °C were $4.7 \pm 1.5$ and $7.3 \pm 1.5\%$. For samples containing 1 wt% BDP the indices were $6.8 \pm 1.5$ and $15.2 \pm 1.5\%$, respectively. In both sets of samples, specimens UV treated above $T_g$ of the precursor show higher intensities in the double-bond region compared with that for the samples UV treated below $T_g$. The presence of photoinitiator plays a bigger role on the extent of the ladder-structure formation on samples UV treated above $T_g$ (7.3 vs. 15.2%) compared with the values obtained for samples UV treated below the $T_g$ of the precursor (4.7 vs. 6.8%).

These results can be explained by the fact that the mobility of the polymer chains is limited below the $T_g$ of the precursor, which reduces the reactivity of the precursor. Based on the similarities shown by their respective FTIR spectra, the presence of photoinitiator has little effect on the type of reactions undergone by the samples UV treated at temperatures below the $T_g$ of the precursor. In both sets of samples, with and without photoinitiator UV treated below $T_g$, a very prominent peak at 1520 cm$^{-1}$ mainly associated with carbon–carbon conjugated double bonds is observed. The only subtle difference observed between these two spectra is the wider shoulder shown by the sample containing photoinitiator reaching the 1600 cm$^{-1}$ region, which is associated with carbon–nitrogen double bonds present in the ladder structure. Although some carbon–nitrogen double-bond formation is observed for samples containing photoinitiator UV treated below $T_g$, the

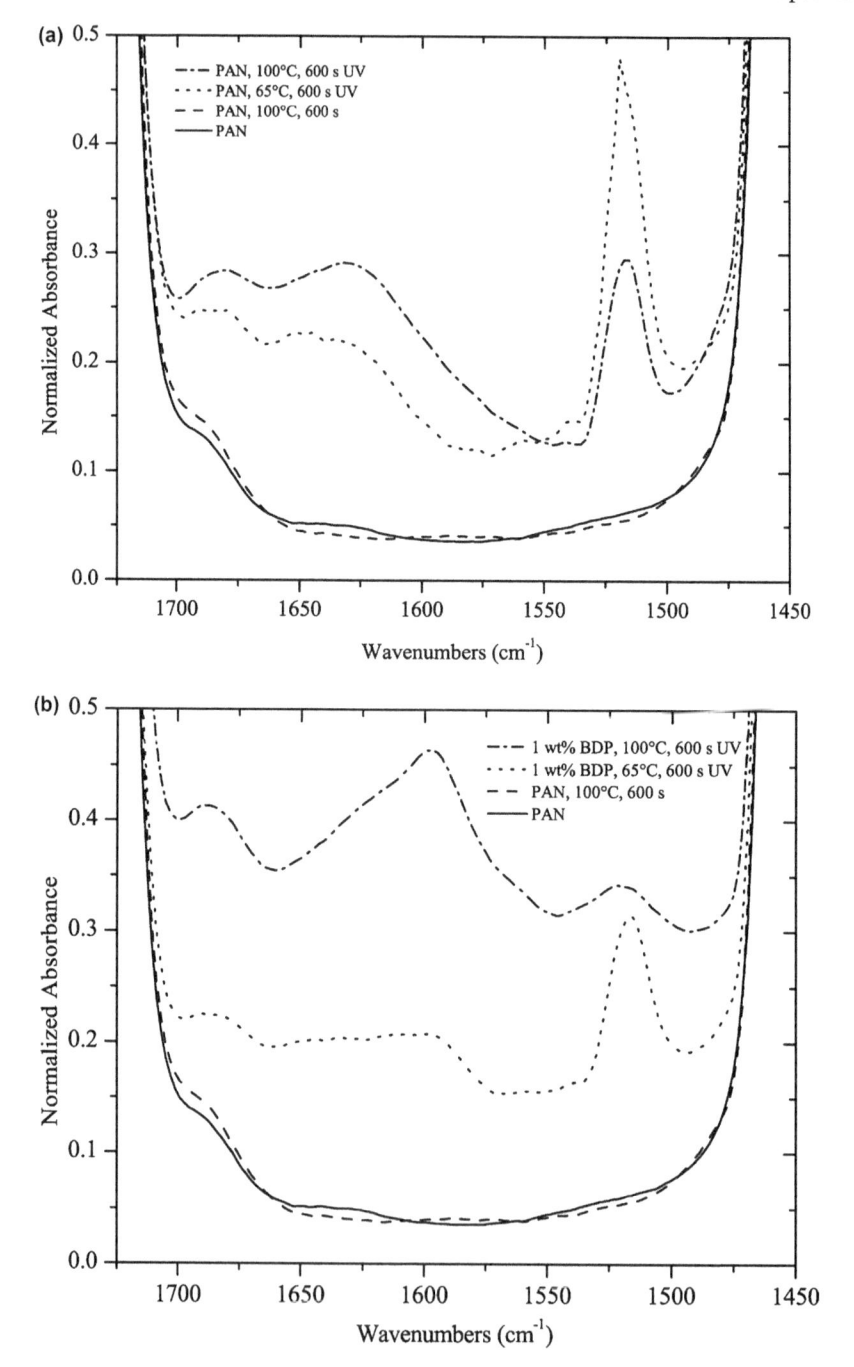

**Figure 13.8** FTIR analysis of PAN-based copolymer UV treated at 65 or 100 °C for 600 s: (a) without photoinitiator, and (b) with 1 wt% BDP; compared against as-produced and thermally treated control samples. (Reproduced from Morales and Ogale.[33])

number of triple bonds attacked by the photoinitiator radicals during the UV treatment has been reduced by the restricted polymer chain mobility. As discussed earlier, samples containing photoinitiator and UV treated above $T_g$ show higher double-bond intensity specially the 1600 cm$^{-1}$ peak related with carbon–nitrogen double bonds. Pure samples are more inclined to form carbon–carbon double bonds when UV treated irrespective of the temperature. The limited mobility of the polymer chains below $T_g$ limits the effect of photoinitiator. Thus, the processing temperature (relative to precursor $T_g$) and the composition (type of photoinitiator) during this photoinduced oxidation process play an important role in achieving the stabilization reaction.

Figure 13.9 displays the FTIR spectra of pure PAN-based copolymer at different stages during the traditional thermal oxidation treatment in air. The overall changes involve the reduction of the C≡N stretching peak (2242 cm$^{-1}$), CH$_2$ stretching peak (2937 cm$^{-1}$), CH$_2$ bending peak (2937 cm$^{-1}$), and carbon backbone bending peak (1051 cm$^{-1}$). On the other hand, primary amines (3390 and 3356 cm$^{-1}$), carbon–oxygen double bond (1680 cm$^{-1}$), carbon–nitrogen double bond (1600 cm$^{-1}$), carbon–carbon double bond (1630 cm$^{-1}$), conjugated carbon–carbon double bond corresponding to the heterocyclic ring structure (1520 cm$^{-1}$), carbon–hydrogen in and out of plane deformations (1370 and 805 cm$^{-1}$, respectively), and carbon–oxygen single bond (1150 cm$^{-1}$) appear. All these peaks are related to the cyclized structure produced during the thermal oxidation treatment.[8,25,31,38,39]

Based on FTIR spectra displayed in Figure 13.9, the FTIR conversion indices for pure, non-UV-treated, thermally stabilized samples are presented in Figure 13.10 as a function of the thermal stabilization time. Also included is the temperature profile followed during these thermal stabilization experiments. These results show that most of the conversion of the precursor occurs during the second half of the thermal treatment process, specifically between 200 and 300 °C. In addition, the dashed line indicates the conversion index of 9.0% for a sample containing 1 wt% BDP UV treated at 100 °C for 300 s. The results show that 300 s (5 min) of UV treatment of the PAN-copolymer containing 1 wt% BDP are approximately equivalent to the first 4800 s (80 min) of the thermal oxidation process, i.e. a reduction of approximately half of the time. In addition, during the UV treatment, the sample was kept at 100 °C; in contrast, the thermal-treated samples have to reach approximately 240 °C to achieve the same conversion index. Thus, the same conversion was achieved in a shorter period of time at a lower temperature. The conversion index for UV-treated samples in the absence of photoinitiator was approximately 2% and not statistically different from the initial indices for pure PAN samples thermally stabilized (therefore, not included in the figure). These kinetic results indicate the potential for developing a novel and more rapid process for stabilization of PAN-based precursors to produce carbon fibers more efficiently.

274  Chapter 13

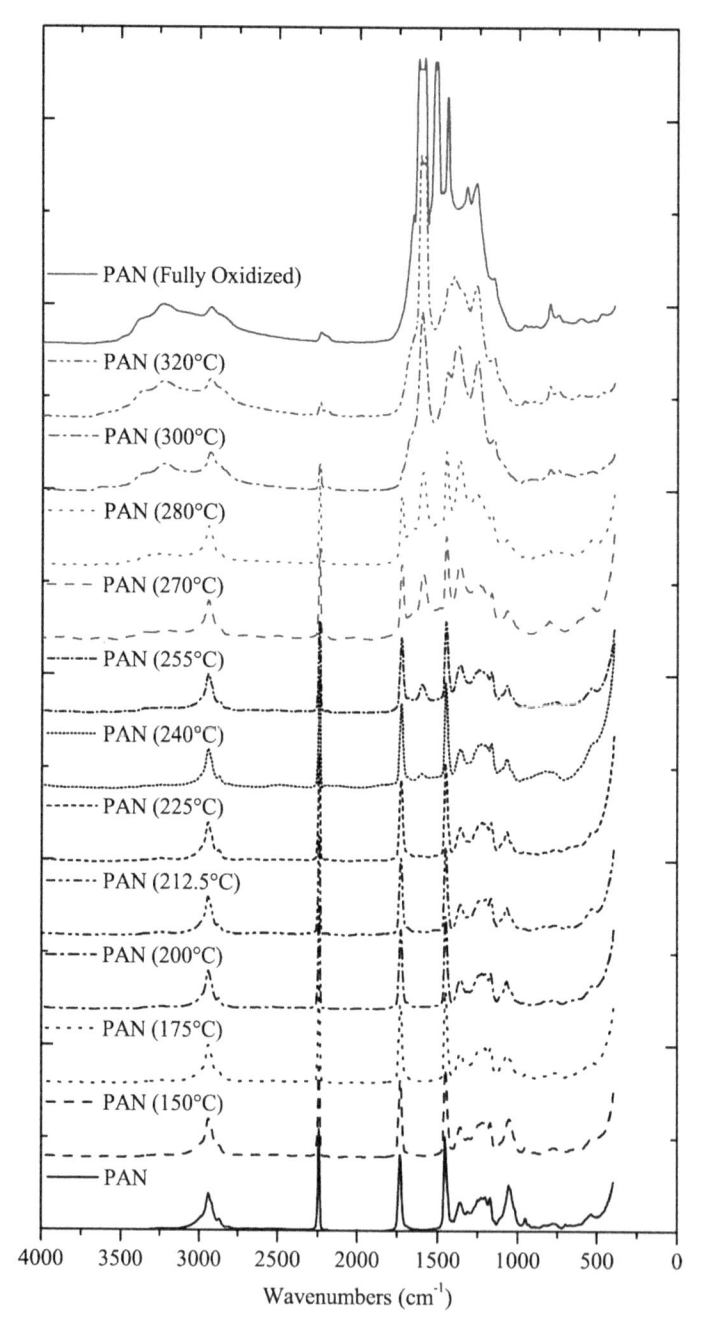

**Figure 13.9** FTIR spectra of pure PAN-based copolymer at different stages during the traditional thermal oxidation treatment in air. (Reproduced from Morales and Ogale.[33])

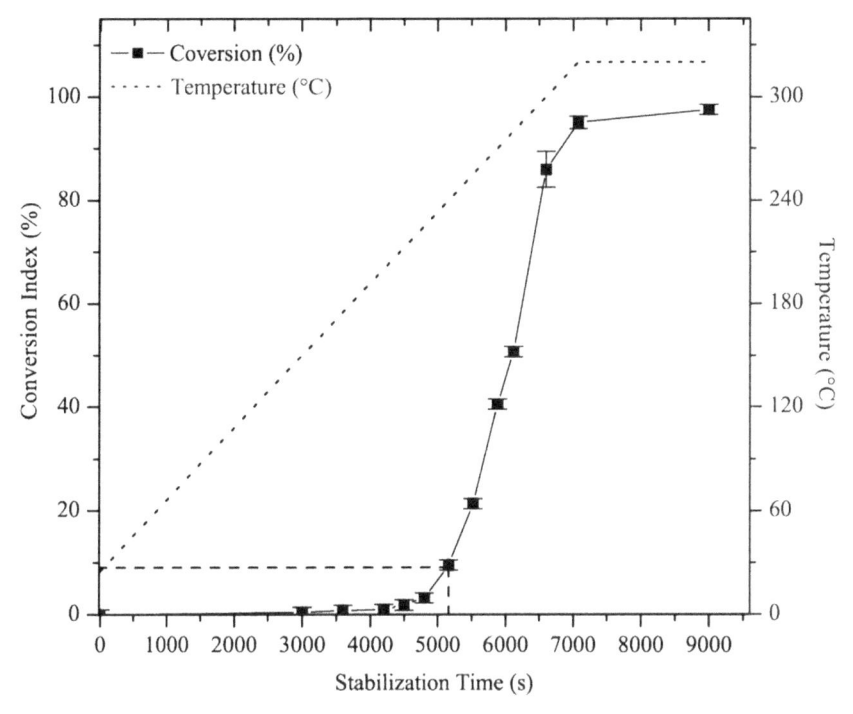

**Figure 13.10** Temperature profile and calculated conversion indices (with trend line) of PAN-based copolymer during the thermal oxidation step. (Reproduced from Morales and Ogale.[33])

## 13.3.3 Influence of UV Radiation on Polyacrylonitrile Fibers

Figure 13.11 shows representative tensile-testing curves of each set of samples. Figure 13.11(a) corresponds to as-produced, thermally exposed, and UV treated with mercury bulb pure PAN fibers. Figure 13.11(b) shows the same treatment conditions but for fibers containing 1 wt% BDP. The limit of linear proportionality was approximately 1% for all samples. Figure 13.11 also shows that the upper and lower yield points seem very close to the proportionality limit. Also, a small necking section is observed just before breakage. Table 13.1 summarizes the tensile-testing results conducted on single filaments. These tensile properties of the PAN fibers are consistent with those reported in the literature: 10–25% for the breaking strain, 0.1–0.7 GPa for strength, and 1–10 GPa for tensile modulus.[15,19,30,50,53–57]

The tensile-testing results shown in Table 13.1 indicate that, at 95% confidence, there is a significant difference in the breaking strain between pure and 1 wt% BDP fibers; the presence of photoinitiator reduces the elongation capabilities of the fibers by ~1%. These results show no significant statistical differences, at 95% confidence, between the ultimate tensile strength among different specimens. On the other hand, at 95% confidence, the two set of fibers containing 1 wt% BDP UV treated for 300 s

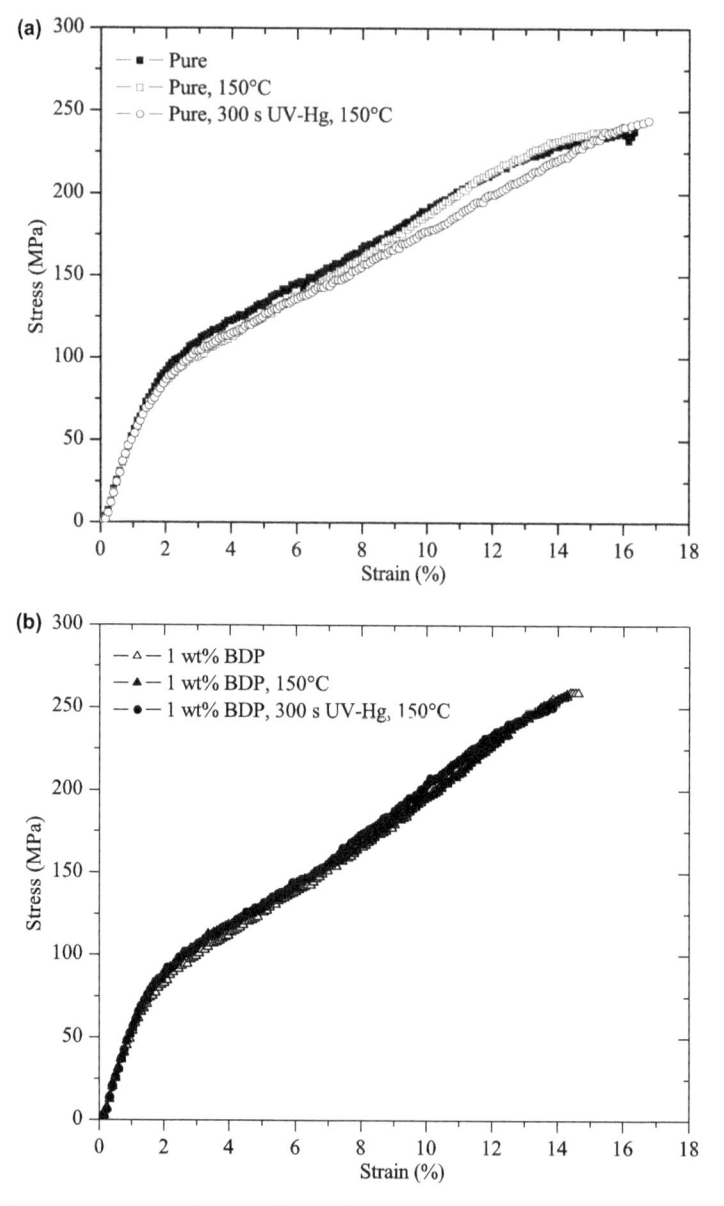

**Figure 13.11** Representative tensile-testing curves of each set of samples: (a) pure PAN, (b) PAN fibers containing 1 wt% BDP. (Reproduced from Morales and Ogale.[34])

(5 min) displayed higher tensile modulus than the other set of samples. In addition, these tensile-testing results show that fibers containing 1 wt% BDP UV treated with a mercury source exhibit the highest tensile modulus of all six set of samples (significant at 95% confidence). This proves the

**Table 13.1** Single filament tensile results with 95% confidence intervals. (Reproduced from Morales and Ogale.[34])

| Sample | Tensile modulus (GPa) | Max stress (GPa) | Break strain (%) |
| --- | --- | --- | --- |
| Pure | 7.0 ± 0.3 | 0.24 ± 0.01 | 16.4 ± 0.5 |
| Pure, 150 °C | 7.1 ± 0.3 | 0.24 ± 0.01 | 14.9 ± 1.2 |
| 1 wt% BDP | 7.0 ± 0.3 | 0.26 ± 0.01 | 14.3 ± 0.5 |
| 1 wt% BDP, 150 °C | 7.1 ± 0.4 | 0.25 ± 0.01 | 14.0 ± 0.8 |
| Pure, 300 s Hg, 150 °C | 7.0 ± 0.3 | 0.24 ± 0.01 | 16.7 ± 0.5 |
| 1 wt% BDP, 300 s Hg, 150 °C | 7.7 ± 0.2 | 0.25 ± 0.01 | 13.8 ± 0.8 |

combined positive effect of the addition of photoinitiator and 300 s of UV treatment on the fibers.

Figure 13.12 shows representative wide-angle X-ray diffractograms of as-spun pure, poststretched pure, and poststretched with 1 wt% BDP PAN fibers. Integrated 2-theta scans, displayed in Figure 13.12(a), show the first peak at ~17° corresponding to the (1 0 0) plane of the PAN precursor. The second peak corresponds to the combination of the amorphous halo at ~26° and the (1 1 0) plane peak of PAN at ~29.5°.[44–48,58–60] A silicon standard was added to each of the different fiber samples as a calibration standard. This silicon standard shows three sharp peaks at 28.44°, 47.3°, and 56.12° corresponding to its (1 1 1), (2 2 0), and (3 1 1) planes, respectively.

As noted before, the orientation of the crystals in the fibers was measured from the azimuthal scans conducted on the (1 0 0) peak of the PAN spectra. Figure 13.12(b) shows the azimuthal scans of (1 0 0) peak of as-spun pure, poststretched pure, and poststretched with 1 wt% BDP PAN fibers. Sharp azimuthal peaks for a given set of planes mean higher orientation, whereas broad peaks indicate low orientation. As expected, poststretched PAN fibers have significantly higher orientation in comparison with that of as-spun fibers. Similar orientation is achieved for pure and 1 wt% BDP PAN fibers after being poststretched. To compare the crystal orientation in the different set of fibers, the full width at half-maximum (FWHM) was measured for each azimuthal scan. The smaller the FWHM value, the more oriented are the crystals within the fibers.

Table 13.2 summarizes the WAXD results obtained from all the different sets of samples. After poststretching, the interplanar spacing between the (1 0 0) plane was reduced and the orientation and crystal size was increased for pure and 1 wt% BDP-PAN fibers. Also, the results show no negative effects (at 95% confidence) for the interplanar spacing and size of the crystals within the fibers containing photoinitiator as compared with pure control fibers. In a previous study, it was shown that PAN-based precursors containing 1% BDP UV treated for 300 s show higher extents of cyclization reaction as compared with other set of samples.[33] Here, the WAXD results show that during the UV treatment (cyclization reaction) these samples are able to retain their molecular orientation. This is not the case for pure UV-treated samples. The retention of the orientation of the polymer chains

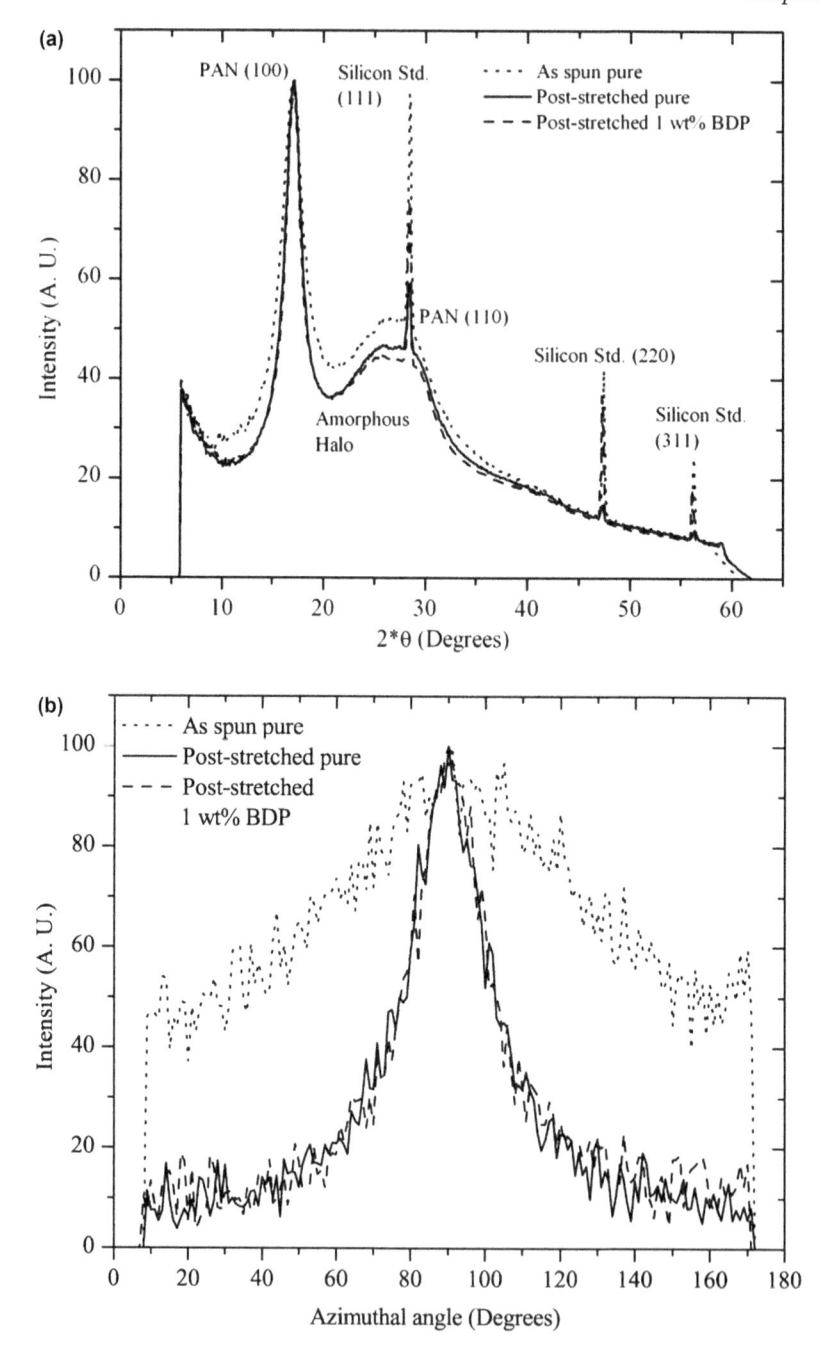

**Figure 13.12** Representative WAXD result of as-spun and poststretched PAN-based fibers: (a) integrated 2-theta scan; (b) azimuthal scan of the PAN (100) peak.
(Reproduced from Morales and Ogale.[34])

Table 13.2 WAXD results with 95% confidence intervals. (Reproduced from Morales and Ogale.[34])

| Sample ID | $l_a$ (Å) | $a$ (Å) | (1 0 0) FWHM (°) |
|---|---|---|---|
| PURE as spun | 31.1 ± 3.2 | 6.048 ± 0.025 | 75.2 ± 2.5 |
| 1 wt% BDP as spun | 29.4 ± 3.3 | 6.064 ± 0.029 | 75.8 ± 2.0 |
| PURE after stretch | 42.3 ± 1.9 | 6.001 ± 0.020 | 26.1 ± 0.9 |
| 1 wt% BDP after stretch | 45.8 ± 2.2 | 6.019 ± 0.019 | 24.6 ± 1.4 |
| PURE (150 °C, NO UV, control) | 41.8 ± 1.8 | 6.028 ± 0.038 | 28.3 ± 0.9 |
| 1 wt% BDP (150 °C, NO UV, control) | 42.4 ± 0.4 | 6.034 ± 0.015 | 25.3 ± 0.2 |
| PURE, 300 s Hg UV, 150 °C | 43.6 ± 1.4 | 6.047 ± 0.025 | 29.7 ± 0.6 |
| 1 wt% BDP, 300 s Hg UV, 150 °C | 46.3 ± 1.5 | 6.036 ± 0.016 | 25.6 ± 1.1 |

within the fibers can lead to more polymer chains undergoing cyclization, leading to higher conversions. These results agree well with the tensile-testing results that show superior mechanical properties for samples containing 1% BDP and UV treated for 300 s due to the fact that they are able to retain molecular orientation. For this system, higher molecular orientation leads to higher mechanical properties of the fibers.[8,15,17,18,20,21]

Figure 13.13 displays the WAXD spectra of pure PAN fibers at different stages during the traditional thermal stabilization in air. Figure 13.13(a) and (b) correspond to the integrated 2-theta scan and the azimuthal scan of the PAN (1 0 0) peak, respectively. Table 13.3 summarizes the lattice parameters calculated from the spectra shown in Figure 13.13. Between room temperature and 150 °C, the (1 0 0) peak becomes sharper and an increase in orientation and crystal size is observed in the fibers. Literature studies have discussed interesting effects of induced anisotropy due to the applied UV radiation.[61] However, as observed in the conventional thermal stabilized samples, most of the induced anisotropy in the current samples can be attributed to the fact that the samples are under tension during processing.

Between 150 and 225 °C, growth of the crystal size is mainly observed. At 225 °C, dramatic changes are observed in the spectra. Between 225 and 250 °C, the crystal size stops increasing, and around 250 °C, the (1 0 0) PAN peak starts broadening and becomes weaker. This indicates an increment on the interplanar spacing and a reduction on the crystal size and orientation of the original PAN structure. In contrast, a strong aromatic peak starts to emerge. This evolution of structure is consistent with that reported in the literature.[20,41,46,57]

Based on the WAXD spectra displayed in Figure 13.13, the conversion indices for pure, non-UV treated, thermally stabilized samples are presented in Figure 13.14 (and Table 13.3) as a function of the thermal stabilization time. Also included is the temperature profile followed during these thermal stabilization experiments. The dashed lines indicate the conversion indices of the samples containing 1 wt% BDP UV treated at 150 °C for 300 s. The results show that 300 s (5 min) of UV treatment of the PAN-based precursor containing 1 wt% BDP are approximately equivalent to the first 5000 s (about 83 min) of the thermal oxidation process, i.e. a reduction of approximately

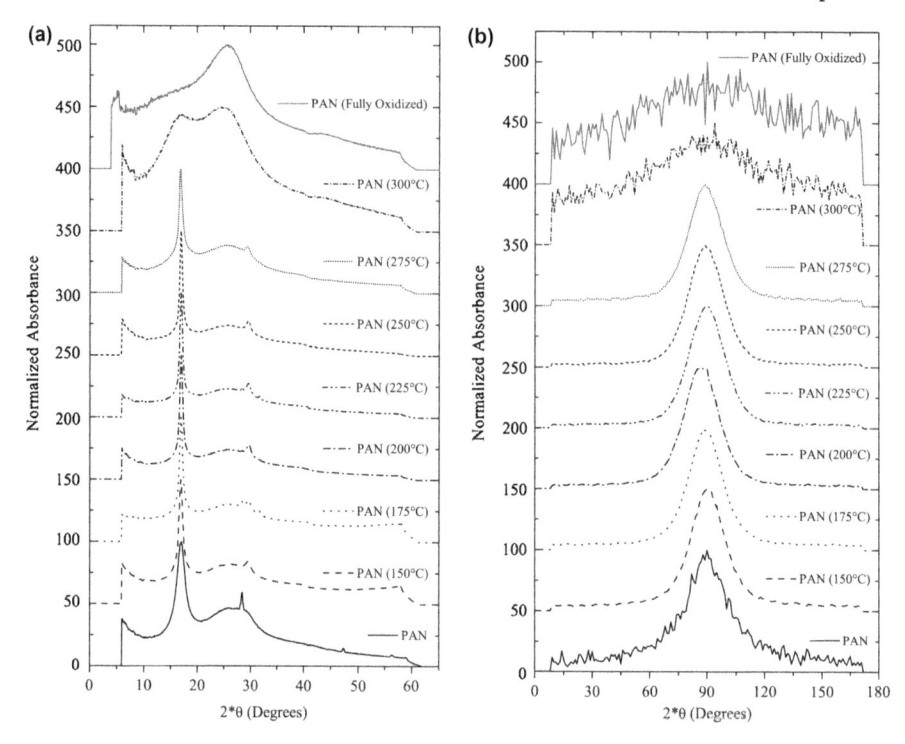

**Figure 13.13** WAXD spectra of pure PAN fibers at different stages during traditional thermo-oxidative stabilization: (a) integrated 2-theta scan; (b) azimuthal scan of the PAN (1 0 0) peak.
(Reproduced from Morales and Ogale.[34])

**Table 13.3** WAXD results with 95% confidence intervals of pure PAN fibers at different stages during traditional thermal stabilization in air. (Reproduced from Morales and Ogale.[34])

| Fiber Temperature (°C) | $l_a$ (Å) | $a$ (Å) | (1 0 0) FWHM (°) | C. I.% |
|---|---|---|---|---|
| 25 | 42 ± 2 | 6.001 ± 0.020 | 26.1 ± 0.9 | 1 ± 2 |
| 150 | 90 ± 2 | 6.025 ± 0.013 | 20.3 ± 1.1 | 1 ± 2 |
| 175 | 107 ± 2 | 6.012 ± 0.020 | 20.1 ± 0.9 | 2 ± 2 |
| 200 | 119 ± 2 | 6.001 ± 0.020 | 21.2 ± 1.0 | 2 ± 2 |
| 225 | 126 ± 4 | 6.003 ± 0.018 | 22.1 ± 0.9 | 1 ± 2 |
| 250 | 134 ± 4 | 6.009 ± 0.017 | 21.3 ± 0.9 | 16 ± 8 |
| 275 | 98 ± 2 | 6.040 ± 0.020 | 21.3 ± 1.2 | 37 ± 5 |
| 300 | 13 ± 1 | 6.198 ± 0.124 | 79.6 ± 5.5 | 52 ± 14 |
| 300 (after 30 min) | 10 ± 1 | 6.438 ± 0.049 | 100.4 ± 2.1 | 70 ± 2 |
| 1 wt% BDP, 300 s Hg UV, 150 °C | 46 ± 2 | 6.036 ± 0.016 | 25.6 ± 1.1 | 14 ± 2 |

half of the time needed for conventional stabilization process. This correlation between conversion index and processing time is consistent with that obtained from our previous study using FTIR spectra.[33] The other conversion

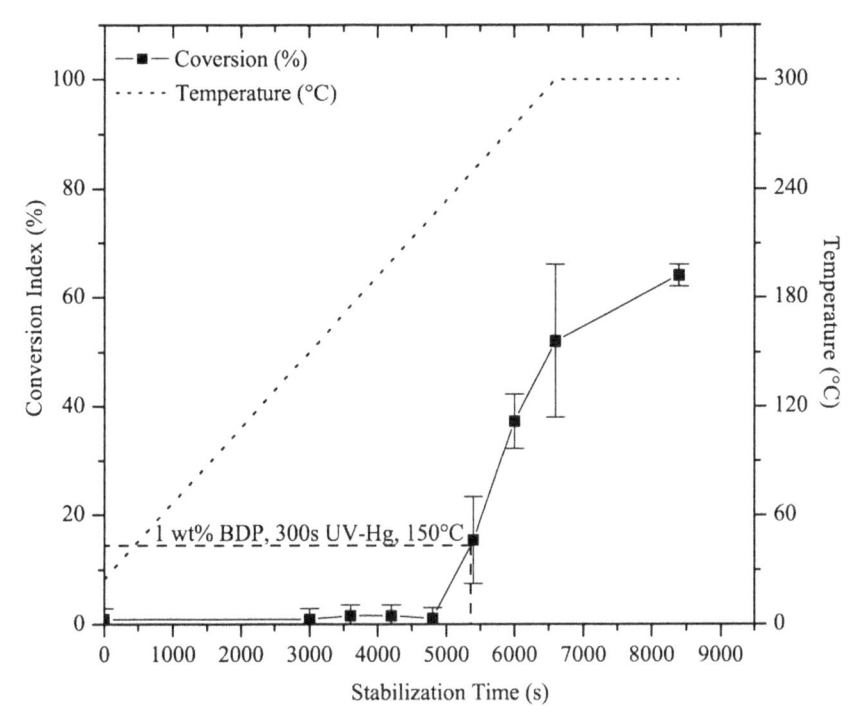

**Figure 13.14** Correlation between stabilization temperature and conversion indices (with trend line) of pure PAN fibers during the thermal stabilization step.
(Reproduced from Morales and Ogale.[34])

indices for UV-treated samples in the absence of photoinitiator were not statistically different from the initial indices for pure PAN samples thermally stabilized, and are not included in the figure. These kinetic results indicate the potential for developing a novel and more rapid process for stabilization of PAN-based precursors to produce carbon fibers more efficiently.

## 13.3.4 Influence of UV Radiation and Photoinitiator on Carbon Fibers

Figure 13.15 displays representative SEM micrographs of carbonized PAN fibers with and without photoinitiator. Figure 13.15(a) and (b) correspond to pure PAN fibers and those containing 1 wt% BDP, respectively. Figure 13.15(c) and (d) are representative micrographs of pure and 1 wt% BDP-PAN fibers, both UV treated with the mercury source, respectively. These first four specimens were conventionally thermal stabilized. In contrast, Figure 13.15(e) and (f) correspond to fast-thermally stabilized non-UV-treated pure and UV-treated 1 wt% BDP-PAN fibers, respectively. In all cases the fibers retained the characteristic kidney shape of wet-spun

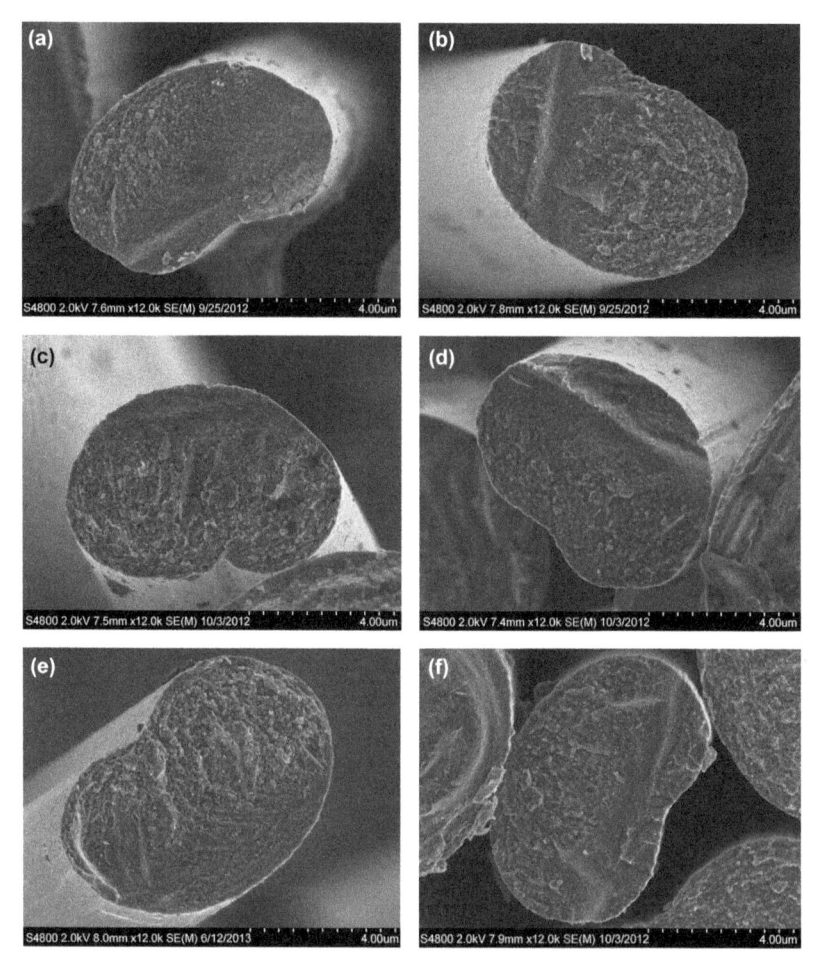

**Figure 13.15** Representative SEM micrographs of carbonized PAN fibers with and without photoinitiator: (a) pure PAN fibers, (b) fibers containing 1 wt% BDP, (c) pure UV-treated PAN fibers, (d) 1 wt% BDP PAN UV-treated fibers, (e) fast-thermal stabilized pure PAN fibers, and (f) fast-thermal stabilized UV-treated 1 wt% BDP PAN fibers.
(Reproduced from Morales and Ogale.[35])

PAN-based fibers.[6,45,62] The effective diameter of pure and 1 wt% BDP PAN fibers were $6.6 \pm 0.1$ and $6.4 \pm 0.1$ μm, respectively. At 95% confidence level, there was no significant difference between the effective diameters among the different samples.

No noticeable deterioration or change in the microstructure was observed on UV-treated fibers containing photoinitiator when compared with pure control fibers. Thus, the presence of photoinitiator and UV treatment did not significantly affect the morphology of the fibers. Another important observation is that the UV treatment of the fibers did not lead to any obvious skin–core structure formation. However, this is not the case for fast-thermally

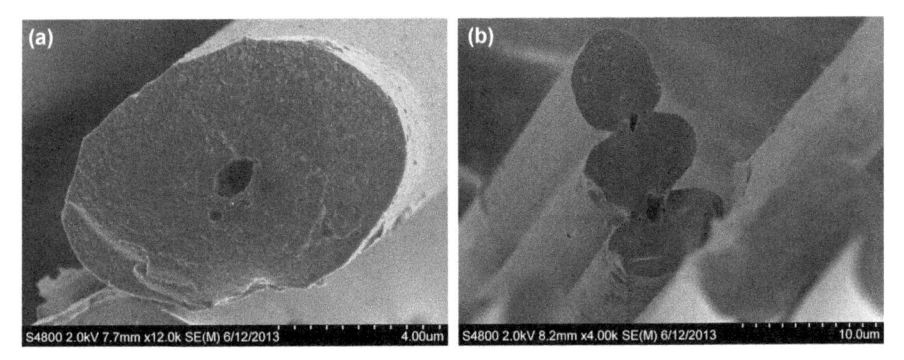

*UV-Based Dual Mechanism for Crosslinking and Stabilization* 283

**Figure 13.16** SEM micrographs of defects observed on fast-thermal stabilized pure PAN carbon fibers attributed to inadequate thermal stabilization: (a) core-hollow fiber, (b) fused hollow fibers. (Reproduced from Morales and Ogale.[35])

stabilized pure fibers. Figure 13.16 shows SEM micrographs of defects observed on fast-thermal-stabilized pure PAN carbon fibers attributed to inadequate thermal stabilization. Figure 13.16(a) and (b) display a core-hollow fiber and fused hollow fibers, respectively. Note that these core-hollow fiber defects were not observed in any of the other sets of samples. This demonstrates skin-core structure formation in fast-thermally stabilized pure PAN fibers. As mentioned before, skin-core structure is undesired during the stabilization of PAN-based fibers because it leads to poor quality (mechanical properties of the) stabilized and carbonized fibers.[11,12,46,54]

### 13.3.5 Mechanical Properties

Figure 13.17 displays representative tensile responses of various carbonized fibers. Figure 13.17(a) corresponds to conventionally thermal stabilized non-UV-treated pure and UV-treated pure as well as fast-thermal stabilized pure PAN fibers. Figure 13.17(b) shows the same treatment conditions but for fibers containing 1 wt% BDP. Table 13.4 summarizes the tensile-testing results conducted on single filaments. These tensile properties of the PAN fibers are consistent with those reported in the literature for commercial and experimental grade PAN-based fibers: 0.3–2.5% for the breaking strain, 0.1–6.5 GPa for strength, and 30–500 GPa for tensile modulus.[2,6,9,15,30,54]

Tensile-testing results shown in Table 13.4 indicate that, at 95% confidence, there are significant differences in the breaking strain and ultimate tensile strength between pure control and pure UV-treated fibers. The UV treatment reduces the elongation capabilities of the pure fibers and the ultimate tensile strength by approximately one third. These reductions could be possibly attributed to the excessive fiber fusion observed on pure UV-treated fibers leading to defects on these fibers. As mentioned before, it is suspected that the main phenomena taking place during the UV treatment at temperatures slightly above the glass transition of the PAN precursor

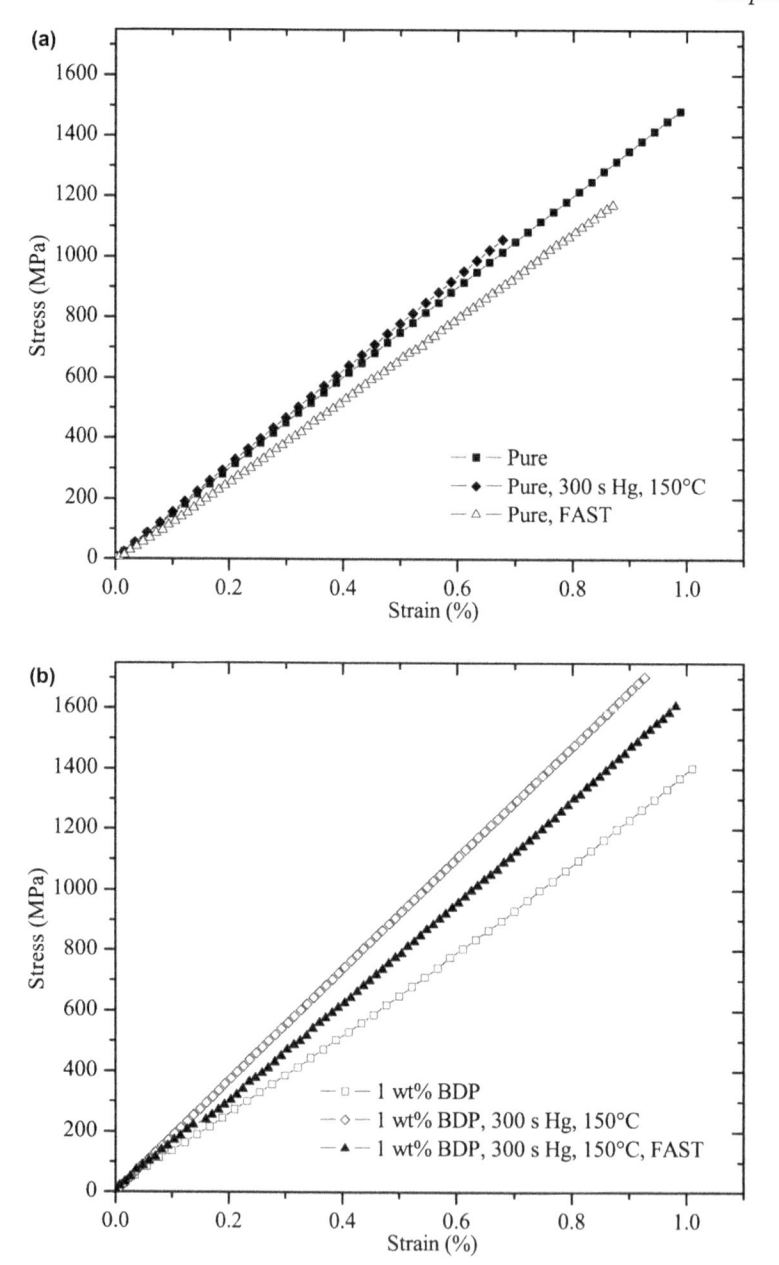

**Figure 13.17** Representative tensile-testing curves of each set of carbonized fibers: (a) pure PAN, (b) PAN fibers containing 1 wt% BDP. (Reproduced from Morales and Ogale.[35])

without photoinitiator are fiber fusion and polymer chain relaxation leading to a significant reduction in the elongation capabilities and ultimate tensile strength of the fibers.

**Table 13.4** Single-filament tensile results of the produced carbon fibers (CF) with 95% confidence intervals. (Reproduced from Morales and Ogale.[35])

| Sample | Tensile modulus (GPa) | Max stress (GPa) | Break strain (%) |
| --- | --- | --- | --- |
| Pure | 153.5 ± 6.6 | 1.47 ± 0.17 | 1.0 ± 0.1 |
| 1 wt% BDP | 150.2 ± 10.1 | 1.40 ± 0.20 | 0.9 ± 0.1 |
| Pure, 300 s Hg, 150 °C | 157.6 ± 6.2 | 1.05 ± 0.13 | 0.7 ± 0.1 |
| 1 wt% BDP, 300 s Hg, 150 °C | 184.8 ± 10.4 | 1.65 ± 0.12 | 0.9 ± 0.1 |
| Pure, FAST | 137.9 ± 4.8 | 1.22 ± 0.18 | 0.9 ± 0.1 |
| 1 wt% BDP, 300 s Hg, 150 °C, FAST | 170.6 ± 6.0 | 1.60 ± 0.11 | 1.0 ± 0.1 |

Pure fast-thermal stabilized fibers also show a significant reduction in the tensile modulus of the fibers by ∼10% and the ultimate tensile strength by ∼20% (at 95% confidence) when compared with pure control fibers. For these specimens considerable fiber fusion is observed as well, similar to that observed on pure UV-treated fibers. Fast heating rates during thermal stabilization lead to the formation of skin-core structure in the fibers confirmed by core-hollow fibers (Figure 13.16) which is not desired because it is harmful and negatively affects the mechanical properties of the produced fibers, as indicated by the tensile-testing results.

For UV-treated fibers containing photoinitiator, the results are significantly different, the two set of fibers containing 1 wt% BDP UV treated for 300 s (5 min) display higher tensile modulus and ultimate tensile strength than all the other set of samples. Specifically, conventionally thermal stabilized UV-treated fibers containing 1 wt% BDP exhibit the highest tensile modulus of all six sets of samples (at 95% confidence). An improvement in the tensile modulus of approximately 20% was observed in conventional thermal-stabilized UV-treated fibers containing 1 wt% BDP when compared with conventional thermal-stabilized control pure PAN fibers. The addition of photoinitiator and further UV treatment of the fibers increase the mechanical properties of the fibers during conventional thermal stabilization.

UV-treated fast-thermal-stabilized fibers containing 1 wt% BDP exhibit higher tensile modulus and ultimate tensile strength when compared with conventional and fast-thermal stabilized pure fibers. UV-treated fast-thermal stabilized fibers containing 1 wt% BDP show around 10%, 15%, and 25% improvement in the tensile modulus when compared with conventional thermal stabilized pure, conventional thermal stabilized 1 wt% BDP fibers, and fast-thermal stabilized pure control fibers, respectively. In addition, UV-treated fast-thermal stabilized fibers containing 1 wt% BDP display approximately 33% improvement in the ultimate tensile strength when compared with fast-thermal stabilized pure control fibers. It is important to note that, besides the demonstrated improvement in the mechanical properties, the fast-thermal stabilization step takes less than half the duration of the conventional thermal stabilization (60 vs. 140 min, respectively). This proves the combined positive effect of the addition of photoinitiator and

300 s of UV treatment on the fibers. The addition of photoinitiator and further UV treatment of the fibers could be used to increase the mechanical properties of the fiber during conventional thermal stabilization or reduce the thermal oxidation time (faster fiber production) while retaining the mechanical properties of the carbon fibers thus produced.

## 13.4 Conclusions

The effect of UV photoinitiators and UV radiation on photoinduced stabilization of polyacrylonitrile was investigated. Hydrogen abstraction was found to be a more efficient mechanism in comparison with homolytic cleavage during the photoinduced cyclization of PAN precursor polymer. FTIR and DSC results confirm that samples containing 4,4'-bis(diethylamino)benzophenone (hydrogen-abstraction type) achieved higher extents of cyclization than did samples containing 1-hydroxycyclohexyl phenyl ketone (hemolytic-cleavage type). Also, the presence of 4,4'-bis(diethylamino)benzophenone enables higher extents of post-UV thermal cyclization. The presence of photoinitiator is more effective in promoting carbon–nitrogen double-bond formation for samples UV treated above the $T_g$ of the precursor in comparison with UV-treated samples without photoinitiator and samples UV treated below $T_g$ where carbon-carbon double-bond formation was more favorable. At temperatures below the $T_g$ of the precursor, the presence of photoinitiator does not play a significant role on the extent of the cyclization due to the limited mobility of the polymer chains. Increasing UV exposure time leads to greater formation and higher FTIR conversion index in the double-bond region. The present results confirm that it is possible to produce UV-induced cyclization in PAN-based precursors by the external addition of photoinitiator into the polymer solution (*i.e.* without necessarily incorporating photoinitiators into the polymer chain itself). Conversion indices calculated from FTIR and WAXD spectra prove that the addition of 1 wt% photoinitiator to PAN and UV treatment for only 5 min increase the rate of the cyclization reaction and reduce the thermal oxidation time by over an hour, which could significantly reduce the conventional stabilization time by half.

PAN fibers containing 1 wt% photoinitiator showed neither deterioration nor significant morphological differences as compared to pure fibers. Further UV treatment for a short period of time (5 min) of the PAN fibers with 1 wt% photoinitiator led to enhancement of the tensile modulus. Tensile-testing results indicated that fibers containing photoinitiator show higher tensile modulus for all set of samples, confirming the positive influence of photoinitiator and further UV treatment in the fibers.

After UV treatment, precursor fibers could be rapidly thermo-oxidatively stabilized and successfully carbonized. SEM micrographs show no noticeable deterioration, skin-core structure, or change in the microstructure of the UV-treated fibers containing photoinitiator after thermal stabilization and carbonization. On the other hand, fast-thermal-stabilized pure PAN carbon fibers show core-hollow fiber defects attributed to inadequate thermal stabilization. Note that these core-hollow fiber defects were not observed in any of the other set of samples.

Tensile-testing results show fibers containing 1 wt% photoinitiator UV treated for 300 s (5 min) display higher tensile modulus and ultimate tensile strength than all the other sets of thermal stabilized and carbonized fibers. Specifically, conventionally thermal stabilized UV-treated fibers containing 1 wt% BDP exhibit the highest tensile modulus and ultimate tensile strength of all six sets of thermal-stabilized and carbonized fibers. This was not the case for pure UV-treated and fast-thermal-stabilized pure PAN fibers that display a reduction in their mechanical properties when compared with control pure conventionally thermal stabilized fibers. These reductions were attributed to the lost molecular orientation and excessive fiber fusion leading to defect formation. Fast thermal stabilization of pure PAN precursors leads to excessive degradation of the PAN precursor, formation of skin-core structure within the fibers, and reduction of the mechanical properties of the fibers. On the other hand, no negative effects were observed on the UV-treated fibers containing photoinitiator as compared with all the other types of carbon fibers.

The results obtained in this study prove the combined positive effect of the addition of photoinitiator and UV treatment on the fibers as prethermal stabilization on the properties of the PAN-based stabilized and carbon fibers. Under the same thermal treatment conditions, it was possible to improve the mechanical properties of the thermal stabilized and carbonized fibers. In addition, with this novel process, it was possible to reduce the thermal stabilization time by more than half while maintaining the final mechanical and physical properties of the carbon fibers. This can be used in the development of a novel and more rapid process for stabilization of PAN-based precursors to produce carbon fibers more efficiently.

## Acknowledgements

As a subcontractor to Cytec Carbon Fibers, LLC under AFRL Prime Contract No. FA8650-05-D-5807 (AFRL/RXBT Program Monitor: Dr Karla Strong), Clemson University received partial financial assistance to support the initial phase of this overall project, which is gratefully acknowledged.

## References

1. W. Zhang and M. Li, *J. Mater. Sci. Technol.*, 2005, **21**, 581.
2. H. A. Katzman, P. M. Adam, T. D. Le and C. S. Hemminger, *Carbon*, 1994, **32**, 379.
3. M. S. Reisch, *Chem. Eng. News*, 2011, **89**, 10.
4. J. D. Buckley, in *Carbon-Carbon Materials and Composites*, ed. J. D. Buckley and D. D. Edie, National Aeronautics and Space Administration, Washington, DC, 1992, pp. 1–17.

5. R. J. Diefendorf, *Engineered Materials Handbook, Vol. 1: Composites*, ASM International, 1987, p. 49.
6. D. D. Edie and R. J. Diefendorf, in *Carbon-Carbon Materials and Composites*, ed. J. D. Buckley and D. D. Edie, National Aeronautics and Space Administration, Washington, DC, 1992, pp. 19–39.
7. A. J. Clarke and J. E. Bailey, *Nature*, 1973, **243**, 146.
8. A. K. Gupta, D. K. Paliwal and P. Bajaj, *J. Macromol. Sci., Rev. Macromol. Chem. Phys.*, 1991, **C31**, 1.
9. P. H. Wang, Z. R. Yue, R. Y. Li and J. Liu, *J. Appl. Polym. Sci.*, 1995, **56**, 289.
10. J. Hou, X. Wang and L. Zhang, *Appl. Phys. Lett.*, 2006, **89**, 152504.
11. Y. Hou, T. Sun, H. Wang and D. Wu, *Text. Res. J.*, 2008, **78**, 806.
12. Y. Hou, T. Sun, H. Wang and D. Wu, *J. Appl. Polym. Sci.*, 2008, **108**, 3990.
13. M. Yu, Y. Bai, C. Wang, Y. Xu and P. Guo, *Mater. Lett.*, 2007, **61**, 2292.
14. D. C. Gupta, R. C. Sharma and J. P. Agrawal, *J. Appl. Polym. Sci.*, 1989, **38**, 265.
15. M. C. Paiva, P. Kotasthane, D. D. Edie and A. A. Ogale, *Carbon*, 2003, **41**, 1399.
16. J. Mittal, O. P. Bahl and R. B. Mathur, *Carbon*, 1997, **35**, 1196.
17. Y. Wang, C. Wang, Y. Bai and Z. Bo, *J. Appl. Polym. Sci.*, 2007, **104**, 1026.
18. L. Tan, J. Pan and A. Wan, *Colloid Polym. Sci.*, 2012, **290**, 289.
19. M. Yu, C. Wang, Y. Bai, Y. Wang and B. Zhu, *J. Appl. Polym. Sci.*, 2006, **102**, 5500.
20. G. Wu, C. Lu, L. Ling, A. Hao and F. He, *J. Appl. Polym. Sci.*, 2005, **96**, 1029.
21. P. H. Wang, *J. Appl. Polym. Sci.*, 1998, **67**, 1185.
22. A. Singh; J. Silverman, *Radiation Processing of Polymers*, Hanser Publisher, Munich, Germany, 1992.
23. I. W. Boyd and J. Y. Zhang, *Nucl. Instrum. Methods Phys. Res., Sect. B*, 1997, **121**, 349.
24. A. Gupta and A. A. Ogale, *Polym. Compos.*, 2002, **23**, 1162.
25. S. M. Badawy and A. M. Dessouki, *J. Phys. Chem. B*, 2003, **107**, 11273.
26. A. Endruweit, M. S. Johnson and A. C. Long, *Polym. Compos.*, 2006, **27**, 119.
27. A. Ledwith, in *Photochemistry and Polymeric Systems*, ed. J. M. Kelly; C. B. Mc Ardle and M. J. de F. Maunder, Royal Society of Chemistry, Cambridge, 1993, pp. 1–14.
28. J. P. Fouassier, *Prog. Org. Coat.*, 1990, **18**, 229.
29. J. Fouassier, D. Burr and F. Wieder, *J. Polym. Sci. Part A*, 1991, **29**, 1319.
30. T. Mukundan, V. A. Bhanu, K. B. Wiles, H. Johnson, M. Bortner, D. G. Baird, A. K. Naskar, A. A. Ogale, D. D. Edie and J. E. McGrath, *Polymer*, 2006, **47**, 4163.
31. A. K. Naskar, R. A. Walker, S. Proulx, D. D. Edie and A. A. Ogale, *Carbon*, 2005, **43**, 1065.
32. C. Decker, in *Photochemistry and Polymeric System*, ed. J. M. Kelly; C. B. McArdle and M. J. de F. Maunder, Royal Society of Chemistry, Cambridge, 1993, pp. 32–46.

33. M. S. Morales and A. A. Ogale, *J. Appl. Polym. Sci.*, 2013, **128**, 2081.
34. M. S. Morales and A. A. Ogale, *J. Appl. Polym. Sci.*, 2013, **130**, 2494.
35. M. S. Morales and A. A. Ogale, *J. Appl. Polym. Sci.*, 2014, **131**(16), DOI: 10.1002/app.40623.
36. F. Khan, *Biomacromolecules*, 2004, **5**, 1078.
37. H. Kubota and Y. Ogiwara, *J. Appl. Polym. Sci.*, 1982, **27**, 2683.
38. I. Shimada, T. Takahagi, M. Fukuhara, K. Morita and A. Ishitani, *J. Polym. Sci. Part A*, 1986, **24**, 1989.
39. T. Sun, Y. Hou and H. Wang, *J. Appl. Polym. Sci.*, 2010, **118**, 462.
40. E. Fitzer and D. J. Müller, *Carbon*, 1975, **13**, 63.
41. Y. Zhu, M. A. Wilding and S. K. Mukhopadhyay, *J. Mater. Sci.*, 1996, **31**, 3831.
42. S. Dalton, F. Heatley and P. M. Budd, *Polymer*, 1999, **40**, 5531.
43. B. D. Cullity, *Elements of X-Ray Diffraction*, Addison-Wesley Publishing Company, Inc., Reading, MA, 1978.
44. X. Dong, C. Wang and C. Juan, *Polym. Bull.*, 2007, **58**, 1005.
45. X. Dong, C. Wang, Y. Bai and W. Cao, *J. Appl. Polym. Sci.*, 2007, **105**, 1221.
46. M. Yu, C. Wang, Y. Bai, Y. Wang and Y. Xu, *Polym. Bull.*, 2006, **57**, 757.
47. G. Peng, X. Zhang, Y. Wen, Y. Yang and L. Liu, *J. Macromol. Sci., Part B: Phys.*, 2008, **47**, 1130.
48. Z. Bashir, *J. Macromol. Sci., Part B: Phys.*, 2001, **40 B**, 41.
49. X.-P. Hu, *J. Appl. Polym. Sci.*, 1996, **62**, 1925.
50. T. Ko, H. Ting and C. Lin, *J. Appl. Polym. Sci.*, 1988, **35**, 631.
51. A. S. Newton, in *Radiation Effects on Organic Materials*, ed. R. O. Bolt and J. G. Carroll, Academic Press, New York, NY, 1963, pp. 35–62.
52. K. L. Hall, R. O. Bolt and J. G. Carroll, in *Radiation Effects on Organic Materials*, ed. R. O. Bolt and J. G. Carroll, Academic Press, New York, NY, 1963, pp. 63–125.
53. Y. Hou, T. Sun, H. Wang and D. Wu, *J. Appl. Polym. Sci.*, 2009, **114**, 3668.
54. L. Jie and Z. Wangxi, *J. Appl. Polym. Sci.*, 2005, **97**, 2047.
55. A. Shiedlin, G. Marom and A. Zilkha, *Polymer*, 1985, **26**, 447.
56. T. Mikolajczyk, G. Szparaga, M. Bogun, A. Fraczek-Szczypta and S. Blazewicz, *J. Appl. Polym. Sci.*, 2010, **115**, 3628.
57. P. Wang, J. Liu, Z. Yuf and R. Li, *Carbon*, 1992, **30**, 113.
58. M. Yu, C. Wang, Y. Bai, B. Zhu, M. Ji and Y. Xu, *J. Polym. Sci. Part B*, 2008, **46**, 759.
59. V. K. Matta, R. B. Mathur, O. P. Bahl and K. C. Nagpal, *Carbon*, 1990, **28**, 241.
60. J. Tsai and H. Hsu, *J. Mater. Sci. Lett.*, 1992, **11**, 1403.
61. B. Sahraoui, I. V. Kityk, J. Kasperczyk, M. Salle and T. T. Nguyen, *Opt. Commun.*, 2000, **176**, 503.
62. A. Sedghi, R. E. Farsani and A. Shokuhfar, *J. Mater. Process. Technol.*, 2008, **198**, 60.

CHAPTER 14

# Analytical Methods for Determining Photoinitiators in Food-Contact Materials

M. A. LAGO, R. SENDÓN AND
A. RODRÍGUEZ-BERNALDO DE QUIRÓS*

Department of Analytical Chemistry, Nutrition and Food Science, Faculty of Pharmacy, University of Santiago de Compostela, E-15782, Spain
*Email: ana.rodriguez.bernaldo@usc.es

## 14.1 Introduction

In the past, the main function of food packaging was as a container. Nowadays, the main function of packaging is to protect food and preserve its quality.[1,2]

In this context, food packaging must form an inert barrier to prevent transfer of mass to the foodstuff from outside. However, mass transfer may also occur from the food-contact material (FCM) to the foodstuff, representing a possible hazard to human health.[3–5]

The packaging sector represents about 2% of the gross national product (GNP) in developed countries, and half of the packaging produced is used for foodstuffs.[6] Any problems associated with the FCM could therefore have serious effects on public health and national economies. The use of printing inks on the external face of food packaging is of particular concern, as many of these are UV-curable inks that contain photoinitiators.

The first reported case of migration of photoinitiators to foodstuffs occurred in September 2005.[7] The Italian authorities detected isopropylthioxanthone (ITX) in baby milk from Spain. Since then, the European Union's Rapid Alert System for Food and Feed (RASFF), (a network involving the European Commission's Directorate-General for Health and Consumer Protection, the European Food Safety Authority [EFSA], the EFTA Surveillance Authority [ESA] and the members states) has reported another 144 notifications (44 alerts and 100 informations) until the end of 2013.[8]

A brief glance at these 144 notifications shows the following: 12 different photoinitiators and one amine synergist were detected in foodstuffs; the notifications originated in 16 different countries; the migration occurred from/through different types of packaging to reach many different kinds of foodstuffs; and finally, not all the products were seized or withdrawn from the markets. The data gathered by the RASFF indicate that there are many different factors involved in the problem of migration of photoinitiators from FCM to foods, making this problem difficult to address.

The European Food Safety Agency (EFSA) considered the problem of photoinitiators in an external scientific report (ESCO)[9] compiled jointly by the food industry and food safety agencies. This report formed the basis of future legislation, and the potential toxicity of some photoinitiators was determined by the Cramer classification,[10] whereby the photoinitiators and other molecules are classified as of potentially high, medium or low toxicity on the basis of their molecular structure.

Some researchers have studied how photoinitiators can reach the foodstuffs and have identified two main routes:[11,12] (a) direct transfer by permeation from the outer printed surface through different layer(s) of packaging, or by offset (transfer of the photoinitiators during storage of the printed material in reels or stacks),[13,14] and (b) indirect transfer by gas phase, for volatile photoinitiators such as benzophenone and derivatives, see Figure 14.1.[15,16]

However, it is not only the starting substances that can migrate to the foodstuffs and contaminate them. Depending on the initial photoinitiator(s) involved, different byproducts (photolysis products and radicals) can be

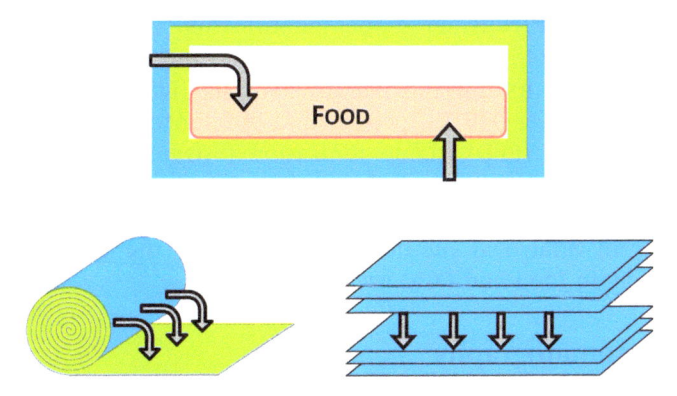

**Figure 14.1** The upper scheme illustrates the indirect transfer and direct transfer of photoinitiators from the external layer of the packaging to the food. In the lower scheme, the set-off effect in reels and stacks is represented.

generated by photopolymerization and photocuring reactions. Foodstuffs can also be contaminated with these products, in a complex process about which very little is known, which makes it very difficult to evaluate the hazards associated with these unknown substances.

The above-mentioned byproducts are included in the so-called non-intentionally added substances (NIAS), which are defined as "impurities in the substances used or reaction intermediates formed during the production process or decomposition or reaction products which can occur in the final product" by European Commission Regulation (EU) No 10/2011 of 14 January 2011 on plastic materials and articles intended to come into contact with food.[17]

Although there is no standard procedure for the determination of photoinitiators and NIAS in FCM or foodstuffs, some authors have developed methods for the detection and/or quantification of one or more photoinitiators. To date, methods have been developed for determining 28 photoinitiators, of which only 19 have been declared as suitable for use in coatings, inks and varnishes for the noncontact side of food packaging by the European Printing Ink Association (EuPIA).[18] This EuPIA list, see Table 14.1, contains more than 100 photoinitiators declared as appropriated for the mentioned uses.

Determination of NIAS is a difficult task, mainly because these products are unknown. The reactions that occur in the process of photopolymerization in FCM have not been widely studied, and it is not generally known which molecules are involved. Thus, the first step required is to identify and quantify the photolysis products generated by each photoinitiator or combination of photoinitiators/synergists.[19]

In accordance with this scenario, the different techniques for determining photoinitiators and their derivatives in FCM and foodstuffs are discussed in this chapter.

## 14.2 Photoinitiators

Several studies have considered the identification and/or quantification of one or more photoinitiators. The technique most commonly used is chromatography, in its different variants: liquid (LC), gas (GC) or high-performance thin-layer (HPTLC), coupled to different detectors: diode array (DAD), fluorescence (FLD), flame ionization (FID), mass (MS), or some combination of these. Other techniques such as voltammetric methods or, the most novel, direct analysis in real time (DART) are also used.

The different analytical methods developed for the determination of photoinitiators in food and FCM are described below.

### 14.2.1 Chromatography

Selection of liquid or gas chromatography should be based on the nature and the properties of the target photoinitiators, *i.e.* liquid chromatography is usually most suitable for photoinitiators of low volatility.

Analytical Methods for Determining Photoinitiators in Food-Contact Materials 293

Table 14.1 Photoinitiators included in the EuPIA list and the analytical methods developed for their determination and quantification.

| Structure | Name | PI type | Mw | CAS Nr | Mp (°C) | Bp (°C) | Log P (o/w) | Water solubility (mg L$^{-1}$) (25 °C) (pH = 7) | Vapor pressure (mmHg) | SML (mg kg$^{-1}$) |
|---|---|---|---|---|---|---|---|---|---|---|
|  | Irgacure 184® | I | 204.26 | 947-19-3 | 48.00 | 339.00[a] | 2.18[a] | 1400.00[a] | 3.75×10$^{-5a}$ | — |
|  | Irgacure 369® | I | 366.50 | 119313-12-1 | — | 528.80[a] | 3.38[a] | 730.00[a] | 2.87×10$^{-11a}$ | 0.15[c] |
|  | Irgacure 651® | I | 256.30 | 24650-42-8 | 64.00 | 361.10[a] | 3.62[a] | 150.00[a] | 1.06×10$^{-05a}$ | — |
|  | Irgacure 907® | I | 279.40 | 71868-10-5 | — | 420.10[a] | 2.44[a] | 500.00[a] | 2.89×10$^{-07a}$ | 0.05[c] |

**Table 14.1** (Continued)

| Structure | Name | PI type | Mw | CAS Nr | Mp (°C) | Bp (°C) | Log P (o/w) | Water solubility (mg L$^{-1}$) (25 °C) (pH = 7) | Vapor pressure (mmHg) | SML (mg kg$^{-1}$) |
|---|---|---|---|---|---|---|---|---|---|---|
| | Irgacure 1173® | I | 164.20 | 7473-98-5 | 164.00 | 260.80$^a$ | 1.49$^a$ | 4400.00$^a$ | 6.08×10$^{-03a}$ | — |
| | Lucirin TPO® | I | 348.37 | 75980-60-8 | — | 519.00$^a$ | 4.62$^a$ | 11.00 | 6.71×10$^{-11a}$ | 0.05$^c$ |
| | ITX | II | 254.35 | 5495-84-1 | 76.75 | 398.20$^a$ | 5.11$^a$ | 28.00$^a$ | 1.43×10$^{-06a}$ | 0.05 |

## Analytical Methods for Determining Photoinitiators in Food-Contact Materials   295

| | | | | | | | | |
|---|---|---|---|---|---|---|---|---|
| DETX | II | 268.37 | 82799-44-8 | — | 427.90[a] | 5.67[a] | 13.00[a] | $1.58 \times 10^{-07a}$ | — |
| Benzo-phenone | II | 182.22 | 119-61-9 | 47.80 | 305.40 | 3.18 | 137.00 | $8.23 \times 10^{-04a}$ | 0.6[b] |
| 2-MBP | II | 196.24 | 131-58-8 | −18.00 | 308.00 | 3.69[a] | 71.00[a] | $6.17 \times 10^{-04a}$ | 0.05 |
| 3-MBP | II | 196.24 | 643-65-2 | 159.50 | 317.00 | 3.69[a] | 65.00[a] | $3.92 \times 10^{-04a}$ | 0.05 |
| 4-MBP | II | 196.24 | 134-84-9 | 59.50 | 328.10[a] | 3.69[a] | 65.00[a] | $1.94 \times 10^{-04a}$ | 0.05 |

**Table 14.1** (*Continued*)

| Structure | Name | PI type | Mw | CAS Nr | Mp (°C) | Bp (°C) | Log P (o/w) | Water solubility (mg L$^{-1}$) (25 °C) (pH = 7) | Vapor pressure (mmHg) | SML (mg kg$^{-1}$) |
|---|---|---|---|---|---|---|---|---|---|---|
| | DEAB | II | 324.46 | 90-93-7 | 96.00 | 475.70$^a$ | 5.91$^a$ | 0.39$^a$ | 3.25×10$^{-09a}$ | — |
| | MBB | II | 240.25 | 606-28-0 | 52.00 | 351.00 | 2.70 | 80.00 | 1.53×10$^{-05a}$ | 0.05 |
| | PBZ | II | 258.31 | 2128-93-0 | 102.00 | 419.50 | 5.14$^a$ | 3.90$^a$ | 3.11×10$^{-07a}$ | — |

| Structure | Abbreviation | Type | MW | CAS | | | | | | |
|---|---|---|---|---|---|---|---|---|---|---|
| | BIS | Cationic | 876.43 | 74227-35-3 | 235.00 | — | — | — | — | — |
| | THIO | Cationic | 516.51 | 8156-13-8 | — | — | — | — | — | — |
| | EDB | Amine synergist | 193.24 | 10287-53-3 | 63.50 | 296.50$^a$ | 2.51$^a$ | 410.00$^a$ | $1.43 \times 10^{-03a}$ | 0.05$^c$ |
| | EHA | Amine synergist | 277.40 | 21245-02-3 | 243.00 | 382.90$^a$ | 5.41$^a$ | 4.70$^a$ | $4.57 \times 10^{-06a}$ | 2.4 |

$^a$No data.
$^b$The sum of BP, 2-MBP, 3-MBP and 4-MBP.
$^c$Under re-evaluation.

## 14.2.1.1 Liquid Chromatography (LC)

This is the most commonly used technique for the determination of photoinitiators in foodstuffs and FCM.[12,14,16,20–39] The sum of these methods determines a total of 20 different photoinitiators of different kinds (cationic, type I, type II and even amine synergists).

The analytical methods can be classified into two main groups depending on the detector used: diode array or fluorescence detectors, and mass spectrometry. The techniques in the first group are those most extensively used for the identification and quantification of photoinitiators, and those in the second group are very useful for confirmation purposes.

**14.2.1.1.1 HPLC-DAD/FLD.** Before the scandal involving the presence of ITX in baby's milk,[7] very few studies involving the determination of photoinitiators had been reported.[20–22] Castle *et al.*[22] determined 3 different photoinitiators (Michler's ketone (MK), 4,4'-bis(diethylamino)-benzophenone (DEAB) and 4-(dimethylamino)benzophenone (DMAB)) in different paper and cardboard FCM, using ethanol as solvent in the extraction. The chromatographic separation was achieved with a Zorbax SC85 (Hichrom®) (i.d., 250 mm×4.6 mm and particle size, 5 µm) reversed-phase column with low carbon loading (6%), in isocratic mode (methanol:water 4:1). Detection was performed with DAD at 350 nm.

MK, DEAB and DMAB were detected in 10–26% of the samples at concentrations of respectively 0.05, 0.1 and 0.1 mg kg$^{-1}$, which are above the limits of detection (LOD), but unlikely to pose a risk to human health.[20]

In a later study, Papilloud and Baudraz[21,22] used a method involving solid-phase extraction (SPE) to determine the following compounds in food simulants: ITX, benzophenone (BP), bis(4-diphenylsulfonium)phenylsulfide-bis(hexafluorophosphate) (BIS); (diphenyl[(phenylthio) phenyl] sulfonium hexafluorophosphate) (THIO); 1-hydroxycyclohexyl-phenyl-ketone (Irgacure 184®), 2,2-dimethoxy-2-phenyl acetophenone (Irgacure 651®), methyl-1-(4-methylthio)phenyl-2-morpholinopropan-1-one (Irgacure 907®) and 2-ethylhexyl-4-dimethylamino benzoate (EHA). Chromatographic separation was performed with a reversed-phase column with cyano groups (CN Nucleosil 100-5; i.d., 250 mm×4.6 mm and particle size, 5 µm), in gradient mode and with a mobile phase composed of buffered water, methanol and acetonitrile. The results obtained indicated good sensitivity, with LOD and LOQ of µg kg$^{-1}$ order of magnitude.

After the above-mentioned scandal, many researchers focused on the health problem of photoinitiators in foodstuffs, and many analytical methods have since been published. Sanches-Silva *et al.*,[23–26] Pastorelli *et al.*,[16] Rodríguez-Bernaldo de Quirós[27] and Koivikko[28] developed some methods of identifying and quantifying the following photoinitiators in different FCM, foodstuffs and food simulants: (BP, ITX, EHA, DEAB, Irgacure 184®, 907®, 2-benzyl-2-dimethylamino-4-morpholino butyrophenone [Irgacure 369®], 4-methyl-benzophenone [4-MBP], 2-hydroxybenzophenone [2-HBP],

4-hydroxybenzophenone [4-HBP], methyl-2-benzoylbenzoate [MBB], 4-phenyl benzophenone [PBZ] and benzophenone acrylate [BPAcr]).

These authors used similar schemes to develop their methods. All achieved the chromatographic separation with a C18 column: Kromasil 100 (Teknocroma®) or Eclipse XDB (Agilent Technologies®) of length 15 cm, i.d. 4.6 mm and particle size 5 µm, except Rodriguez-Bernaldo de Quirós et al.[27] and Pastorelli et al.,[16] who used columns of length 25 cm. The mobile phase used was a gradient of acetonitrile and water.

The photoinitiators and amine synergist were detected with DAD at different wavelengths: 246 nm for Irgacure 184®, 254 or 256 nm for BP, 256 nm for Irgacure 651®, MBB, DEAB, BPAcr, 2-HBP and 4-HBP, 256 or 260 nm for 4-MBP, 290 nm for PBZ, 306 nm for Irgacure 907®, 310 nm for EHA and 386 or 256 nm for ITX.

Regarding the extraction procedure, single-solvent extraction with acetonitrile was used for FCM, while for the foodstuffs, the procedure depended on the nature of the samples. Thus, the single-solvent extraction procedure was used for samples such as powdered milk and cakes; for beverages, the samples were simply diluted with acetonitrile, and for milk (because of its acidic character), the extraction was carried out with ammoniac and two successive extractions with hexane. Samples of food simulants were injected directly into the HPLC system.

These methods showed low LODs of between 17–30 µg $L^{-1}$ for Irgacure 184®, 651®, 907®, ITX, BP and EHA and slightly higher values (46 µg $L^{-1}$) for 4-MBP and PBZ. For all samples, the LOQ was around 0.14 mg $L^{-1}$. Because of the nature of BPAcr, a polymeric photoinitiator, the chromatographic behavior was different; this compound yielded a sum of peaks rather than a single peak in the chromatograms, and the LOD and LOQ were therefore higher than those obtained for the other photoinitiators (LOD = 1.22 mg $L^{-1}$ and LOQ = 3.67 mg $L^{-1}$). Benzophenone was common in the FCM analyzed, and some other photoinitiators were also detected. Photoinitiators were detected in foodstuffs but to a lesser extent than in FCM.

More recently, Rothenbacher et al. developed an HPLC-DAD/FLD method for the determination of 2-ITX in different types of packaging and foodstuffs purchased in German markets.[14] The FCM extraction was carried out with hexafluoro-2-propanol and ethanol. However, for foodstuffs, the extraction was based on the "quick, easy, cheap, effective, rugged, and safe" (QuEChERS) approach.[40] The QuEChERS method is based on 4 steps (with shaking of the mixture between each): extraction with acetonitrile; addition of magnesium sulfate and sodium chloride; addition of the internal standard; and, finally, mixing of the supernatant with magnesium sulfate (to remove the residual water) and primary secondary amine (PSA) (to remove sugars and fatty acids).

The chromatographic method was carried out with a Supelcosil® LC-PAH (Agilent Technologies®) reversed-phase C18 column (i.d., 250 mm × 4.6 mm and particle size, 5 µm) and acetonitrile and water as mobile phase in isocratic mode. 2-ITX was detected with DAD, at 260 nm, and with FLD, at

273 nm (excitation) and 440 nm (emission). In this study, ITX was detected in 26% of the packages at concentrations higher than the LOD and LOQ (2 and 5 µg L$^{-1}$), but lower than those obtained by Sanches-Silva et al. in the previously mentioned studies.[23–26] Regarding the food, the concentrations of 2-ITX varied up to 357 mg kg$^{-1}$ in orange juice.

Jung et al. first developed an HPLC-DAD-FLD method[12] and more recently another HPLC-DAD method for determining different photoinitiators.[29] They used the first method to determine Irgacure 907®, ITX and ethyl-4-(dimethylamino)benzoate) (EDB) in yoghurt samples. The extraction procedure was the same as described by Rothenbacher et al.,[14] with slight differences to improve extraction. The analysis was performed with a gradient of acetonitrile and water as mobile phase, and a Luna C18 column (i.d., 100 mm×2 mm and particle size, 3 µm) was used as the stationary phase. Quantification was carried on with DAD, at 260 nm, and with FLD, at 264 and 440 nm as excitation and emission wavelengths. The limits of detection and quantification of the compounds in yoghurt were as follows: 0.4 and 2.5 µg kg$^{-1}$ for ITX; 3.5 and 10.7 µg kg$^{-1}$ for EDB, and 3.8 and 11.9 µg kg$^{-1}$ for Irgacure 907®. These data confirm better results for ITX than in the previous studies and are similar to those obtained for Irgacure 907®.

In a recent study, Jung et al. analyzed eleven photoinitiators (BP, 4-MBP, PBZ, Irgacure 184®, 4-HBP, DEAB, MK, EDB, EHA, DMAB and Methyl-2-benzoylbenzoate [MBB]) in cartonboard packaging, confirming the tendency of the development of multimethods for ever greater numbers of photoinitiators.[29] The method used to extract these photoinitiators from the FCM was that proposed by Pastorelli et al.[16]

Chromatographic separation was achieved with a reversed-phase C18 column (Purospher® Star RP 18e: i.d., 250 mm×4 mm and particle size, 5 µm) and a mobile phase composed by water and acetonitrile, in gradient mode. This method yielded good resolution, but the time required for the analysis was 70 min. The photoinitiators were detected at 245 nm (Irgacure 184® and MBB), 255 nm (BP and 4-MBP), 290 nm (4-HBP and PBZ), 310 nm (EHA and EDB) and finally at 365 nm (DEAB, MK and DMABP).

The results showed good sensitivity, with LODs and LOQs in the different cartonboards ranging from 2.8 µm dm$^{-2}$ (for BP) to 29.0 µm dm$^{-2}$ (for PBZ). The LOQs ranged from 9.4 µm dm$^{-2}$ (for BP) to 95.0 µm dm$^{-2}$ (for PBZ). BP was detected and quantified in 49% of the 310 samples analyzed, and the other photoinitiators, except 4-HBP, were detected in at least one of the samples. This data complemented and enhanced the survey of these photoinitiators initiated by Koivikko et al.,[28] in which only BP was detected in paper and cartonboard.

Only one other HPLC-DAD method for the determination of photoinitiators in beverages has been reported. Sagratini et al. determined 5 different photoinitiators (ITX, BP, EHA, EDB and Irgacure 184®) in beverages and packaging.[30] To determine these photoinitiators, the beverages were centrifuged and the resulting supernatant was extracted with n-hexane in the presence of Na$_2$SO$_4$ and then subjected to SPE.

The results, obtained with a Luna C18 reversed phase column (i.d., 250 mm×4.6 mm and particle size, 5 µm) and a mobile phase composed by methanol and water in gradient mode, showed LODs ranging between 20 and 50 µg L$^{-1}$ and LOQs between 100 and 300 µg L$^{-1}$, *i.e.* similar values to those obtained by other authors for the same matrices.[24]

In summary, on the basis of the methods reported in the literature reviewed, the following is a suitable and standardized HPLC-DAD method for the determination of photoinitiators in FCM and food: chromatographic separation with a general reversed-phase C18 column and a mobile phase composed by water and acetonitrile/methanol. By way of example, an HPLC-DAD chromatogram of fourteen photoinitiators is illustrated in Figure 14.2.

Regarding the extraction procedure for FCM, a single-step extraction with acetonitrile 24 h at 70 °C is appropriate. However, the procedure is more complex for foodstuffs, and several methods have been developed. The current tendency is to follow the QuEChERS approach, see Figure 14.3, with some variations depending on the complexity of the food matrix.

**14.2.1.1.2 HPLC-MS and HPLC-MS/MS.** These techniques can be used for confirmatory purposes or for both identification and confirmation. Previously, HPLC-MS was only used to confirm the results obtained by other methods because of the low availability of mass detectors; however, HPLC-MS and MS/MS are now commonly used to detect various substances.

Sanches-Silva *et al.* used HPLC-MS-TOF with electrospray ion source in positive mode (ESI+) as a confirmatory technique under the same chromatographic conditions as in the HPLC-DAD method,[23] although acidification of the mobile phases was required, in this case with formic acid (0.1% and 0.2%, respectively).

Confirmation was carried out by comparison of mass spectra of samples and standards. The most-abundant fragments were protonated molecules $[M+H]^+$ in the case of Irgacure 907® and EHA; a protonated molecule and a sodium adduct $[M+Na]^+$ for BP and ITX, only a sodium adduct for Irgacure 184®; and, finally, a sodium adduct and a molecule without a methoxy group $[M-OMe]^+$ for Irgacure 651®.

Benetti *et al.* used HPLC-MS (ESI +) to determine ITX in dairy products.[31] The extraction procedure was single-solvent extraction with acetonitrile, and the chromatographic separation was achieved with a Gemini reversed-phase C18 column (i.d., 100 mm×2.0 mm and particle size, 5 µm) (Phenomenex®) and a gradient of methanol and 20 mM ammonium formate.

This method showed a good detection capability (7.2 µg kg$^{-1}$). The protonated molecule $[M-H]^+$ $m\,z^{-1} = 255$ was selected as identification and quantification ion, and $m\,z^{-1} = 213$ was chosen as the confirmation ion. This method was tested with 50 samples of milk and yoghurt, and ITX was detected in 20% of the samples.

HPLC-MS and HPLC-MS/MS methods were developed and compared in three different studies.[29,30,32] Jung *et al.* first developed a HPLC-MS method

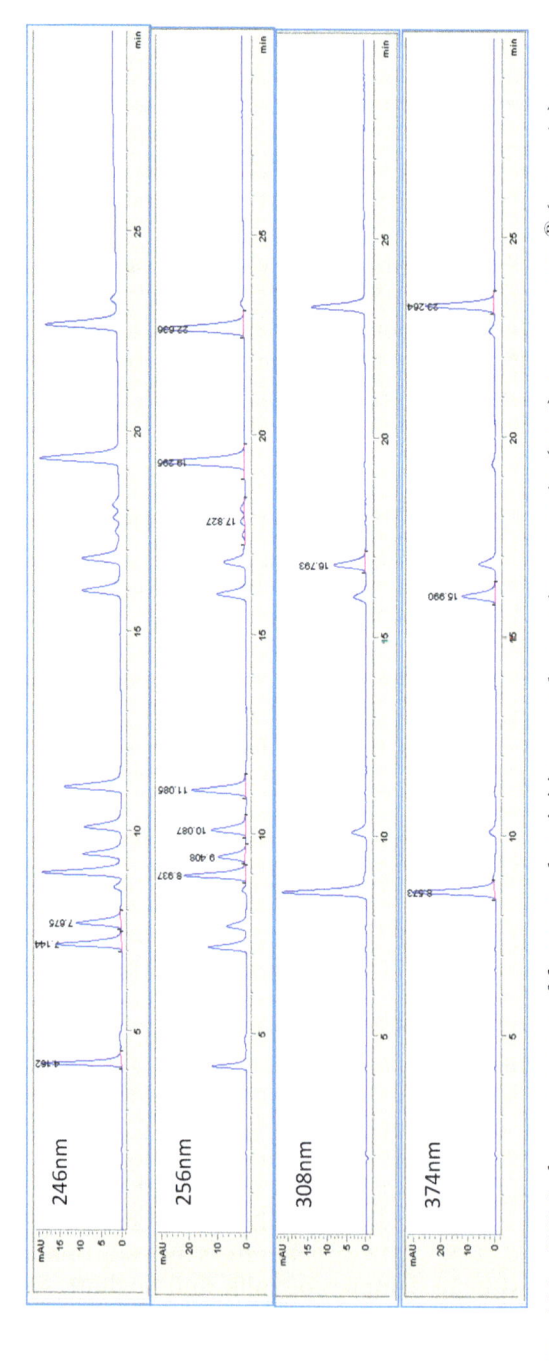

**Figure 14.2** HPLC chromatograms of fourteen photoinitiators and an amine synergist (EHA): Irgacure 1173® (4.16 min), Irgacure 184® (7.14 min), MBB (7.67 min), EDB (8.57 min), BP (8.94min), 2-HBP (9.41 min), Irgacure 651® (10.09 min), 4-MBP (11.09 min), PBZ (15.99 min), DEBP (16.79 min), BPAcr (17.83 min), ITX (19.29 min), DETX (22.69 min) and EHA (23.26 min).

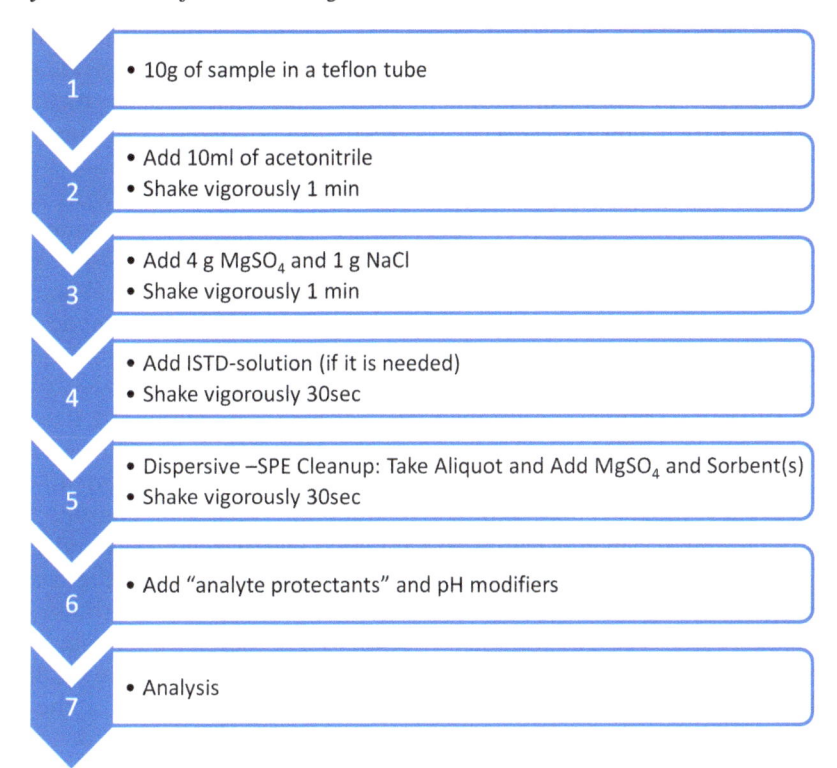

Analytical Methods for Determining Photoinitiators in Food-Contact Materials 303

```
1 • 10g of sample in a teflon tube

2 • Add 10ml of acetonitrile
  • Shake vigorously 1 min

3 • Add 4 g MgSO₄ and 1 g NaCl
  • Shake vigorously 1 min

4 • Add ISTD-solution (if it is needed)
  • Shake vigorously 30sec

5 • Dispersive –SPE Cleanup: Take Aliquot and Add MgSO₄ and Sorbent(s)
  • Shake vigorously 30sec

6 • Add "analyte protectants" and pH modifiers

7 • Analysis
```

**Figure 14.3** Steps included in the "quick, easy, cheap, effective, rugged, and safe" (QuEChERS) approach.[40]

with a single quadrupole and an ESI source in positive mode to determine photoinitiators in foodstuffs after extraction with acetonitrile.[29] For separation, these authors used a chromatographic Hypersil® GOLD reversed-phase column C18 (i.d., 100 mm×2.1 mm and particle size, 3 μm) (Thermo® Scientifics) and a gradient of methanol and water, both modified with 5 mM ammonium formate and 0.1% formic acid.

In this study, the protonated molecules $[M-H]^+$ were selected as quantifier ions and the qualifier ions were selected on the basis of the most-abundant fragments. The LODs ranged between 2.3 and 14 μm kg$^{-1}$, and LOQs ranged between 8.1 and 47 μg kg$^{-1}$. The BP limits were higher than for the other photoinitiators: 195 and 625 μg kg$^{-1}$. In the case of HPLC-MS/MS, the LODs and LOQs were lower: 2.5 and 7.5 μg kg$^{-1}$ for all the photoinitiators except BP, for which the values were again higher (38 and 113 μg kg$^{-1}$).

The HPLC-MS/MS method was performed with a triple quadrupole mass detector and the same source. The photoinitiators were separated on a Synergi® MAX-RP 80A reversed-phase C12 column (i.d., 150 mm×2.0 mm and particle size, 4 μm) (Thermo® Scientifics) and a gradient of methanol and water, both modified with 2 mM ammonium formate and 0.05% formic acid as mobile phase, in gradient mode. Most of the precursor and product ions were the same as the qualifier and quantifier ions in the HPLC-MS method based on the most abundant fragments. Both of these methods yielded positive results in 33% of the foodstuffs analyzed, and the legal limits were exceeded in five samples.[17,41–43]

Sagratini et al. developed three methods for determining ITX in juice samples – one with a single quadrupole mass detector, other with an ion trap and another one with a triple quadrupole detector. For all methods, they used a Luna reversed-phase C18 column (i.d., 150 mm×4.6 mm and particle size, 5 μm) and methanol and water as mobile phase, in isocratic mode. The ITX was extracted from fruit-juice samples with acetone/hexane (50:50) and pressurized-liquid extraction.[32]

For HPLC-MS (single quadrupole), the selected ion monitored was the sodium adduct of ITX. In the ion trap method, a transition was chosen from the protonated molecular ion. The same transition was selected for the HPLC-MS/MS method and another product ion was also chosen. The triple quadrupole detector exhibited very low LOD (0.01 μg L$^{-1}$) and LOQ (0.05 μg L$^{-1}$), and for single quadrupole and ion-trap detectors the corresponding values were 3 and 10 μg L$^{-1}$, respectively. With these methods, ITX was detected and quantified in eleven of the thirty samples analyzed.

Sagratini et al. reported some improvements to the above-mentioned methods.[30] In this study, an HPLC-MS method was developed for the determination of five photoinitiators in beverages, and HPLC-MS/MS was used as a confirmatory technique. The HPLC-MS method was basically the same as in the previous study,[32] except that the mobile phase was operated in gradient mode. The ions selected for determination were the sodium adducts formed with the different photoinitiators.

The results were confirmed by HPLC-MS/MS, and two different sources were tested: ESI and ambient-pressure photoionization (APPI). Chromatographic separation was achieved with the same parameters as in the HPLC-MS method, except for the mobile phases. For the ESI source, the mobile phase comprised methanol and water modified with 0.1% formic acid and 5 mM ammonium acetate in gradient mode, and for APPI, the mobile phase was methanol water, although in isocratic mode.

The results obtained showed that the sensitivity of the technique depended on the photoinitiator used, but that in general the LODs were lower (2–20 μg L$^{-1}$) with the HPLC-MS method. Better results were obtained with the APPI source than with ESI source.

Sun et al. developed an HPLC-MS/MS method for the determination of 2-ITX in beverages and packaging.[33] The equipment used was provided with a turbo ion-spray source operated in positive mode and a triple quadrupole/linear ion-trap mass detector. Chromatographic separation was achieved with two different reversed-phase C18 columns: BDS Hypersil (Thermo Scientifics®) (i.d., 100 mm×3 mm and particle size, 5 μm) and Inertsil ODS-3 (G.L. Sciences®) (i.d., 100 mm×3 mm and particle size, 3 μm), and methanol and 0.1% formic acid were used as mobile phase, in gradient mode.

The extraction procedure was carried out in FCM with acetonitrile; for the foodstuffs, extraction with acetonitrile and Carrez clarification were performed, followed by SPE. The MS parameters were the same as those reported by Sagratini et al.[32] This method showed good sensitivity, with LOD = 0.15 µg kg$^{-1}$ and LOQ = 0.50 µg kg$^{-1}$. Detectable amounts of ITX were found in all packaging, and in 7 of the 39 beverages analyzed.

Bagnati et al. developed another HPLC-MS/MS method for determining both isomers of ITX (2- and 4-ITX) in milk.[34] The extraction was carried out with acetonitrile. The chromatographic method was performed with a triple quadrupole mass detector, with hyperbolic rods, equipped with a turbo ionspray source operated in positive mode. Two columns were tested: Luna C8 (i.d., 50 mm×2 mm and particle size, 5 µm) (Phenomenex®), and a Discovery ZR-PS (i.d., 150 mm×2.1 mm and particle size, 3 µm ) (Sigma-Aldrich®), with zirconium-polystyrene as the bonded phase. The mobile phase was acetonitrile and 0.05% acetic acid, and gradient mode was used.

The mass parameters were the same as in the last two methods for the determination of ITX. Under those conditions, the method yielded a lower LOQ with the Luna C8 column (2.5 µg L$^{-1}$) but only Discovery ZR-PS yielded acceptable separation of the two isomers. The results showed 43% of positive cases in the milk samples tested.

Gallart-Ayala et al. also studied both of the isomers of ITX by using highly selective reaction monitoring (H-SRM).[35] Two columns were tested for the chromatographic separation: SunFire® C18 (i.d., 150 mm×2.1 mm and particle size, 3.5 µm) (Waters®) and a Discovery® HS F5 (i.d., 150 mm×2.1 mm and particle size, 3 µm) (Supelco®). The liquid chromatograph was equipped with a triple quadrupole mass detector and an ESI source operated in positive mode.

The extraction procedure and mass parameters were the same as reported by Sun et al.[33] For analysis of foodstuffs, the column selected was the pentafluorophenylpropyl Discovery® HS F5, because separation of both isomers was not achieved with the Sunfire column. The method yielded excellent LODs of 12.0 (for 2-ITX) and 13.0 ng kg$^{-1}$ (for 4-ITX) in milk samples and LODs of 2.0 (for 2-ITX) and 3.6 ng kg$^{-1}$ (for 4-ITX) in multifruit purée samples; 2-ITX was detected in 4 of the 18 analyzed samples and 4-ITX in only one sample.

Gil-Vergara et al. used HPLC-MS/MS to detect 2-ITX and the amine synergist EHA in milk samples.[36] In this study, two different extraction procedures were tested: pressurized-liquid extraction and liquid–liquid extraction (successive extractions with citrate buffer, acetonitrile and a mixture of tert-butyl methyl ether and hexane). Both extraction procedures yielded the same recoveries.

Chromatographic separation was achieved with a SunFire reversed phase C18 column (i.d., 50 mm×2.1 mm and particle size, 3.5 µm), in isocratic mode, and with methanol and water and 10 mM ammonium formate. The analysis was performed in a liquid chromatograph with triple quadrupole mass detector and an ESI source (+). The 2-ITX mass transitions were the same as in the previously mentioned studies; for EHA, the transitions were selected from the protonated molecule. For EHA and 2-ITX, the LODs obtained with this method were 0.01 and 0.03 µg L$^{-1}$ and the LOQs were 0.03 and 0.1 µg L$^{-1}$, respectively. 2-ITX and EHA were detected in 15% of the samples.

Shen et al. developed another multimethod for the determination of seven photoinitiators (BP, ITX, EHA, Irgacures 184® and 369® and Diphenyl-(2,4,6-trimethylbenzoyl) phosphine oxide (TPO)) in milk samples and packaging.[37] Extraction was performed with acetonitrile for the packaging and with acetonitrile followed by SPE for the foodstuffs.

This method was carried out with a triple quadrupole mass detector and two different sources were tested: ESI and APCI. The best results were obtained with the ESI source operated in positive mode. Chromatographic separation was achieved with a Gemini reversed-phase C18 column (i.d., 150 mm×2.0 mm and particle size, 3 µm) (Phenomenex®) and a gradient of methanol and 20 mM ammonium formate with 0.1% formic acid. The precursor ions selected were the protonated molecules of each photoinitiator, and the product ions were chosen on the basis of the most-abundant fragments originated from each precursor ion. The LODs of the method ranged from 0.05 µg kg$^{-1}$ for TPO and Irgacure 369® to 2.5 µg kg$^{-1}$ for Irgacure 184®, and the LOQs ranged between 0.1 and 5 µg kg$^{-1}$. In this study, it was concluded that ITX and BP were the photoinitiators most commonly used in milk packaging.

In 2011, Gallart-Ayala et al.[38] improved a method that they had developed three years before[35] for the determination of eleven photoinitiators (BP, 2-ITX, 4-ITX, PBZ, DEAB, EDB, EHA, Irgacure 184®, 651®, 2-hydroxy-2-methyl propiophenone (Irgacure 1173®) and 2,4-diethyl-thioxanthone (DETX)) in different foodstuffs and packaging. The equipment and chromatographic parameters were the same in both methods,[38,35] and the precursor ions of the photoinitiators were the protonated molecules $[M+H]^+$, except in the case of Irgacure 651®, and the precursor ion was the molecular ion without a methoxy group $[M-CH_3O]^+$. The product ions were selected on the basis of the most-abundant fragments generated from the precursor ions.

The extraction solvent used for the packages was dichloromethane. Two different methods were tested for the foodstuffs: the QuEChERS method and another based on a method used in a previous study,[30] but no differences were found between the results obtained. The LOQ ranged between 0.2 and 2.3 µg kg$^{-1}$, except for Irgacure 184® and Irgacure 1173®, for which the LOQ ranged between 500 to 710 µg kg$^{-1}$ depending on the extraction method. Positive results were obtained for the foodstuffs and packaging.

The most recent HPLC-MS/MS method reviewed was that used by Biedermann et al. to determine EHA, EDB, BP, 4-MBP, MBB, 2-ITX, PBZ, Irgacure 184® and Irgacure 651® in different foodstuffs and their respective recycled paperboard packages.[39] The QuEChERS method was used to extract the photoinitiators from foodstuffs, and the packages were extracted with acetonitrile. The method yielded a LOD = 0.15 mg kg$^{-1}$.

From the studies reviewed, it can be concluded that HPLC-MS or HPLC-MS/MS are suitable techniques for the analysis and/or confirmation of the presence of photoinitiators in foodstuffs and/or FCM. A standard method could achieve the chromatographic separation with a C18 column. In the case of ITX isomers, columns containing pentafluorophenylpropyl and zirconium-polystyrene, as active compounds of the matrix, proved to be good options.

Use of an ESI source in positive mode and a triple quadrupole mass detector seem to be the best options, with LODs and LOQs as low as the ng kg$^{-1}$ order of magnitude for both FCM and foodstuffs.

A chromatogram obtained from the extraction of a beer can, under the suggested conditions is represented in Figure 14.4. In the same sample analyzed by HPLC-DAD, only BP and EDB were identified, whereas ITX, DEAB, MBB, IRG 184 and EHA were also detected by the LC-MS/MS method.

### 14.2.1.2 Gas Chromatography (GC)

This technique is very common for determining semivolatile photoinitiators such as benzophenone and derivatives. As in liquid chromatography, different detectors can be used with this technique, and the mass detector is the most commonly used (GC-MS),[16,22,24,28,36,44–49] although in this case the use of mass-mass tandem is not very common.[50] A flame ionization detector (FID) was also used.[5,51]

**14.2.1.2.1 GC-FID.** Wang et al.[5] and Huang et al.[51] developed a GC-FID method to study the stability of Irgacure 184® and Irgacure 651® in food simulants. Dichloromethane was used for sample extraction.

The chromatographic separation was achieved with a low polar capillary column: HP-5 (Agilent®) ((5%-phenyl)-methylpolysiloxane) (i.d., 30 m×0.32 mm and film thickness, 0.25 µm). The system was operated in splitless mode, with the oven temperature programmed to increase from 100 to 300 °C, and the detector temperature was set at 250 °C. The photoinitiators were identified by comparison of the retention time with the corresponding peak in a standard solution.

This methodology may be a good option for daily analysis because of the simplicity of sample processing and the low cost of the equipment.

**14.2.1.2.2 GC-MS and GC-MS/MS.** Three GC-MS methods were developed before the first scandal involving ITX was reported.[20–22,44] In the first method, developed by Castle et al., three different photoinitiators (DMAB, MK and DEAB) were confirmed in food packages.[22] The extraction procedure was the same as in the HPLC-DAD method described in Section 14.2.1.2.2. The source used was electron impact (EI) and the injection was in splitless mode. The oven temperature was programmed to increase from 60 °C to 300 °C. The chromatographic separation was achieved with a dimethylpolysiloxane column: CP-Sil 5CB (Chrompack®) (i.d., 17 m×0.25 mm

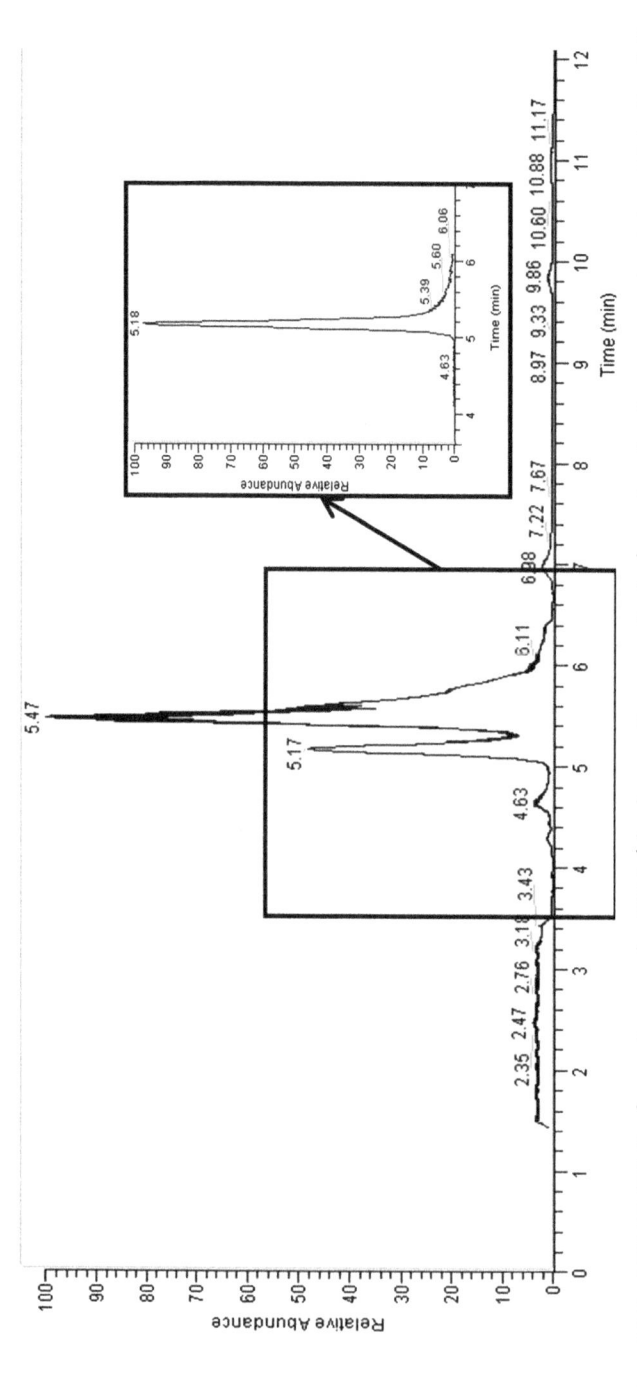

**Figure 14.4** HPLC-MS/MS chromatogram in SRM (single reaction monitoring) mode of a beer can extract in which the following photoinitiators and amine synergist (EHA) were detected: MBB (4.29 min), IRG 184 (4.64 min), EDB (5.17 min), BP (5.48 min), MBB (9.84 min), ITX (13.52 min) and EHA (17.62 min).

and film thickness, 0.12 μm). The LODs, which ranged from 0.14 to 0.6 mg kg$^{-1}$, indicated a lower sensitivity for MK determination than with the HPLC-DAD method developed in the same study.

Another method for the determination and quantification of the same three photoinitiators in food samples is also reported in the same study.[22] In this case, the same equipment was used, but the oven temperature was increased from 100 °C to 300 °C. Sample (food) extraction was carried out (twice) with ethanol and triethylamine. The LOD was 2 μg kg$^{-1}$. The results revealed quantities of photoinitiators that are unlikely to pose a risk to human health.

Papilloud and Baudraz developed a method for determining BP, ITX, EHA and Irgacures 184®, 651® and 907® in food simulants extracted by SPE.[20,21] The system was a gas chromatograph coupled to an ion-trap mass spectrometric detector and equipped with a Optima Delta-6 (Mackerey-Nagel®) medium polar column of i.d. 30 m×0.25 mm. The oven temperature was programmed to increase from 50 °C to 250 °C. The method showed good linearity and reproducibility.

Anderson and Castle developed a GC-MS method for the determination of residues of BP in food and paperboard packaging.[44] The chromatographic method used a nonpolar Rtx-1 column (i.d., 60 m×0.25 mm and film thickness, 0.25 μm) (Restek®). The oven temperature was programmed to increase from 50 to 280 °C and the injection was in splitless mode.

The results showed a good sensitivity in food, with a LOD = 0.01 mg kg$^{-1}$ and a LOQ = 0.05 mg kg$^{-1}$. BP was found in 41% of the packaging and in the foodstuff contained in the contaminated packaging, 71% of the samples contained appreciable amounts of this photoinitiator.

Since the first case of milk contaminated with ITX was reported in Italy, many other GC-MS methods have been developed. Sanches-Silva et al.[24] determined the same photoinitiators as Papilloud and Baudraz[20,21] in beverage packaging. Chromatographic separation was achieved with a low-polarity column Rtx®-5MS (i.d., 30 m×0.25 mm and film thickness, 0.25 μm) (Restek®) operated in splitless mode. The oven temperature was programmed to increase from 120 to 300 °C.

Extraction was performed with acetonitrile. The results showed that this method was suitable for the determination of the six above-mentioned photoinitiators in FCM.

Pastorelli et al.[16] used GC-MS as confirmatory technique for the determination of BP in food and FCM. The equipment was the same as in the previous study,[24] but with the split mode of operation (1 : 20), and the extraction was achieved with acetonitrile. The method showed good results with an analytical run time of only eleven minutes.

Sagratini et al. developed a GC-MS method combined with HPLC-DAD, HPLC-MS and HPLC-MS/MS methods discussed in Section 14.2.1.2.2.[30] In this method BP, ITX, EHA, EDB and Irgacure 184® were determined in beverages and packaging. The FCM was first extracted with dichloromethane and then by SPE; the foodstuffs were extracted three times with n-hexane, sodium sulfate was added and extracted by solid-phase extraction.

Chromatographic separation was achieved with an HP 5 MSI column (i.d., 30 m×0.25 mm and film thickness, 0.25 µm) (Agilent Technologies®) operated in split mode (1:40). The oven temperature was programmed to increase from 80 °C to 300 °C. The lowest LODs and LOQs were obtained using GC/MS (0.2–1 and 1–5 µg L$^{-1}$).

Koivikko et al. developed another GC-MS method to confirm the presence of BP, 4-MBP and various derivatives in FCM.[28] A low-polar column DB-5-HT (Agilent®) (i.d., 30 m×0.25 mm and film thickness, 0.1 µm) was used to achieve for chromatographic separation of all photoinitiators. Injection was in split mode (1:40). The oven temperature was programmed to increase from 100 °C to 250 °C. The results confirmed the data obtained with the HPLC-DAD method discussed in Section 14.2.1.2.2.

Gil-Vergara et al. compared LC-MS/MS and a GC-MS methods for the determination of EHA and ITX in milk and milk-based beverages.[36] For this purpose, the gas chromatograph was equipped with a nonpolar column DB 5 MS (i.d., 30 m×0.25 mm and film thickness, 0.25 µm) (J&W Scientific®). The oven temperature was programmed to increase from 50 to 270 °C and injection was in split mode.

The results showed LODs of 0.1 for ITX and 0.3 µg L$^{-1}$ for EHA and LOQs of 0.5 and 1.0 µg L$^{-1}$, respectively. These values were higher than those reported for HPLC-MS/MS, but the values were lower than other HPLC-MS/MS methods,[33] and this method proved to be a good alternative to HPLC-MS methods.

To determine 7 different photoinitiators (Irgacure 184®, 651®, BP, 4-MBP, ITX, EHA and EDB) in packaged milk, Negreira et al. developed two GC-MS methods, the first for determination and the second to confirm the results.[45] In the first method, chromatographic separation was achieved with a low-polar column HP 5 MS (i.d., 30 m×0.25 mm and film thickness, 0.25 µm) (Agilent Technologies®). The oven temperature was programmed to range from 70 °C to 280 °C, and the injection mode was splitless. In the confirmatory method, an ion trap was used, and the column was a low-polar column DB-5-HT (Agilent Technologies®) (i.d., 30 m×0.25 mm and film thickness, 0.1 µm).

SPME was used for the milk samples and acetonitrile was used for the packaging. The LOQs ranged from 0.6 and 3.5 µg L$^{-1}$. Only benzophenone was detected in the milk samples.

Bentayeb et al. developed a GC-MS method to confirm the results of a DART/TOF-MS method (discussed in section 14.2.3) in FCM.[46] For confirmation, the packages were extracted with methylene chloride and injected into the gas chromatograph in splitless mode. The chromatographic conditions (source and column) were the same as those described by Negreira et al., but the oven temperature was programmed to range from 50 to 295 °C.[45] The results obtained suggest that GC-MS is a good technique for confirmation of detection of photoinitiators in FCM.

Van Hoeck et al. determined BP and 4-MBP in breakfast cereals.[47] For this purpose, the cereals were extracted using a mixture of dichloromethane and acetonitrile, followed by SPE. The chromatographic method used a nonpolar 5% phenyl-polysilphenylene-siloxane SGE BPX-5 column (i.d., 25 m×0.22 mm and film thickness, 0.25 μm) (SGE®). The injection mode was splitless, and the oven temperature was programmed to increase from 50 to 280 °C. The detector was an ion trap. This method was able to detect low concentrations of BP and 4-MBP (LODs = 2 μg kg$^{-1}$) and LOQs = 6–8 μg kg$^{-1}$.

Bradley et al. developed a multimethod for analysing 17 photoinitiators (BP, PBZ, 2-HBP, 4-HBP, MBB, Irgacure 184, 651, 907, 2-ITX, 4-ITX, DETX, EHA, EDB, 4-MBP, 3-methyl benzophenone (3-MBP), 2-methylbenzophenone (2-MBP) and 4-(4-methylphenylthio) benzophenone (BMS)) in 350 different foodstuffs and for determining whether the food packaging was the origin of contamination.[48,49] In this study, the food was extracted twice with acetonitrile/dichloromethane (1:1), followed by two more extractions with hexane and acetonitrile. The paper/cardboard samples were extracted with acetonitrile.

Chromatographic separation was performed with a low-polarity column: Phenomenex® ZB-5ms (i.d., 30 m×0.25 mm and film thickness, 0.25 μm), with injection in splitless mode. The oven temperature was programmed to range from 100 °C to 320 °C. The results showed positive results for nine different photoinitiators in foodstuffs. The LODs ranged between 0.44 and 17 μg kg$^{-1}$ and the LOQs between 1.5 and 57 μg kg$^{-1}$ for the different food samples.

Finally, only one reference to a gas chromatography method with a MS/MS detector was found. Allegrone et al. developed a method for determining ITX in milk.[50] For the extraction procedure, two alternative methods were assayed. In the first method, the milk was mixed with ethanol and water and extracted with SPE; in the second option, an alkaline hydrolysis was carried out previous to the SPE. Chromatographic separation was achieved with a low-polarity column Rtx®-5MS (i.d., 30 m×0.25 mm and film thickness, 0.25 μm) (Restek®) operated in splitless mode. The chromatograph was equipped with an ESI source operated in positive mode and an ion-trap mass spectrometer.

With this method, the LOD was 0.1 μg L$^{-1}$ and the LOQ, 0.5 μg L$^{-1}$, and appreciable amounts of ITX were detected in all of the milk samples.

Gas chromatography can be recommended for the analysis and confirmation of several photoinitiators, e.g. Bradley et al.[48] analyzed 20 different photoinitiators. However, the method is not suitable for some photoinitiators because of their physicochemical properties, e.g. Benzophenone acrylate or Irgacure 369®. Generally, for analysis of these compounds, a low-polar column, an EI source and a MS or MS/MS are recommended, yielding LODs in the order of μg kg$^{-1}$ in most cases.

By way of example, Figure 14.5 shows a chromatogram for sixteen photoinitiators obtained with the proposal suggestions.

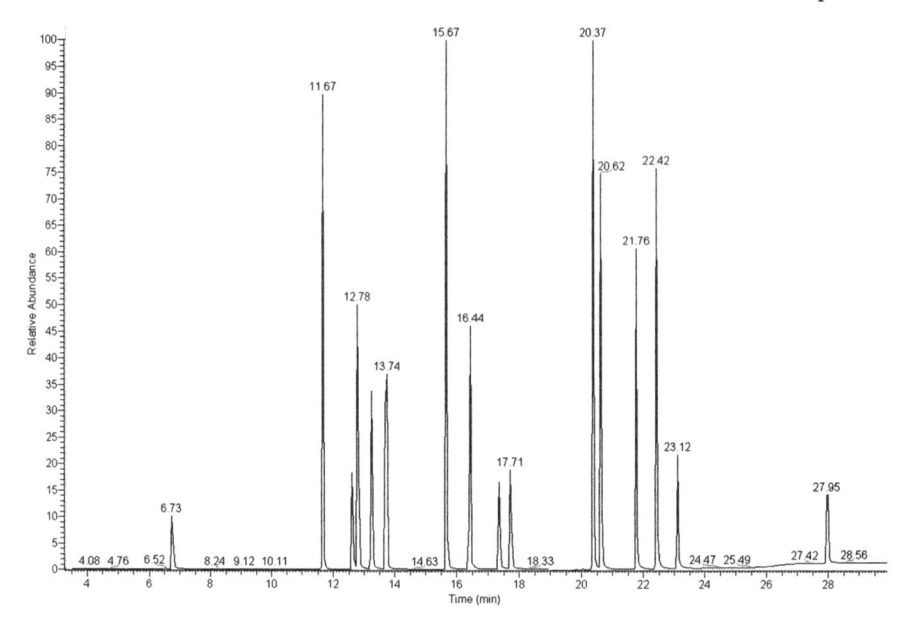

**Figure 14.5** GC-MS chromatogram of sixteen photoinitiators and an amine synergist (EHA): Irgacure 1173® (6.73 min), BP (11.67 min), Irgacure 184® (12.65 min), EDB (12.81 min), 2-HBP (13.19 min), 4-MBP (13.75 min), Irgacure 651® (15.67 min), MBB (16.45 min), 4-HBP (17.29 min), Chimassorb 90® (17.75 min), EHA (20.39 min), Irgacure 907® (20.64 min), ITX (21.78 min), PBZ (22.40 min), DETX (23.10 min) and DEAB (27.95 min).

## 14.2.1.3 High-Performance Thin-Layer Chromatography (HPTLC)

HPTLC was one of the first techniques developed for the identification and quantification of photoinitiators. Morlock and Schwack developed two methods for quantifying ITX in milk, yoghurt, margarine and soybean oil.[52] The extraction procedure was conducted with various solvents, and the best results were obtained with hexane/ethyl acetate. Two stationary phases were tested: silica gel 60 and RP18 HPTLC. For quantification, FLD was used at 254/>400 nm.

To confirm the results, two different approaches were tested: ESI/MS and DART/MS, and good robust results were obtained with LOD = 64 pg and a LOQ = 128 pg in ESI/MS mode.

## 14.2.2 Voltammetry

Ranganathan *et al.* developed cyclic voltammetry (CV) and differential pulse voltammetry (DPV) methods for the determination of ITX in wine.[53] For this purpose, they used an electrochemical workstation with three-electrode cell: a glassy-carbon solid electrode as working electrode, the counterelectrode was a Pt wire and as reference electrode Ag/AgCl. Lithium perchlorate was used as supporting electrolyte. A small amount of LiClO$_4$ was added to the samples.

Haidong *et al.* developed an amperometric sensor for the determination of benzophenone in FCM.[54] The sensor was developed by electropolymerization of o-phenylenediamine (o-PD) on a glassy-carbon electrode (GCE) in the presence of BP. As BP is nonelectroactive, potassium ferricyanide was used. The BP molecules enter the holes of the imprinted film, blocking electron transfer and the concentrations of BP are indicated by the resulting decrease in the current.

The equipment used to test this sensor was the same as that used by Ranganathan *et al.*[53] The extraction procedure was carried out with ethanol. The results were compared with an HPLC-DAD method developed in the same study with a Diamonsil C18 column (i.d., 250 mm×4.6 mm and particle size, 5 µm), with acetonitrile and water as mobile phase, in gradient mode. The data obtained revealed similar LODs as the HPLC-DAD method and proved to be a good alternative to HPLC-DAD and GC-MS methods.

### 14.2.3 DART

This new technique is presented as the most suitable for screening photoinitiators in FCM by offset transfer because of the huge advantage that no sample preparation is required, and the analytical run time is therefore minimal. Bentayeb *et al.* developed a DART/TOF-MS method to determine photoinitiators in FCM.[46] The DART parameters were 500 V of exit grid voltage and the helium current flow was carried out at a temperature of 530 °C in positive-ion mode. However, this methodology has the drawback that it is not expected to be highly quantitative: ion intensities were quite variable, and the LODs were therefore presented as a range: 0.20 µg dm$^{-2}$ < LOD < 20 µg dm$^{-2}$.

## 14.3 Nonadditionally Intended Substances (NIAS)

Two main approaches can be used for the determination of compounds related with the use of photoinitiators can be followed: study of possible byproducts or performance of a nontargeted screening analysis of the printed materials.

### 14.3.1 Study of Byproducts

Photochemistry studies can demonstrate the cleavage routes that each photoinitiator can follow and the intermediate and final molecules generated during the process. Although some of these routes have been described,[55,56] this information is not available for all photoinitiators. Nonetheless, many molecules can be generated from a single photoinitiator.

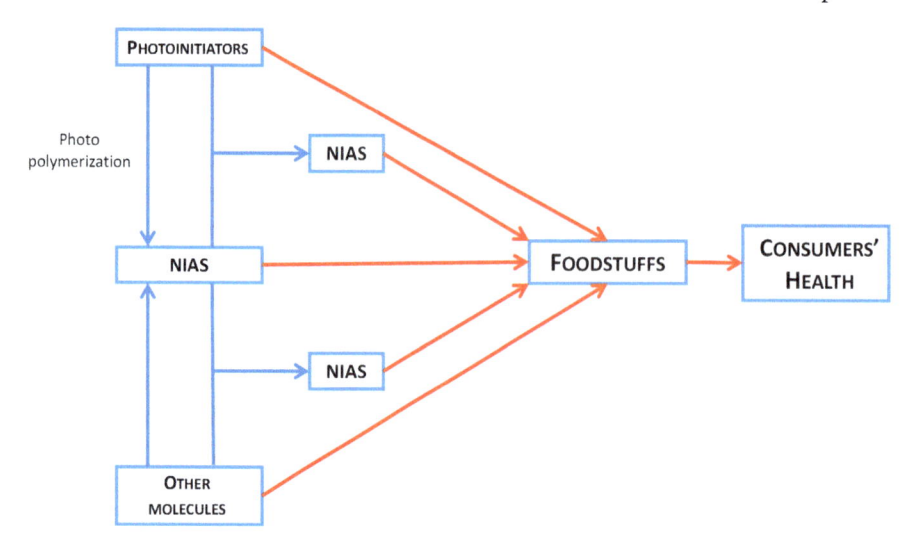

**Figure 14.6** Scheme of the different molecules (and possible byproducts/NIAS) that can reach foodstuffs and pose a risk to consumer health.

As these products can also react with the environment, the number of byproducts increases exponentially, (see Figure 14.6).

As observed in these studies, some hazardous products can be generated, *e.g.* benzene is a photolysis product of the cationic photoinitiator triphenylsulfonium hexafluorophosphate.[56] The International Agency for Research on Cancer (IARC) has classified this molecule as Group I (carcinogenic to humans) on the basis of the results of scientific studies.[57]

If all byproducts, impurities and other photoproducts were known, analytical methods could be developed for determination of these substances in FCM and foodstuffs. However, as there is a huge number of such compounds, this task is almost impossible. Nevertheless, some potentially hazardous molecules should be identified (*e.g.* benzene) and monitored to prevent risks to consumer health.

## 14.3.2 Nontargeted Screening Analysis

The number of possible byproducts is enormous because the byproducts of the photoinitiators are added to others derived from adhesives, plasticizers, and also from many other components of the FCM, foodstuffs and even the environment.

To analyze such molecules, some authors have proposed some strategies for overcoming the above-mentioned problem.[58,59] These can be summarized in 4 different steps: sample work-up, selection of the analytical technique, determination and/or quantification of the compounds, and interpretation of the results.

First, the food or FCM should be extracted using any of the techniques previously discussed in this chapter. For this purpose, it is necessary to consider the compounds of interest, because volatile compounds may readily disappear, and many transformations may occur, such as oxidation processes or other reactions with the environment. The extraction procedure should therefore differ in relation to the sample and the compounds of interest. According to Coulier *et al.*, the best option is to avoid this step, although this was not possible until the appearance of the DART technique, which enables direct analysis of samples.[59,60]

Selection of the analytical technique is critical because nontargeted analysis does not distinguish compounds, which is a problem for the chromatographic separation. GC and LC are therefore the most commonly used techniques in this field, even in two-dimensional separations (GC×GC or LC×LC). Use of MS or TOF detectors is the most convenient option with these techniques.[59]

A pure standard should be used for the quantitative determination of the possible migrating molecules; however, in many cases such standards are not available and comparison of the response of the unknown with another molecule of similar structure is the best solution.

The final step in this strategy is to analyze the data obtained. Whether the quantity of migrant is relevant or not will depend on the results and on the LODs and LOQs of the method. However, if a component does not appear in the chromatograms, this does not mean that it does not represent a risk to consumer health, because the LODs may be higher than the toxicological thresholds of this unknown compound.

In the absence of toxicological data on the migrant, it is imperative to follow some criteria to decide whether the quantities of the molecules represent potential risks to consumer health. Under this premise, the threshold of toxicological concern (TTC) can be applied to the evaluation of these substances in FCM based on the Cramer classification of the structure of each of the molecules.[10,61] With this methodology, molecules are classified as of low toxicity (Class I) or significant toxicity (Class III), with class II being of intermediate toxicity. This strategy has been recommended since the last decade and followed by most researchers and authorities, see Figure 14.7.[62–64]

## 14.4 Conclusions and Future Overview

The lack of specific European legislation for the use of printing inks in food packaging potentially creates a risk to consumer health. General standardized protocols must be developed to harmonize procedures for the determination of photoinitiators and their byproducts. The methods proposed by different authors are shown in Figure 14.1, which can be used as a starting point for standardizing the procedure.

From the literature reviewed, HPLC-MS/MS-MS appears to be the analytical tool of choice because the limits of detection are the lowest obtained and this may be the most versatile of all of the methods tested.

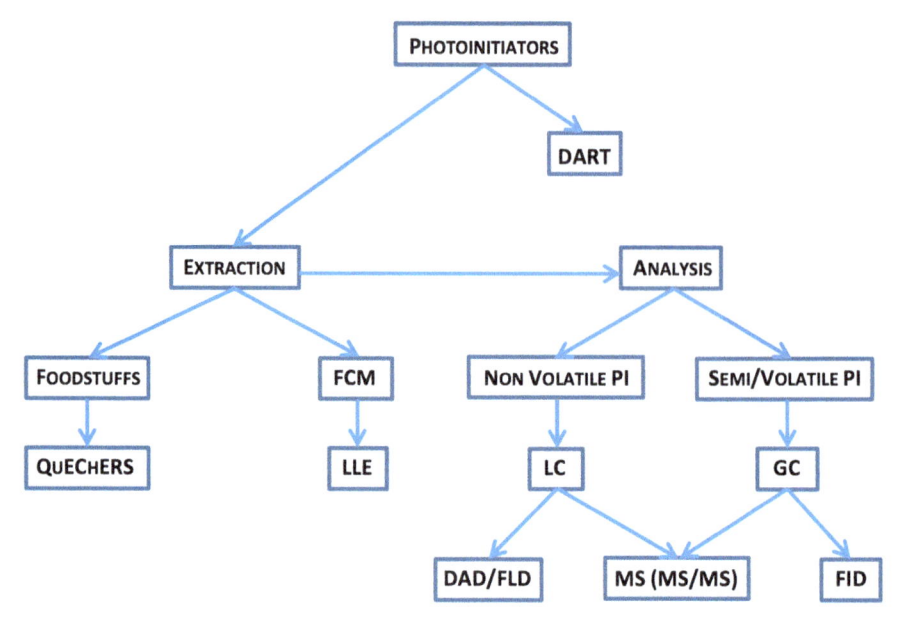

**Figure 14.7** Proposed method for determining and/or quantifying photoinitiators in FCM/foodstuffs based on the predominant extractions and analytical techniques in use at present.

However, the most difficult challenge is to use analytical methods to determine NIAS. It is therefore imperative to carry out more detailed studies of the photoproducts derived from each photoinitiator used in FCM to identify as yet unknown products.

Moreover, legislation must also be developed in relation to NIAS. Although efforts are being made in this field, they are not yet sufficient and the following years will be decisive as regards protection of consumer health.

## Acknowledgements

The study was financially supported by the "Ministerio de Economía y Competitividad", Ref. No. AGL/2011-26531 "MIGRATIN". The authors are grateful to "Ministerio de Economía y Competitividad" for the FPI predoctoral fellowship (Ref. BES-2012-051993) awarded to Miguel Ángel Lago. Raquel Sendón is grateful to the "Parga Pondal" program financed by "Consellería de Innovación e Industria, Xunta de Galicia" for a postdoctoral contract.

## References

1. N. M. De Kruijf, R. R. Van Beest, T. Sipilinen-Malm, P. Paseiro and B. De Meulenaer, *Food Addit. Contam.*, 2002, **19**, 144.
2. S. J. Risch, *J. Agric. Food Chem.*, 2009, **57**, 8089.

3. M. F. Poças and T. Hogg, *Trends Food Sci. Technol.*, 2007, **18**, 219.
4. H. Widén, A. Leufvén and T. Nielsen, *Food. Addit. Contam.*, 2004, **21**, 993.
5. Z. W. Wang, X. L. Huang and C. Y. Hu, *Packag. Technol. Sci.*, 2009, **22**, 151.
6. G. L. Robertson, *Food Packaging Principles and Practice*, CRC Press, Boca Raton FL US, 3rd edn, 2013, ch. 1, pp. 1–8.
7. RASFF. 2005. Isopropyl thioxanthone in milk for babies from Spain. Alert notification 2005.631, 8 September 2005. Available: http://ec.europa.eu/food/food/rapidalert/reports/week37-2005_en.pdf.Accessed 29 November 2013.
8. RASFF, 2013. Rapid Alert System for Food and Feed (RASFF). Available from: http://ec.europa.eu/food/food/rapidalert/rasff_portal_database_en.html. Accessed 29 November 2013.
9. EFSA, 2012. European Food Safety Authority. External Scientific Report. Report of ESCO WG on non-plastic Food Contact Materials. 2011:EN-139.
10. G. M. Cramer, R. A. Ford and R. L. Hall, *Food Cosmet. Toxicol.*, 1978, **16**, 255.
11. J. O. Choi, F. Jitsunari, F. Asakawa, H. J. Park and D. S. Lee, *Food Addit. Contam.*, 2002, **19**, 1200.
12. T. Jung, T. J. Simat and W. Altkofer, *Food Addit. Contam. A*, 2010, **27**, 1040.
13. G. Morlock and W. Schwack, *Anal. Bioanal. Chem.*, 2006, **385**, 586.
14. T. Rothenbacher, M. Baumann and D. Fügel, *Food. Addit. Contam.*, 2007, **24**, 438.
15. European Food Safety Authority (EFSA). 2009. EFSA statement on the presence of 4-methylbenzophenone found in breakfast cereals. EFSA J. RN-243:1–19. Available from: http://www.efsa.europa.eu/en/efsajournal/doc/243r.pdf. Accessed 2 December 2013.
16. S. Pastorelli, A. Sanches-Silva, J. M. Cruz, C. Simoneau and P. Paseiro-Losada, *Eur. Food Res. Tech.*, 2008, **227**, 1585.
17. Commission Regulation (EU) No 10/2011 of 14 January 2011 on plastic materials and articles intended to come into contact with food. Off J Eur Union. L12/1.
18. European Printing Ink Agency (EuPIA). EuPIA Suitability List of Photoinitiators for Low Migration UV Printing Inks and Varnishes. February, 2013.
19. Screening Tests for Visible and Non-Visible Set Off: Final Report. Project code: FS231076. April 2012. Pyra International. Lord, A.W.; Berry, C.; Hutchings, J.; Obhrai, S.; Callan, N. and Sullivan, D. Available on: http://www.foodbase.org.uk//admintools/reportdocuments/759-1-297_FS231076__PART_1.pdf. Accessed 2 December 2013.
20. S. Papilloud and D. Baudraz, *Food Addit. Contam.*, 2002, **19**, 168.
21. S. Papilloud and D. Baudraz, *Prog. Org. Coat.*, 2002, **45**, 231.
22. L. Castle, A. P. Damant, C. A. Honeybone, S. M. Johns, S. M. Jickells, M. Sharman and J. Gilbert, *Food. Addit. Contam.*, 1997, **14**, 45.

23. A. Sanches-Silva, S. Pastorelli, J. M. Cruz, C. Simoneau, I. Castanheira and P. Paseiro, *J. Dairy Sci.*, 2008, **91**, 900.
24. A. Sanches-Silva, S. Pastorelli, J. M. Cruz, C. Simoneau and P. Paseiro, *J. Food Sci.*, 2008, **73**, C92.
25. A. Sanches-Silva, S. Pastorelli, J. M. Cruz, C. Simoneau, I. Castanheira and P. Paseiro-Losada, *J. Agric. Food Chem.*, 2009, **56**, 2722.
26. A. Sanches-Silva, C. Andre, I. Castanheira, J. M. Cruz, S. Pastorelli, C. Simoneau and P. Paseiro-Losada, *J. Agric. Food Chem.*, 2009, **57**, 9516.
27. A. Rodríguez-Bernaldo de Quirós, R. Paseiro-Cerrato, S. Pastorelli, R. Koivikko, C. Simoneau and P. Paseiro-Losada, *J. Agric. Food Chem.*, 2009, **57**, 10211.
28. R. Koivikko, S. Pastorelli, A. Rodríguez-Bernaldo de Quirós, R. Paseiro-Cerrato, P. Paseiro-Losada and C. Simoneau, *Food. Addit. Contam. A*, 2010, **27**, 1478.
29. T. Jung, T. J. Simat, W. Altkofer and D. Fügel, *Food. Addit. Contam. A*, 2013, **30**, 1993.
30. G. Sagratini, G. Caprioli, G. Cristali, D. Giardina, M. Ricciutelli, R. Volpini, Y. Zuo and S. Vittori, *J. Chromatogr. A*, 2008, **1194**, 213.
31. C. Benetti, R. Angeletti, G. Binato, A. Biancardi and G. Biancotto, *Anal. Chim. Acta*, 2008, **617**, 132.
32. G. Sagratini, J. Mañes, D. Giardina and Y. Picó, *J. Agric. Food Chem.*, 2006, **54**, 7947.
33. C. Sun, S. H. Chan, D. Lub, H. M. W. Lee and B. C. Bloodworth, *J. Chromatogr. A*, 2007, **1143**, 162.
34. R. Bagnati, G. Bianchi, E. Marangon, E. Zuccato, R. Fanelli and E. Davoli, *Rapid Commun. Mass Sp.*, 2007, **21**, 1998.
35. H. Gallart-Ayala, E. Moyano and M. T. Galceran, *J. Chromatogr. A*, 2008, **1208**, 182.
36. A. Gil-Vergara, C. Blasco and Y. Picó, *Anal. Bioanal. Chem.*, 2007, **389**, 605.
37. D. Shen, H. Lian, T. Ding, J. Xu and C. Shen, *Anal. Bioanal. Chem.*, 2009, **395**, 2359.
38. H. Gallart-Ayala, O. Núñez, E. Moyano and M. T. Galceran, *J. Chromatogr. A.*, 2011, **1218**, 459.
39. M. Biedermann, J. E. Ingenhoff, M. Zurfluh, L. Richter, T. Simat, A. Harling, W. Altkofer, R. Helling and K. Grob, *Food Addit. Contam. A*, 2013, **30**, 885.
40. M. Anastassiades, S. J. Lehotay, D. Stajnbaher and F. J. Schenck, *J. AOAC Int.*, 2003, **83**, 412.
41. Council of Europe. 2009. Policy statement concerning paper and board materials and articles intended to come into contact with foodstuffs [Internet]. Version 4–12.02.2009. Available from: http://www.coe.int/t/e/social_cohesion/soc-sp/public_health/food_contact/PS%20PAPER%20AND%20BOARD%20Version%204%20E.pdf. Accessed 2 December 2013.

42. FDHA. Ordinance of the FDHA on articles and materials (RS 817.023.21) of 23 November 2005. Annex 6. Lists of permitted substances from 1 May 2011 for the manufacture of packaging inks, subject to the requirements set out therein. Available from: http://www.bag.admin.ch/themen/lebensmittel/04867/10015/index.html?lang=en. Accessed 2 December 2013.
43. Standing Committee. 2009. Standing Committee on the Food Chain and Animal Health Section Toxicological Safety – Conclusions of the meeting of 06 March 2009 [Internet]. http://ec.europa.eu/food/food/chemicalsafety/foodcontact/docs/conclusions_20090306statement.pdf. Accessed 2 December 2013.
44. W. A. C. Anderson and L. Castle, *Food Addit. Contam.*, 2003, **20**, 607.
45. N. Negreira, I. Rodríguez, E. Rubí and R. Cela, *Talanta*, 2010, **82**, 296.
46. K. Bentayeb, L. K. Ackerman, T. Lord and T. H. Begley, *Food Addit. Contam. A*, 2013, **30**, 750.
47. E. Van Hoeck, T. De Schaetzen, C. Pacquet, F. Bolle, L. Boxus and J. Van Loco, *Anal. Chim. Acta*, 2010, **663**, 55.
48. E. L. Bradley, J. S. Stratton, J. Leak, L. Lister and L. Castle, *Food Addit. Contam. B*, 2013, **6**, 73.
49. FSIS Number 03/11. 2011. A 4-year rolling programme of surveys on chemical migrants from food-packaging materials and articles survey 4: migration of selected ink components from printed packaging materials into foodstuffs and screening of printed packaging for the presence of mineral oils. Food Standards Agency, Food Survey Information Sheet Number 03/11, December 2011. Available from: http://www.food.gov.uk/multimedia/pdfs/fsis0311.pdf. Accessed on 2 December 2013.
50. G. Allegrone, I. Tamaro, S. Spinardi and G. Grosa, *J. Chromatogr. A*, 2008, **1214**, 128.
51. X. L. Huang, Z. W. Wang, C. Y. Hu, Y. Zhu and J. Wang, *Packag. Technol. Sci.*, 2012, **26**, 23.
52. G. Morlock and W. Schwack, *Anal. Bioanal. Chem.*, 2006, **385**, 586.
53. D. Ranganathan, S. Zamponi, P. Conti, G. Sagratini, M. Berrettoni and M. Giorgetti, *Anal. Lett.*, 2011, **44**, 2335.
54. L. Haidong, G. Huaimin, D. Hong, T. Yuejin, Z. Xianen, Q. Wenjing, M. Saadat and X. Guobao, *Talanta*, 2012, **99**, 811.
55. W. Arthur Green, *Industrial Photoinitiators. A Technical Guide*, CRC Press, Boca Raton, FL, US, 1st edn, 2010, ch. 6, pp. 139–148.
56. J. L. Dektar and N. P. Hacker, *J. Am. Chem. Soc.*, 1978, **112**, 6004.
57. IARC. IARC monographs: Benzene. http://monographs.iarc.fr/ENG/Monographs/vol100F/mono100F-24.pdf.
58. A. Feigenbaum, D. Scholler, J. Bouquant, G. Brigot, D. Ferrier, R. Franz, L. Lillemark, A. M. Riquet, J. H. Petersen, B. Van Lierop and N. Yagoubi, *Food Addit. Contam.*, 2002, **19**, 184.
59. L. Coulier, S. Koster, B. Muilwijk, L. van Stee, R. Peters, E. Zondervan-van den Beuken, R. Rijk, M. Rennen, W. Leeman, G. Houben and W. D. van Dongen, in *Comprehensive Analytical Chemistry: Food Contaminants &*

*Residue Analysis*, ed. Y. Picó, Wilson & Wilson's, Jordan Hill, Oxford, UK, 1st edn, 2008, ch. 22, vol. 51, pp. 775–794.
60. J. Hajslova, T. Cajka and L. Vaclavik, *TrAC-Trend Anal. Chem.*, 2011, **30**, 204.
61. I. C. Munro, R. A. Ford, E. Kennepohl and J. G. Sprenger, *Food Chem. Toxicol.*, 1996, **34**, 829.
62. R. Kroes, C. Galli, I. Munro, B. Schilter, L. Tran, R. Walker and G. Würtzen, *Food Chem. Toxicol.*, 2002, **38**, 255.
63. R. Kroes, A. G. Renwick, M. Cheeseman, J. Kleiner, I. Mangelsdorf, A. Piersma, B. Schitter, J. Schalatter, F. van Schothorst, J. G. Vos and G. Würtzen, *Food Chem. Toxicol.*, 2004, **42**, 65.
64. R. Pinalli, C. Croera, A. Theobald and A. Feigenbaum, *Trends Food Sci. Technol.*, 2011, **22**, 523.

CHAPTER 15

# Methacrylate and Epoxy Resins Photocured by Means of Visible Light-Emitting Diodes (LEDs)

CLAUDIA I. VALLO* AND SILVANA V. ASMUSSEN

Institute of Materials Science and Technology (INTEMA), Universidad Nacional de Mar del Plata-National Research Council (CONICET), Av. Juan B. Justo 4302, (7600), Mar del Plata, Argentina
*Email: civallo@fi.mdp.edu.ar

## 15.1 Introduction

Photopolymerization has found numerous applications in films and coatings, graphic arts, adhesives, dentistry, contact lenses, and semiconductor fabrication.[1,2] The photoinitiation step usually requires the presence of a molecule (photoinitiator), which absorbs the exciting light and leads to the production of a reactive species, which initiate the polymerization.[3,4] Based on the mechanism of photoinitiation, photopolymerization reactions can be broadly divided into free-radical and cationic systems. In free-radical systems, light-sensitive photoinitiator molecules present within a liquid monomer (typically an acrylate or a methacrylate) react with photons of light to generate highly reactive free radicals. These radicals initiate the polymerization process, attacking reactive double-bond groups on monomer molecules and converting them to a polymer.[4] In cationic polymerization, the use of cationic photoinitiators, such as diaryliodonium and triarylsulfonium salts, provides a convenient method of generating powerful Brønsted acids under irradiation, which is the primary species that initiates polymerization.[5-9] A wide variety of monomers and oligomers can be used to tailor the structure and properties of the cured materials, while pigments, additives, *etc.*, may be included to confer color, stability, or chemical functionality, leading to a variety of different applications. In comparison to thermal polymerization, photopolymerization has the added environmental benefits of eliminating the use of solvents and high temperatures. The high rate of crosslinking achieved on irradiation is ideal for high-speed applications.

Traditional curing units emit in a wide wavelength range from UV to IR. Conversely, the spectral output of light-emitting diodes (LEDs) is concentrated in a comparatively narrow wavelength range. In practice, the light absorption of photoinitiators should correlate with the spectral emission profiles of light-curing units compared on an equivalent basis. Only those wavelengths where the photosensitizer absorbs are useful for photopolymerization, the rest of the spectrum is unused and has to be filtered. Since all of the spectral output of the LEDs is absorbed by the photoinitiator, more efficient curing should be possible with current LED lights, resulting in reduced curing time compared with conventional lamp light sources. In addition, it yields the positive aspect of eliminating the IR and UV components. As a result, when compared to conventional curing units, there is less heat transfer to the substrate (no IR) and no harmful UV rays. Through this chapter, the photopolymerization of typical methacrylate and epoxy monomers carried out with LEDs will be described.

## 15.2 Light-Emitting Diodes

Light is created by LEDs units *via* electroluminescence, which is the phenomenon of a semiconductor material emitting light when electric current or an electric field is passed through it.[10] By controlling the chemical composition of the semiconductor material, one can control the emission of light in a specific wavelength range. For example, red and infrared LEDs are made with gallium arsenide, while bright blue is made with gallium nitride. There are also orange, green, blue, violet, purple, ultraviolet LEDs. Table 15.1 shows the available colors of LEDs and corresponding semiconductor material.

Appearing as practical electronic components in 1962,[10] early LEDs emitted low-intensity red light, but modern versions are available across the visible, ultraviolet, and infrared wavelengths, with very high brightness. The first commercial LEDs were commonly used as replacements for incandescent and neon indicator lamps in expensive equipment such as laboratory and electronics test equipment, then later in such appliances as TVs, radios, telephones, and calculators. Red LEDs based on gallium arsenide phosphide (GaAsP), suitable for indicators, were produced in 1968. Later, other colors became widely available and appeared in appliances and equipment. The innovative processing and packaging methods developed resulted in important cost reductions. The development of LED technology

Table 15.1 Wavelength ranges and materials of light-emitting diodes (LEDs).

| Color | Wavelength [nm] | Semiconductor material |
|---|---|---|
| Infrared | $\lambda > 760$ | Gallium arsenide (GaAs) <br> Aluminum gallium arsenide (AlGaAs) |
| Red | $610 < \lambda < 760$ | Aluminum gallium arsenide (AlGaAs) <br> Gallium arsenide phosphide (GaAsP) <br> Aluminum gallium indium phosphide (AlGaInP) <br> Gallium(III) phosphide (GaP) |
| Orange | $590 < \lambda < 610$ | Gallium arsenide phosphide (GaAsP) <br> Aluminum gallium indium phosphide (AlGaInP) <br> Gallium(III) phosphide (GaP) |
| Yellow | $570 < \lambda < 590$ | Gallium arsenide phosphide (GaAsP) <br> Aluminum gallium indium phosphide (AlGaInP) <br> Gallium(III) phosphide (GaP) |
| Green | $500 < \lambda < 570$ | Traditional green: <br> Gallium(III) phosphide (GaP) <br> Aluminum gallium indium phosphide (AlGaInP) <br> Aluminum gallium phosphide (AlGaP) <br> Pure green: <br> Indium gallium nitride (InGaN)/Gallium(III) nitride (GaN) |
| Blue | $450 < \lambda < 500$ | Zinc selenide (ZnSe) <br> Indium gallium nitride (InGaN) <br> Silicon carbide (SiC) as substrate <br> Silicon (Si) as substrate |
| Violet | $400 < \lambda < 450$ | Indium gallium nitride (InGaN) |
| Purple | multiple types | Dual blue/red LEDs, blue with red phosphor, or white with purple plastic |
| Ultraviolet | $\lambda < 400$ | Diamond (235 nm) <br> Boron nitride (215 nm) <br> Aluminum nitride (AlN) (210 nm) <br> Aluminum gallium nitride (AlGaN) <br> Aluminum gallium indium nitride (AlGaInN)—down to 210 nm |
| Pink | multiple types | Blue with one or two phosphor layers: yellow with red, orange or pink phosphor added afterwards, or white phosphors with pink pigment or dye over top |
| White | Broad spectrum | Blue/UV diode with yellow |

has caused their efficiency and light output to rise exponentially, with a doubling occurring approximately every 36 months since the 1960s. This trend is generally attributed to the parallel development of other semiconductor technologies and advances in optics and material science.

During the last decade, LEDs have replaced incandescent, fluorescent, and neon lamps due to their ability to produce high luminosity at low currents and voltages.[11] LEDs have many advantages over incandescent light sources including lower energy consumption, comparatively longer service lives, improved physical robustness, smaller size, and faster switching. LEDs are used in applications as diverse as aviation lighting, automotive lighting, advertising, large video screens, general lighting and traffic signals. LEDs have allowed new video displays, and sensors to be developed, while their high switching rates are also useful in advanced communications technology. Infrared LEDs are also used in the remote control units of many commercial products including televisions, DVD players and other domestic appliances.[10]

The majority of modern resin-based oral restorative biomaterials are cured *via* photopolymerization processes. A variety of light sources is available for this. Quartz-tungsten-halogen light-curing units have dominated light curing of dental materials for decades and now are almost entirely replaced by modern light-emitting diode light-curing units.[12,13] LEDs used in dentistry are blue because that is the wavelength range for activating the photoinitiator camphorquinone (CQ), which is present in light cure dental resins. Tungsten-halogen-curing lights have been the most frequently used polymerization source in dentistry. The advantage of halogen curing lights is that they are derived from relatively low-cost technology. However, they have low efficiency and present several drawbacks. The light from a halogen curing light is produced by an electric current flowing through a thin tungsten filament.[12,13] The filament functions as a resistor, and when it is heated by the current, it becomes incandescent and emits electromagnetic radiation in the form of visible light, as well as a large amount of infrared radiation. The light emitted is not selective; therefore, when blue light is given off, the rest of the spectrum is unused and has to be filtered out. Because the filament generates high temperatures, the curing light has to be cooled by a ventilating fan that forces airflow through slots in the casing. This feature causes the hand piece to become cumbersome and noisy. The high temperatures attained can also cause the bulb components to have limited lifetimes and requires frequent monitoring and replacement of the curing light's bulb. Newer types of curing lights have been introduced to photopolymerize dental materials. Plasma arc curing, or PAC, lights have been introduced with the claim that they can decrease curing times significantly without a concomitant reduction in mechanical properties and performance of the cured materials.[12,13] In PAC lights, a high voltage is applied between two electrodes, resulting in a light arc between them. Like halogen curing lights, PAC lights also have low efficiency, and their power consumption is higher than that of halogen curing lights. PAC lights have high operating temperatures, which makes the use of ventilating fans necessary. In addition, the light must be filtered to provide the useful blue light. More recently, LEDs that produce blue light have been introduced for the photopolymerization of dental materials.[12] Highly bright blue LEDs have been available since the mid-1990s. A new semiconductor material system – the gallium nitride – forms the basis for the blue emission, as well as for the

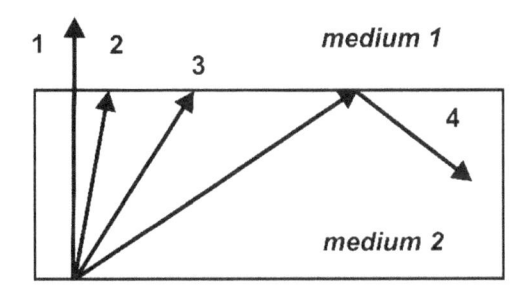

**Scheme 15.1** Loss mechanism that decreases photoemission of LEDs. If the angle of incidence is greater than the critical angle the light beam will be totally reflected back internally or trapped inside the die causing low light output.

high efficiency of devices that use it. The earlier versions of blue LEDs were low in intensity and required the use of a large number of LEDs to provide adequate performance. Improvements in LED technology introduced during the last decade have resulted in the development of several types of commercial LED light-curing units that have improved intensity output, which results in increased performance. Further advancements in technology have made it possible for manufacturers to produce high-powered LED lights. LED light-curing units that are capable of delivering power densities of about 1000 mW cm$^{-2}$ have been marketed recently.[13]

There are three loss mechanisms that decrease the efficiency photoemission of LEDs: absorption of the photons within the LED material itself, critical angle loss, and Fresnel loss.[11] When a light beam crosses a boundary between media of different refractive indices, the light beam will be partially refracted at the boundary surface and partially reflected. However, if the angle of incidence is greater than a certain value known as the critical angle, then the light beam will be totally reflected back internally or trapped inside the die, causing low light output (Scheme 15.1).[11] This can only occur when light travels from a medium with a higher to one with a lower refractive index. The value of the critical angle, $\theta_c$, is given by the Snell's law.[11]

$$\theta_c = \sin^{-1}\left(\frac{n_2}{n_1}\right) \quad (15.1)$$

where $n_1$ and $n_2$ are the refractive index of each medium and $n_1 > n_2$. According to Snell's law, the larger the refractive index of the encapsulant, the higher the light extraction from the LED device. In addition to critical-angle loss, a portion of the light emitted from the LED chip is reflected back at discontinuities of refractive index. Light that is reflected back into the LED chip is lost, and, therefore, the efficiency of light extraction is reduced. The light reflection is caused by a change in the refractive index that occurs at the interface between the LED chip and the encapsulant material. This reflection loss is referred to as the Fresnel reflection loss. The portion of incident light

(light of normal incidence) that is reflected back into the source can be approximated from the following relationship:

$$R = \left(\frac{n_1 + n_2}{n_1 + n_2}\right)^2 \quad (15.2)$$

$R$ is the fraction of the incident light reflected, $n_1$ is the refractive index of the LED chip, and $n_2$ is the refractive index of the encapsulant material. It is clear from eqn (15.1) and (15.2) that the higher the refractive index of the encapsulant, the more efficient the light extraction from the LED. The progress of light-emitting diodes has been impressive. In fact, it has been reported that in the past thirty years the light intensity has increased twenty-fold, while the cost has decreased ten-fold each decade.[11] Still, considerable efforts are being devoted now to reduce the aforementioned losses that decrease the efficiency photoemission.

## 15.3 Light-Curable Methacrylate Resins

In general, a photopolymerizable formulation contains three components:

- A monomer or oligomer, which is generally a multifunctional molecule, that polymerizes to give a crosslinked network.
- A diluent, to reduce the viscosity of the resin.
- A photoinitiator system, which absorbs light and initiates the polymerization.

### 15.3.1 Methacrylate Monomers

Scheme 15.2 shows typical dimethacrylate monomers used in light-cured formulations.[14]

The backbone of the monomers has a chemical structure, which can be varied and confers to the polymerized material its special mechanical and physical properties. These monomers, exhibit different chemical structures that are expected to result in different intra- and intermolecular interactions. The presence of these interactions can cause a decreased mobility of the medium during polymerization and also increased rigidity of the corresponding polymeric network.[14] For BISGMA, weaker hydrogen bonds are formed between the OH group and methacrylate carbonyls, while stronger hydrogen-bonding interactions arise between BISGMA hydroxyl groups.[14] In the UDMA monomer, NH hydrogen-bonded interactions can be differentiated into weaker associations with the methacrylate carbonyls and stronger bonds formed with the more electron-rich urethane carbonyl. In addition, a combination of intra- and intermolecular associations is possible with the more flexible aliphatic UDMA structure. The lack of hydrogen bonding in BISEMA allows a greater degree of mobility and is considered responsible

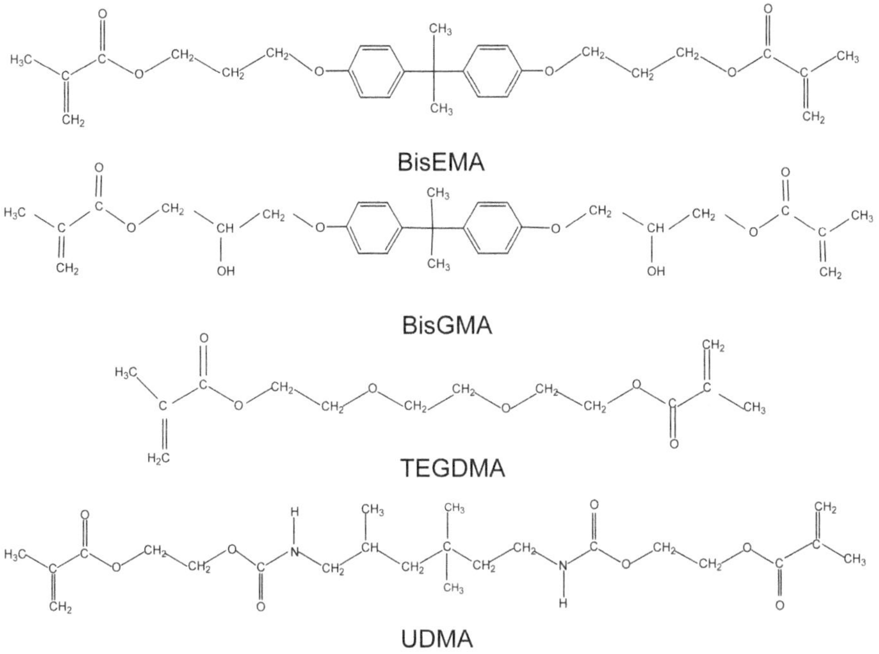

**Scheme 15.2** Structure of methacrylate monomers used in light-curable formulations.

of its comparatively low viscosity. TEGDMA is commonly used as reactive diluent for BISGMA.

## 15.3.2 Photopolymerization Reactions of Methacrylate Resins

Photopolymerization reactions are ubiquitous in the production of materials such as adhesives, films, coatings, dental restorative composites, contact lenses, photolithographic resists and optoelectronic coatings. The main steps of the mechanism of radical polymerization are an initial photochemical event, which leads to the first active species, the classical propagation and then a termination reaction.[15]

$$\text{Initiation} \quad I \xrightarrow{h\nu} R^{\bullet} \xrightarrow{M} RM^{\bullet}$$
$$\text{Propagation} \quad RM_n^{\bullet} + M \rightarrow RM_{n+1}^{\bullet}$$
$$\text{Termination} \quad RM_n^{\bullet} + RM_t^{\bullet} \rightarrow RM_{n+t}R$$
$$\text{Termination} \quad RM_n^{\bullet} + RM_t^{\bullet} \rightarrow RM_n + RM_t$$

Free-radical polymerization has many benefits, including a broad array of versatile monomers with excellent mechanical properties, readily controllable reaction kinetics, and rapid reactions with low energy demand. However, because of highly efficient radical termination reactions, these systems require a continuous production of initiating radicals, and thus, in

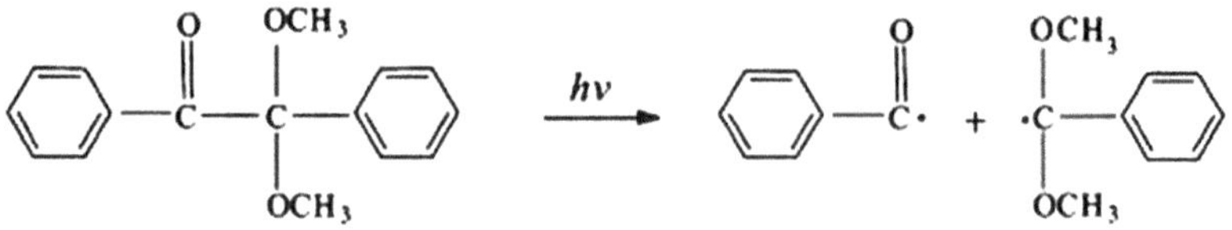

**Scheme 15.3** Photolysis of DMPA under irradiation with a LED centred at 365 nm.

photopolymerization processes, the reaction rapidly ceases if illumination is interrupted. In addition, free-radical photocuring is not able to proceed in shadow regions or at increasing depths where significant light attenuation occurs.

The photoinitiation step usually requires the presence of a molecule (photoinitiator, I), which absorbs the exciting light and leads to the production of a reactive species.[3] Based on the mechanism by which initiating radicals are formed, photoinitiators are generally divided into two classes. Norrish type-I photoinitiators undergo a unimolecular bond cleavage upon irradiation to yield free radicals. Scheme 15.3 shows the decomposition of 2,2-Dimethoxy-2-phenylacetophenone (DMPA) under irradiation with a LED (peak wavelength at 365 nm).[16]

Norrish type-II photoinitiators undergo a bimolecular reaction where the excited state of the photoinitiator interacts with a second molecule (a coinitiator) to generate free radicals. The behavior of a Norrish type-II photoinitiator, such as the ketone/amine (AH) pair, in the presence of monomer (M) can be represented by the following simplified reaction scheme:

$I + h\nu \rightarrow {}^1I \rightarrow {}^3I$
${}^3I \rightarrow$ physical deactivation
${}^3I \rightarrow$ radicals
${}^3I + M \rightarrow$ products
${}^3I + AH \rightarrow$ charge transference

where $I$, ${}^1I$ and ${}^3I$ are the ground state, the excited singlet and the excited triplet, respectively. It is generally accepted that upon absorption of light the carbonyl group of the ketone is promoted to an excited singlet state. This excited state may return to the ground state or it may decompose to another species. The excited singlet may also undergo intersystem crossing to the triplet state. The excited triplet then forms an exciplex with the electron donor (in this case the amine AH) by charge transfer to the carbonyl, thus producing two radical ions. In some cases, this charge-transfer complex may be deactivated to the starting species or may form a degradation product. However, if proton abstraction occurs, then two free radicals are formed. It is generally considered that the amine radical is responsible for initiating the polymerization and that the radical formed from the ketone is not an efficient initiator and dimerizes. Scheme 15.4 shows the decomposition

**Scheme 15.4** Photodecomposition of CQ/amine systems under visible irradiation.

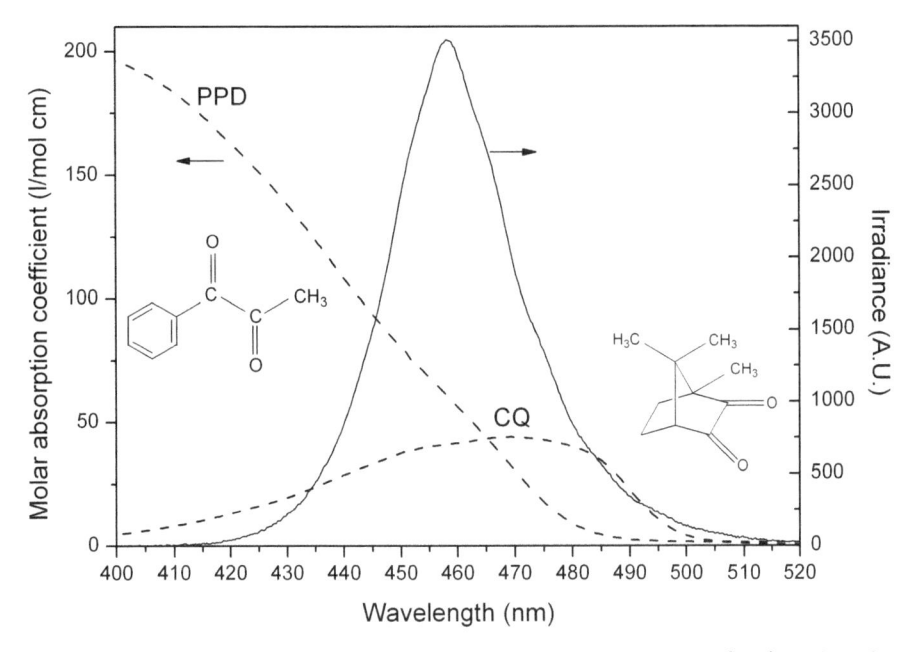

**Figure 15.1** Light irradiance of the LED unit in arbitrary units (a.u.) and molar extinction coefficients of the photoinitiators CQ and PPD.

of the camphorquinone (CQ)/amine pair under irradiation with blue LED (467 nm).[17–19]

### 15.3.3 Photon-Absorption Efficiency

In practice, the light absorption of photoinitiators should correlate with the spectral emission profiles of light-curing units compared on an equivalent basis. Only those wavelengths where the photosensitizer strongly absorbs are useful for the photopolymerization. Figure 15.1 shows the molar absorptivity of CQ and 1-phenyl-1,2-propanedione (PPD) in conjunction with the emission spectrum of a LED curing lamp.

It is seen that CQ is activated in the wavelength range 400–520 nm with an absorption peak at 470 nm and PPD is activated from below 350 nm to 480 nm with a maximum at around 400 nm. In comparison, the LED light source emits in the wavelength range 420–510 nm with a peak around 470 nm. Although the absorption peak for PPD is in the UV region, its absorption curve extends into the visible range and, in fact PPD exhibits a molar absorbance at 470 nm equivalent to that of CQ. The efficiency of a given curing unit to excite a given photoinitiator, can be quantified from the spectral irradiance of the curing unit $(I(\lambda))$ and the molar absorption coefficient distribution of the photoinitiator $(\varepsilon(\lambda))$.[20] This quantity is termed the photon-absorption efficiency (PAE).[20] The PAE can be assessed as follows. If $(I(\lambda)$, in W m$^{-2})$ is the energy incident to an area per unit time, then, the number of photons of a given wavelength $(N(\lambda))$ are equal to the irradiance of the source at the wavelength $\lambda$ divided by the energy of one photon:

$$N(\lambda) = \frac{I(\lambda)\lambda}{hc} \quad \frac{\text{photons}}{m^2 s} \tag{15.3}$$

where $h$ is the Planck constant, $c$ is the speed of light and $hc/\lambda$ is the energy of one photon. The total number of photons of the incident light $(N_t)$ is given by the summation of the photons produced at each wavelength.

$$N_t = \sum_{\lambda_1}^{\lambda_2} N(\lambda)\Delta\lambda = \int_{\lambda_1}^{\lambda_2} \frac{I(\lambda)\lambda}{hc} d\lambda \tag{15.4}$$

where $\lambda_1$ and $\lambda_2$ are the limits of the wavelength emission of the source. For radiation passing through a layer of sample of thickness dL, containing a concentration of photoinitiator $c$ (molecules per cm3), the fraction of absorbed photons $(f_{abs})$ is:

$$f_{abs} = \frac{\text{Number of photons absorbed}}{\text{Number of photon incident}} = 1 - 10^{-\varepsilon(\lambda)cdL} \tag{15.5}$$

assuming that the material obeys the Beer–Lambert model. For an infinitely thin layer, this equation may be simplified to

$$f_{abs} = 2.303 \ \varepsilon(\lambda) \, c \, dL \tag{15.6}$$

Hence, the theoretical number of photons available to be absorbed by the sample per unit time in the volume $A$ dL is obtained by combining eqn (15.4) and eqn (15.6) to give eqn (15.7) where $K$ is a constant. This equation provides a means of measuring the efficiency that a particular photoinitiator can utilize the radiation of a given light source to produce an excited state. If these excited states all produce radicals then this is the photocuring efficiency (PE) of the source–initiator pair.

$$PE = K \int_{\lambda_1}^{\lambda_2} I(\lambda)\varepsilon(\lambda)\lambda d\lambda \tag{15.7}$$

Figure 15.2 shows the product of the spectral distribution of the light source times the molar absorption coefficient distribution of CQ and PPD.

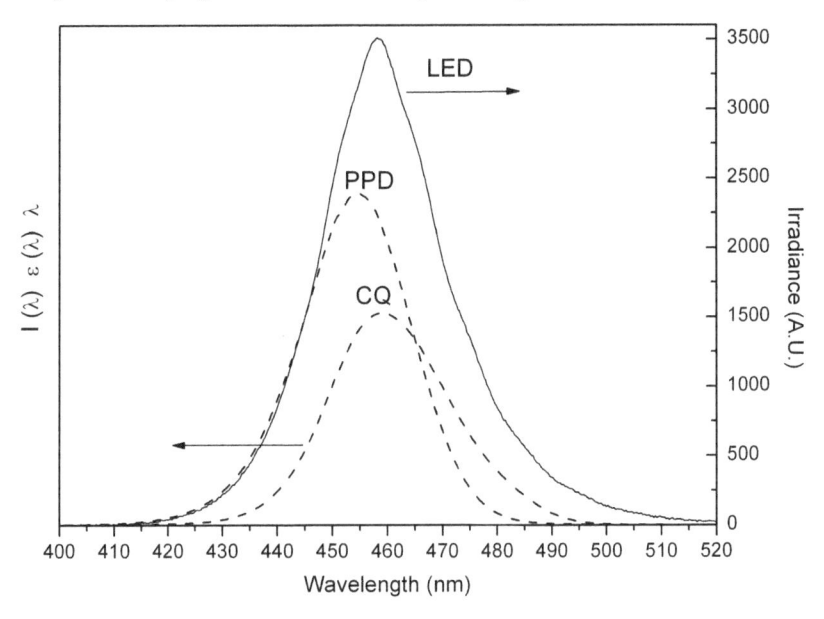

*Methacrylate and Epoxy Resins Photocured by Means of Visible LEDs* 331

**Figure 15.2** Product of the spectral distribution of the light source times the molar absorption coefficient distribution of CQ and PPD.

The PAE for PPD, from the integrated curves, is about 40% higher than the PAE for CQ. The knowledge of the absorption spectrum of a photosensitizer in conjunction with the emission spectrum of the curing lamp permits one to assess the photon-absorption efficiency, which measures the amount of photons actually absorbed by the photosensitizer when irradiated by a specific light unit. It is worth mentioning that resins photoactivated with PPD can be photopolymerized with commercially available LEDs sources with its irradiance centered at 390 nm, which is the absorption peak of PPD.[21]

From Figures 15.1 and 15.2 it is seen that the spectral output of the LED is concentrated in a comparatively narrow wavelength range. Since all of the spectral output of the LED is absorbed by the photoinitiator, more efficient curing should be possible, resulting in reduced curing time compared with conventional lamp light sources.

## 15.3.4 Light Attenuation in Thick Layers

Photopolymerization of thick systems (*ca.* 1–2 mm) is less common than radiation-induced polymerization of thin films and coatings (*ca.* 20–100 μm). This can be attributed to the reduced cure rate and depth of cure caused by the effect of the initiator absorptivity on the optical attenuation of the radiation with increasing sample thickness. On the other hand, in many photoinitiation systems, the photoinitiator is photobleached during the irradiation, producing moieties, which are transparent or have lower absorption coefficients at the wavelength of interest. Because the absorbing photoinitiator is destroyed upon radiation exposure, the incident radiation progressively penetrates deeper into thick samples.[18,19,22] Thus, the study of photobleaching characteristics of photoinitiators used for the polymerization of thick layers is an essential step for a proper understanding of the photopolymerization process.

CQ photoinitiator displays an intense dark yellow color due to the presence of the conjugated diketone chromophore that absorbs at 470 nm. During irradiation of CQ and reduction of one of the carbonyl groups, the conjugation is destroyed, causing a blue shift of the remaining ketone's absorption and loss of the yellow color. Thus, the photodecomposition rate of CQ can be assessed by measuring the decrease in absorbance as a function of the irradiation time by UV-vis spectroscopy. Figure 15.3 shows typical spectral changes during irradiation of solutions of CQ/amine in BisGMA/TEGDMA 70/30 (wt%) under irradiation with a LED centered at 470 nm.

It is seen that the absorption band of CQ at 470 nm decreases monotonically under irradiation. Because of the photobleaching of CQ with exposure time, the radiation intensity will increase with depth. Initially, the initiator concentration is uniform, and the light intensity will decrease with depth according to the Beer–Lambert law. Immediately after irradiation, the initiator will be photobleached at a rate proportional to the local light

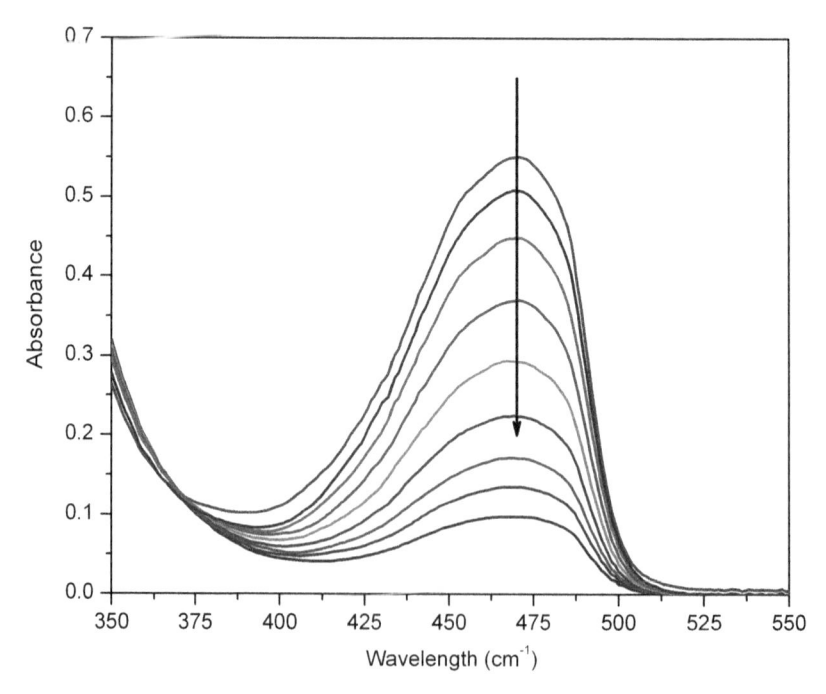

**Figure 15.3** Typical spectral changes during irradiation of solutions of CQ/amine in BisGMA/TEGDMA 70/30 (wt%) under irradiation with a blue LED (peak at 470 nm).

intensity, thereby leading to an initiator concentration gradient along the beam direction. The photobleaching of the photoinitiator is accompanied by a deeper penetration of the light through the underlying layers. This leads to temporal variation of local light intensity – depending on the nature of the photolysis products, this consumption of the photoinitiator can either lead to a increase in light intensity in the underlying lays (if the photolysis product is more transparent at the irradiating wavelengths) or, in some cases, a reduction in light intensity (if the photolysis product is strongly absorbing or in light scattering media).[22] Thus, a mathematical description of a photoinitiation process with a photobleaching photoinitiator must take into account the coupling of the effect of light intensity gradient and the initiator concentration gradient. The attenuation of the radiation as a function of depth into the material not only affects the cure kinetics but also often limits the depth of cure. Figure 15.4 shows the dimensionless CQ concentration as a function of irradiation time and depth.

At a given time and position within the sample, the degree of light attenuation is determined by the absorbance of the layers nearest to the irradiation source. The photobleaching rate is proportional to the intensity of the irradiation source. Consequently, current LED units of high intensity result in a fast photobleaching of the photoinitiator and, thus, in efficient polymerization of thick layers.

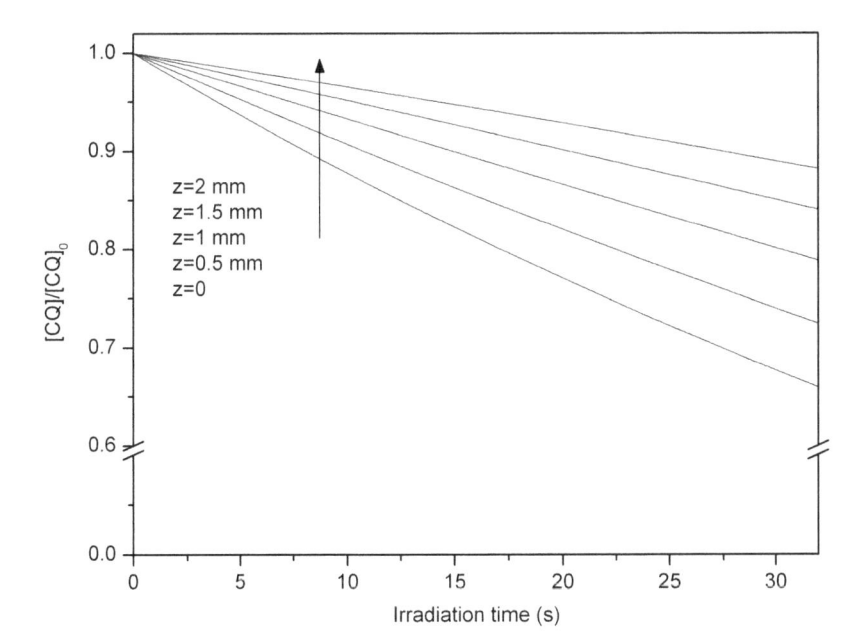

**Figure 15.4** Dimensionless concentration of CQ *vs.* irradiation time at different depth $z$. The sample was 2 mm thick. The resin was BisGMA/TEGDMA 70/30 (wt%) photoactivated with the CQ/amine pair. Irradiation was carried out with a blue LED.

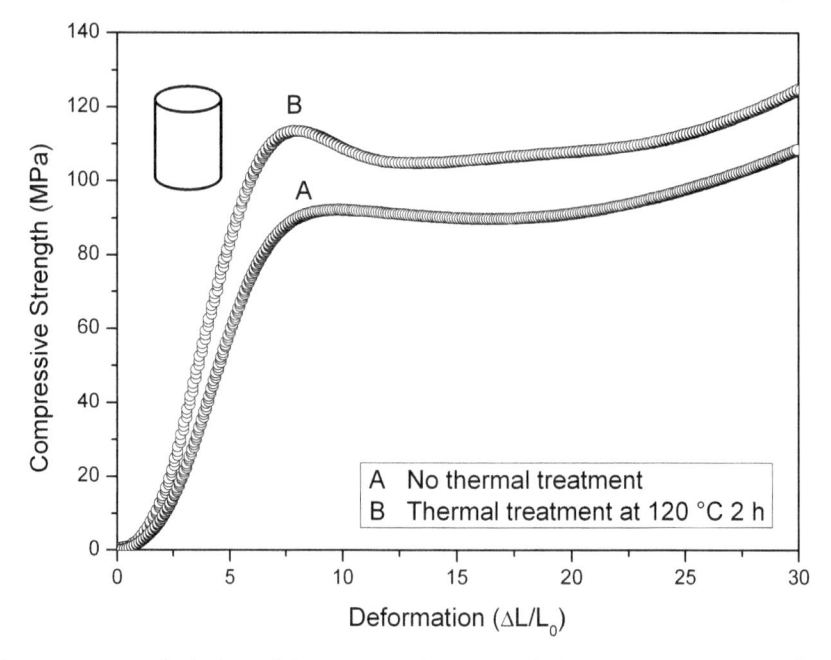

**Figure 15.5** Typical plots of the compression tests of BisGMA/TEGDMA 70/30 (wt%) resin. Cylindrical specimens having 6 mm internal diameter and 9 mm length were deformed.

The capacity of LEDs sources to cure thick layers is illustrated in Figure 15.5, which shows typical stress–deformation curves obtained from compression tests.

Cylindrical specimens were made by injecting the resins into polypropylene cylindrical molds. Samples were irradiated for 60 s on each side. After removal from the molds, the compression specimens were machined to reach their final dimensions. Cylindrical specimens having 6 mm internal diameter ($D$) and 9 mm length ($L$) were deformed between metallic plates. All test specimens were photopolymerized in ambient at room temperature ($20 \pm 2$ °C). A set of test specimens was also postcured at 120 °C for 2 h. The polymerization of dimethacrylates in the absence of external heating leads to glassy resins in which only some of the available double bonds are reacted. Before the completion of conversion, the vitrification phenomenon decelerates the reaction to a hardly perceptible rate.[17] The presence of nonreacted monomer can have a plasticizing effect on the polymer, thereby altering the physical and mechanical properties of the hardened material.[23] The postcuring treatment at 120 °C, which increases the chain mobility, increases monomer conversion and reduces the plasticizing effect of the nonreacted monomer on the mechanical behavior of the nanocomposites.[17] Thus, the higher value of compressive strength of samples subjected to a postcuring treatment are attributed to a reduced amount of nonreacted monomer.[23]

The LED unit used in measurements of Figures 15.3–15.5 produces a narrow spectrum of blue light in the 400 to 500 nm range (with a peak wavelength of about 467 nm),[18,19] which is the useful energy range for activating the CQ molecule. Since a narrow band of light is emitted, all of the spectral output of the LED is absorbed by the photoinitiator. In addition, because there is no infrared emission, the curing lights have low amounts of wasted energy, leading to minimum heat generation, which obviates the need for cooling fans. Results reported in Figures 15.3–15.5 demonstrate the potential of the blue LED technology for curing thick specimens.

### 15.3.5 Oxygen Inhibition in Free-Radical Polymerization

Oxygen inhibition is known to be one of the major problems in light-induced free-radical polymerization. Three inhibition mechanisms involving oxygen in photoinduced radical polymerization are: quenching of the photoinitiator triplet state, reaction with primary radicals or reaction with propagating radicals.[24] Photoinitiating systems that involve tertiary aliphatic amines can be useful in suppressing inhibition by the reaction of oxygen with the C-based radicals through the radical chain-transfer process which regenerates the initiation reaction:[24]

$$R^{\bullet} + O_2 \rightarrow ROO^{\bullet} \quad (15.8)$$

$$ROO^{\bullet} + A\text{-}H \rightarrow ROOH + A^{\bullet} \quad (15.9)$$

The radical, $R^{\bullet}$, in eqn (15.8) may be an amine or monomer radical.

Oxygen is known to inhibit free-radical polymerizations by reacting with initiator, primary, and growing polymer radicals to form peroxy radicals.[24] The peroxy radicals are more stable and do not readily reinitiate polymerization, and thus, the oxygen essentially terminates or consumes radicals. The atmospheric oxygen inhibits photopolymerization by scavenging initiator radicals and so competing with the reaction between initiator radicals and monomer. Consequently, the photocuring reaction is delayed until nearly all the oxygen is consumed. Another variable that dramatically affects the extent of oxygen inhibition of free-radical photopolymerization is the initiation rate. If the initiation rate (determined by the initiator concentration and activity and the radiation intensity) is sufficiently low, all of the radicals generated could be consumed in the inhibition process. Thus, polymerization reactions carried out with LED sources of high intensity will result in reduced induction periods because the inhibiting species will be consumed faster.

### 15.3.6 Conversion of Methacrylate Groups under Irradiation

The most widely used technique to assess the extent of polymerization of methacrylate monomers has been the measurement of the heat of reaction liberated during polymerization by differential scanning calorimetry (photo-DSC). The photo-DSC technique has been extensively used to follow the reaction kinetics of many photoinitiated polymerization reactions of dimethacrylate monomers and comonomer mixtures. However, photo-DSC suffers from rather long response time, which makes it impossible to monitor very fast reactions. In addition, because reaction rate rather than conversion is monitored, the technique is not adequate to monitor accurately the slow increase in monomer conversion when the vitrification process decelerates the reaction rate. Alternatively, a variety of infrared spectroscopic techniques, which enable the direct measurement of monomer conversion, have been an extremely valuable tool for the characterization of methacrylate resins. Among them, the near-IR spectroscopy (NIR) has been recently applied to the measurement of conversion in methacrylate resins and composites.[17,25] The nondestructive nature of the NIR technique provides a simple means to follow conversion on individual specimens as a function of time. The absorption of C–H is located at about 4743 cm$^{-1}$ while the first overtone is at 6165 cm$^{-1}$. Because the base line drops sharply near the 4743 cm$^{-1}$ band, reliable peak-area measurements are difficult in this region. Conversely, the region at 6165 cm$^{-1}$, is very stable and there is no ambiguity in baseline construction. Figure 15.6 shows methacrylate conversion *versus* irradiation time of a 70 : 30 BisGMA/TEGDMA resin activated for visible light polymerization with the CQ/amine pair.[20] The light source

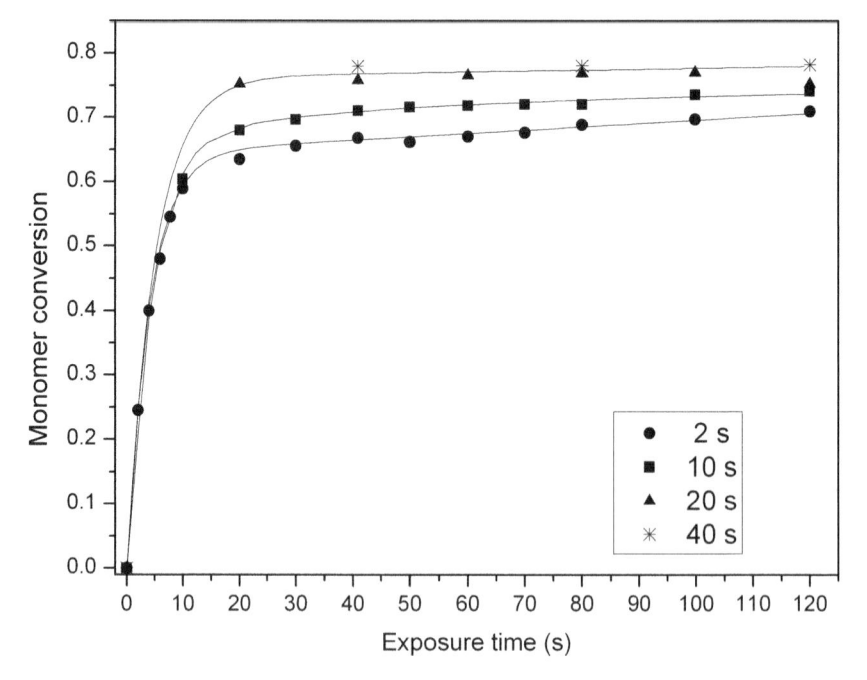

**Figure 15.6** Conversion profiles of BisGMA/TEGDMA 70/30 (wt%) resin photoactivated with CQ in combination with ethyl-4-dimethylaminobenzoate (EDMAB) for visible-light polymerization.

was a LED unit with a wavelength range 420–520 nm and irradiance equal to 400 mW cm$^{-2}$.

The specimens were irradiated at regular time intervals. Spectra were collected immediately after each exposure interval. The conversion profiles were calculated from the decay of the absorption band located at 6165 cm$^{-1}$. The effect of the radiation dose on the extent of polymerization was examined by irradiating four samples for different intervals. One sample was irradiated in 2 s intervals up to 10 s and then it was irradiated in 10 s increments. The other three samples were irradiated using 10 or 20 or 40 s irradiation periods. In each of these experiments, there was a 2 min data acquisition and software data processing interval between two successive irradiations periods. For the 2 and 10 s exposure times, the samples achieved lower conversion than samples cured by more continuous irradiation. However, for the 20 and 40 s exposure intervals, the final conversion was almost the same. This influence of the exposure period on the monomer conversion can be attributed to the temperature rise in the sample due to the exothermic reaction. The maximum temperature reached during polymerization depends on the relative rate of heat generation to heat transfer between the sample and the surroundings. As described above, the cure schedule of the samples consisted of the illumination period followed by the 2 min data acquisition period. During the latter period, the sample is not irradiated and no exotherm is generated, but heat transfer to the environment still occurs, which contributes to a decrease in the sample temperature. Thus, the temperature of samples irradiated for shorter periods does not rise as high as samples with longer irradiation periods. A higher sample temperature increases the mobility of the reaction environment (*i.e.* monomer, radical and polymer) and consequently increases the reaction-rate parameters. Thus, the greater double-bond conversion in samples subjected to relatively continuous illumination is attributed to a combination of both photo and thermal effects. In addition, polymerization of these monomers causes the glass-transition temperature ($T_g$) to rise. Once the $T_g$ approaches the photocuring temperature, the material will vitrify and the reaction will stop, thus limiting maximum conversion. Therefore, the samples photoirradiated for longer periods attained a higher temperature and therefore reached a higher limiting conversion. Studies under irradiation in steps (Figure 15.6) can be carried our accurately with LEDs sources because they do not require warm-up period, *i.e.* they reach their full intensity in nanoseconds.

Polymerization of multifunctional dimethacrylates exhibit complex behavior due to the fact that the kinetics depends on the mobility of the medium, which changes during the process.[26,27] Figure 15.7 shows the characteristic curing behavior of a 70:30 BisGMA/TEGDMA blend.[28]

The data illustrates autoacceleration immediately after irradiation commences so that the reaction rate reaches a maximum value at a double-bond conversion of about 10%,[26,27] As the polymerization progresses, the polymer molecular weight increases and gelation occurs so that the mobility of the

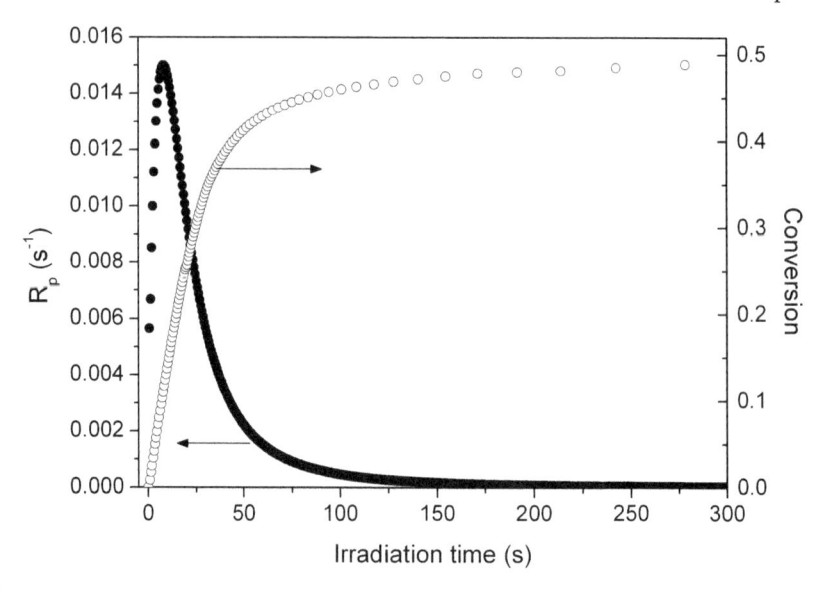

**Figure 15.7** Double-bond conversion and polymerization rate *versus* irradiation time during polymerization of BisGMA/TEGDMA 70/30 (wt%) resin containing 1 wt% CQ/EDMAB.

environment decreases resulting in a reduction in the termination rate constant and the initiation efficiency and an increase in reaction rate as shown in Figure 15.7. The photoinitiator in the present system is not totally consumed even at the end of the reaction. Thus, the observed rapid drop in the cure rate and the failure of the system to attain full conversion (Figure 15.7) is due to a vitrification effect where the propagation reaction also becomes diffusion controlled because the temperature of the system approaches the conversion-dependent glass-transition temperature ($T_g$). At this stage, the reaction rate approaches zero within the time scale of the experiment.

An alternative technique to assess the extent of polymerization of methacrylate resins in a continuous way is the measurement of the shrinkage development during irradiation.[29] The volume shrinkage during polymerization should be related to the number of functional groups that have reacted before the system enters the glassy state. A linear correlation, between volume contraction and mole of converted double bonds, was first proposed in 1953 by Loshaek and Fox[30] and more recent literature[31] still refers to this early work. Figure 15.8 shows typical plots of shrinkage *vs.* irradiation time of a methacrylate resin photoactivated with CQ/amine.[29]

Specimens were irradiated with a blue LED (11.5 mW cm$^{-2}$, peak 467 nm) for 5, 10 and 20 s. The shrinkage was monitored during and after exposure to complete 300 s. The lines show the times at which the LED was turned off. Only the first 120 s of test are shown to make the plot clearer. It is seen that a considerable amount of shrinkage took place after the LED was turned off.

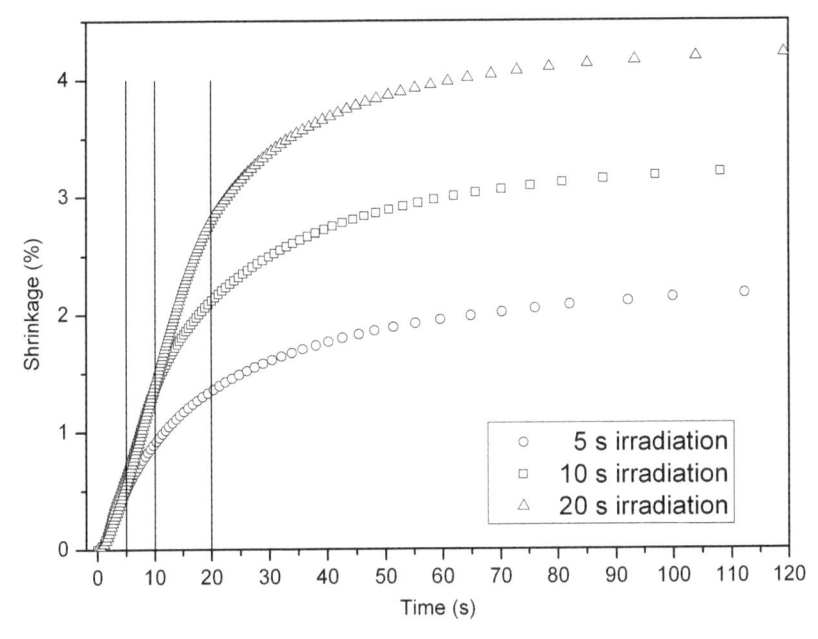

**Figure 15.8** Typical plots of shrinkage vs. time of samples irradiated for 5, 10 and 20 s. The lines show the times at which the LED was turned off. A considerable amount of shrinkage took place after the LED was turned off. The samples were 1 mm thick.

Results of shrinkage evolution presented in Figure 15.8 show that a considerable amount of shrinkage took place after the light was turned off. This additional "dark cure" is attributed to the presence of free radical in the network after irradiation had ceased. Although no new radicals are generated from initiation when the irradiation is terminated, the remaining free radicals continue to propagate and terminate. The extent of post-polymerization is greater in samples with lower initial conversion due to a greater mobility of the network compared to samples that were more highly converted. The intensity of the LEDs sources can be easily set by varying the electrical voltage through the semiconductor: In studies of Figure 15.8, measurements were carried out at low values of irradiance in order to focus on the initial stages of the polymerization, *i.e.* the time before which the double-bond consumption reaches a plateau.

## 15.4 Light-Curable Epoxy Resins

### 15.4.1 Epoxy Monomers

Epoxide-containing monomers are among the most advantageous substrates for photoinitiated cationic polymerization. Epoxide-based coatings, adhesives, and composites are widely used in industry and are noted for their

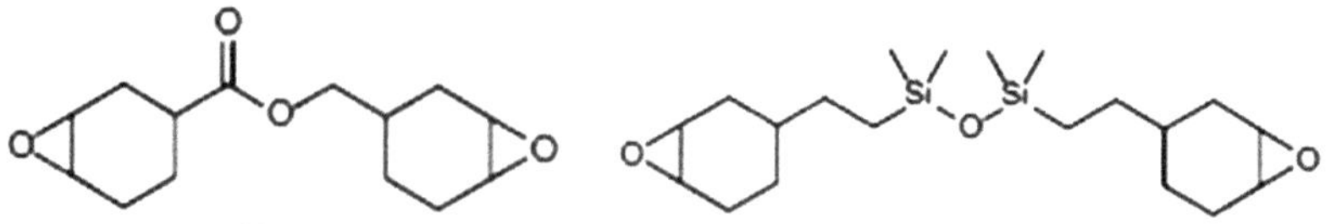

**Scheme 15.5** Structure of epoxy monomers used in light-curable formulations.

outstanding chemical resistance, adhesion, and mechanical properties. However, the photopolymerization of commercially available epoxy monomers is generally regarded as proceeding at slower rates than the corresponding free-radical polymerizations. This is due to the fact that such monomers were designed primarily for use with various amine and anhydride hardening agents and not for photoinduced cationic ring-opening polymerizations. Accordingly, research work carried out by Crivello et al.[32–34] has been directed to the preparation of new epoxy-containing monomers designed specifically for photoinitiated polymerization. Scheme 15.5 shows the structure of typical biscycloaliphatic difunctional epoxy monomers: 3,4-epoxycyclohexylmethyl 3′,4′-epoxycyclohexane carboxylate (E-I) 1,3-bis(3,4-epoxycyclohexyl-2-ethyl),1,1,3,3-tetramethyldisiloxane (E-II). Since the monomer molecules contain two epoxy groups, chain propagation results in a highly crosslinked polymer network.

## 15.4.2 Cationic Photoinitiators

Cationic photopolymerization has become an important method of crosslinking epoxy and vinyl ether monomers, and onium salts are well known as very effective photoinitiators.[5–9] The use of cationic photoinitiators, such as diaryliodonium and triarylsulfonium salts, provides a convenient method of generating powerful Brønsted acids *in situ*, which is the primary species that initiates polymerization, and obviates the difficulties of preparing and handling these materials. The cationic photoinitiators can be dissolved into a multifunctional monomer, and then the polymerization can be triggered on demand by irradiating the mixture with light. The high rate of crosslinking achieved on irradiation is ideal for high-speed applications such as printing inks, coatings and adhesives. However, the major absorption bands of most of these initiators fall in the deep UV region (210–250 nm),[7] and the absence of overlap of these bands with the emission spectra of medium and low-pressure mercury lamps limited their application. Thus, photosensitizers are used in order to broaden the spectral sensitivity of onium salts to longer wavelengths.[32] On irradiation with UV light (210–250 nm), diaryliodonium salts undergo fragmentation to yield a variety of reactive intermediates, including radicals, cations and radical cations.[5–9] The cationic species interact with a proton source, usually the monomer, impurity or solvent to generate strong Brønsted acids. Initiation of polymerization proceeds by protonation of the monomer and is followed by polymerization *via* a dark, *i.e.* nonphotochemical, process. By using dyes as photosensitizers, it is possible to extend the spectral sensitivity of diaryliodonium salts into the visible region of the spectrum.[32] The mechanisms of photosensitization of onium salts have been summarized by Yagci and Reetz,[7] and Crivello and Sangermano.[33] It is worth noting that the number of photosensitizers operating in the long-wavelength UV and visible regions is limited. Crivello and Lam[32] reported attempts to use visible-light-absorbing dyes to sensitize the decomposition of iodonium salts. After screening nearly 75 different dyes, they reported that only five dyes possessed appreciable sensitizing ability. The use of camphorquinone (CQ) as a visible-light photosensitizer in the iodonium-initiated cationic polymerization of epoxy and vinyl ethers has been reported in recent studies. Crivello and Sangermano studied polymerization photoinitiated by a diaryliodonium salt in combination with CQ of monomers that can serve as hydrogen donors.[33] In addition, Crivello developed an efficient three-component visible-light-sensitive photoinitiator system, based on CQ/benzyl alcohol/diaryliodonium salt, for the cationic photopolymerization of epoxide monomers.[34] Oxman *et al.*[35] studied the free-radical/cationic hybrid photopolymerizations of acrylates and epoxides using CQ as sensitizer in a three-component photoinitiator system. Vallo and coworkers reported results on visible-light polymerization of epoxy monomers (E-I and E-II, Scheme 15.5) using a photoinitiator system based on an iodonium salt, a ketone (CQ), and amine (ethyl-4-dimethyl aminobenzoate, EDMAB).[36] The function of the aromatic amine in the cationic polymerization is dictated by its basicity and nucleophilicity toward the propagating cationic centre. If the amine has low basicity and the carbocation is less electrophilic, then the amine functions mainly as coinitiator. Otherwise, it will function as a terminator.

### 15.4.3 Cationic Photopolymerization of Epoxy Monomers

One of the advantages of cationic systems is the absence of air inhibition, a property that eliminates the need for an inert atmosphere during curing, and distinguishes cationic from radical polymerization. Cationic polymerization, once initiated, may continue to proceed after the light source has been removed. This process, called the "dark reaction", is the result of the ability of the relatively long-lived protonic acid or Lewis acid species to continue the polymerization. Free radicals, on the other hand, are extinguished by a variety of termination steps, and no new radicals are formed from the photoinitiators in the absence of light. Furthermore, the cationic photopolymerizable systems are characterized by the absence of toxicity or irritation properties and for this reason are good alternative to acrylates and methacrylates systems usually employed in radical processes.

The cationic polymerization of epoxy resins can be carried out by the use of three-component photoinitiator systems that generates reactive species that initiates polymerization under irradiation with visible light. In particular, the three-component system based on iodonium salt/CQ/EDMAB is an efficient photoinitiator of monomers E-I and E-II (Scheme 15.5). Measurements of epoxy groups conversion showed a rapid polymerization reaction of monomers (E-I) and (E-II) under irradiation with a blue LED (peak at 460 nm). Moreover, almost complete conversion of (E-I) occurs in the absence of external heating.[36] The extent of reaction is markedly affected by sample thickness. For example, the conversion of epoxy groups in thin films (~50 µm) was 48% after 100 s irradiation while in 2 mm thick samples the conversion was 98% after 5 s irradiation. This is attributed to different temperatures reached during polymerization in samples of different thicknesses. The polymerization reaction of epoxy monomers is highly exothermic, which results in an increase in the sample temperature during the process and this effect becomes more important as the sample thickness increases.[36] A higher sample temperature induces an increase in the mobility of the reaction environment (*i.e.* monomer, propagating species and polymer) and consequently increases the reaction rate. Therefore, the greater conversion of epoxy groups in thick samples is attributed to thermal effects. It is worth mentioning that the values of conversion measured at the irradiated and nonirradiated surfaces in 2 mm thick specimens were similar. In contrast to free-radical photopolymerization, the only part of a photoinitiated cationic polymerization that is dependent on light is the photolysis of the photoinitiator. Once the active species are formed, the polymerization itself proceeds by a normal cationic process. Thus, the similar values of conversion at the irradiated and nonirradiated surfaces in 2 mm thick specimens are the result of the well-known "dark cure" or "post-polymerization" effect. This contrasts with most free-radical photopolymerizations, which cease after the irradiation is extinguished because of rapid termination.

The possible initiating mechanism in systems containing a diaryliodonium salt (Ph$_2$ISbF$_6$) in combination with CQ has been proposed by previous researchers.[33] Irradiation of the CQ with visible light results in the formation of its singlet state, which is rapidly converted to its triplet state by intersystem crossing (Scheme 15.6). The excited CQ molecule is initially reduced by a hydrogen donor (in this case the amine) to the ketyl radical, which in turn is oxidized back to CQ by the iodonium salt. The resulting strong Brønsted acid derived from this process initiates the cationic ring-opening polymerization. In the case of the CQ/amine pair (Scheme 15.7). The mechanism is analogous to that proposed by Bi and Neckers for the visible-light cationic polymerization of cyclohexene oxide in the presence of a diaryliodonium salt.[37] These authors developed a system based on a xanthene dye in combination with an aromatic amine to sensitize the diaryliodonium salt. Although the initiating mechanism is apparently complex, it is assumed to involve the generation of α-amino radicals by the abstraction of a hydrogen atom from the amine by the photoexcited dye (Scheme 15.7).

The diaryliodonium salt oxidizes the α-amino radicals to the respective cations, which initiate the polymerization. Subsequently, these free radicals

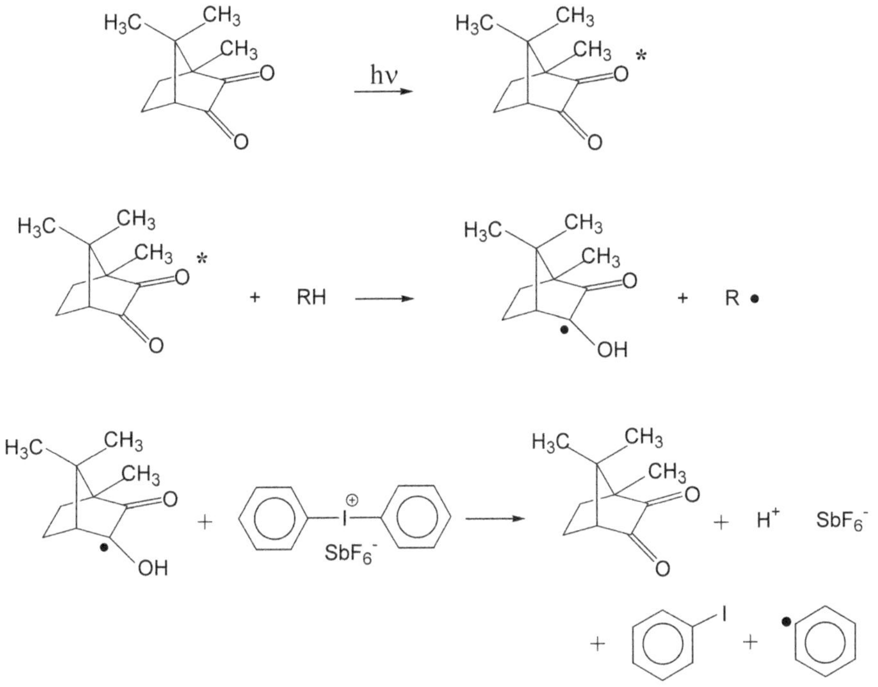

*Methacrylate and Epoxy Resins Photocured by Means of Visible LEDs*     343

**Scheme 15.6** Photosensitization of diaryliodonium salt (Ph$_2$ISbF$_6$) by CQ.

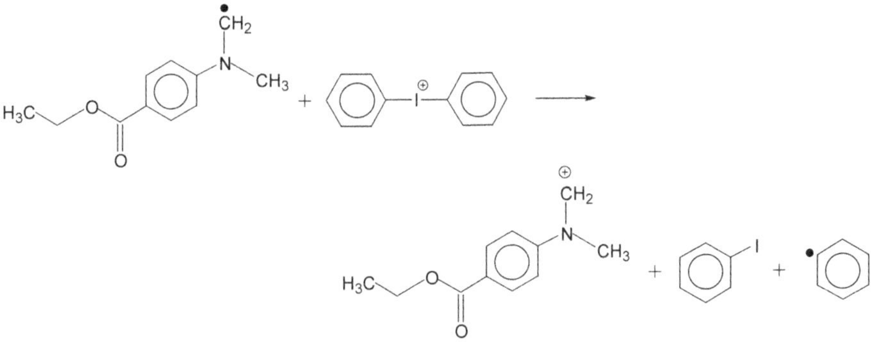

**Scheme 15.7** Photosensitization of diaryliodonium salt (Ph$_2$ISbF$_6$) by CQ/amine.

induce the chain decomposition of the diaryliodonium salt, producing a dramatic rate enhancement in the polymerization.

Crivello developed a photoinitiator system based on CQ in combination with a benzyl alcohol to generate free radicals by the absorption of visible light.[34] Subsequently, the radicals participate in the free-radical chain-induced decomposition of a diaryliodonium salt. A very effective photopolymerization of the monomer E-I (Scheme 15.5) was observed using the three-component photoinitiator. In contrast, no photopolymerization was detected even after visible-light irradiation for 5 min when only CQ and diaryliodonium salt were present. The absence of polymerization of monomer E-II containing iodonium salt and CQ reported by Crivello can be attributed to a low irradiance of the light source used. At high light intensity, the polymerizations generally proceed to high conversion, whereas at low irradiation intensity the conversions are reduced. This is due to the presence of termination reactions that are more important when fewer reactive species are generated at low light intensities.

In the case of polymerization of thin films, the conversion of monomer E-I containing the three-component photoinitiator increased monotonically with time, reaching 48% after 100 s irradiation while monomer E-II with the same photoinitiator system reached 70% conversion after 10 s irradiation. No change in temperature was recorded during irradiation of the thin films and therefore, in the absence of thermal effects, the different degrees of reaction reached by monomers E-I and E-II are explained in terms of the reactivity of each monomer. Crivello and colleagues reported a detailed investigation of the reactivity of a wide variety of epoxy monomers.[38,39] Based on the results of those studies, the authors concluded that epoxide ring strain is one of the most important factors in determining their reactivity. Monomers containing the highly strained epoxycyclohexane and epoxycyclopentane rings were found to be the most reactive and considerably more reactive than open-chain epoxy compounds. On the other hand, Crivello and Varlemann found that epoxide monomers containing ester carbonyl groups undergo polymerization at substantially lower rates than monomers that do not contain those groups.[38] Further studies showed that this effect is due to the higher nucleophilic and basic character of the ester carbonyl, which leads to a different mechanism of propagation in these compounds. The ester carbonyl group can directly interact with the protonated epoxide group in either an intermolecular or intramolecular fashion to produce a comparatively low reactivity dialkoxycarbenium ion. Alternatively, the carbonyl group may reduce the activity of epoxide monomers through hydrogen bonding with the protonated epoxide group produced during the initiation process. Conversely, Jang and Crivello[39] concluded that silicone epoxide monomers such as E-II (Scheme 15.5) that incorporate both epoxycyclohexane groups and siloxane linkages are among the most reactive epoxide monomers. They proposed that the outstanding reactivity of this monomer could be attributed to two major factors. First, the ring strain inherent in the epoxycyclohexane ring system is probably the largest contributing factor. Secondly, there are no basic groups in the structure of E-II that can compete with the epoxide oxygen for either the initiating photogenerated acid or the positively charged growing chain ends. In addition, siloxane (Si–O–Si) bonds are not ether-like and are not readily protonated.

The high intensity provided by current visible LEDs permits the rapid cure of biscycloaliphatic epoxy monomers. This is due to the high rate generation of reactive species compared with termination reactions. Results of epoxide groups conversion in resins irradiated with a LED source of very high irradiance (700 mW cm$^{-2}$) highlight the inherent interlinking of light intensity and extent of reaction in visible-light polymerization of epoxy resins.

## 15.5 Conclusion

Improvements in LED technology introduced during the last decade have resulted in the development of several types of commercial LED light-curing units that have improved intensity output, which results in increased performance. It has been reported that in the past thirty years the light intensity has increased twenty-fold while the cost has decreased ten-fold each decade. LED light-curing units that are capable of delivering power densities of about 1000 mW cm$^{-2}$ have been marketed recently. Still, considerable efforts are being devoted now to reduce optical losses that decrease the efficiency photoemission.

The spectral output of LED sources is concentrated in a comparatively narrow wavelength range. Since all of the spectral output of the LED is absorbed by the photoinitiator, more efficient curing is possible, resulting in reduced curing time and increased depth of cure compared with conventional light sources. In addition, it yields the positive aspect of eliminating the IR and UV components. As a result, when compared to conventional curing units, there is less heat transfer to the substrate and no harmful UV rays.

In the particular case of epoxy monomers, the high intensity provided by current LEDs, permits the rapid cure of biscycloaliphatic epoxy monomers when irradiated with visible light. This is due to the high generation rate of reactive species compared with termination reactions. Results of epoxide groups conversion in resins irradiated with a LED source of very high irradiance (700 mW cm$^{-2}$) highlight the inherent interlinking of light intensity and extent of reaction in visible-light polymerization of epoxy resins.

## References

1. *Photopolymerization Fundamentals and Applications*, ed. A. C. Scranton, C. N. Bowman and R. W. Peiffer, ACS Symp. Series 673, American Chemical Society, Washington, 1997.
2. E. Andrzejewska, *Prog. Polym. Sci.*, 2001, **26**, 605.
3. R. S. Davidson, *J. Photochem. Photobiol., A*, 1993, **73**, 81.
4. S. Jonsson, P. E. Sundell, J. Hultgren, D. Sheng and C. E. Hoyle, *Prog. Org. Coat.*, 1996, **27**, 107.
5. Y. Yagci, S. Jovkusch and N. J. Turro, *Macromolecules*, 2010, **43**, 6245.
6. J. Crivello, *J. Polym. Sci., Part A: Polym. Chem.*, 1999, **37**, 4241.
7. Y. Yagci and I. Reetz, *Prog. Polym. Sci.*, 1998, **23**, 1485.
8. C. Decker, T. N. Viet and H. P. Thi, *Polym. Int.*, 2001, **50**, 986.
9. J. Crivello and R. Acosta Ortiz, *J. Polym. Sci., Part A: Polym. Chem.*, 2001, **39**, 2385.

10. E. F. Schubert, *Light-Emitting Diodes*, Cambridge University Press, Cambridge, UK, 2nd edn, 2006.
11. A. I. Zhmakin, *Phys. Rep.*, 2011, **498**, 189.
12. K. M. Wiggins, M. Hartung, O. Althoff, C. Wastian and S. B. Mitra, *J. Am. Dent. Assoc., JADA*, 2004, **135**, 1471.
13. K. D. Jandt and R. W. Mills, *Dent. Mater.*, 2013, **29**, 605.
14. M. T. Lemon, M. S. Jones and J. W. Stansbury, *J. Biomed. Mater. Res.*, 2007, **83A**, 734.
15. G. Odian, *Principles of Polymerization*, John Wiley, New York, 2010.
16. V. Mucci and C. I. Vallo, *J. Appl. Polym. Sci.*, 2012, **123**, 418.
17. W. F. Shroeder and C. I. Vallo, *Dent. Mater.*, 2007, **23**, 1313.
18. S. Asmusen, G. Arenas, W. D. Cook and C. I. Vallo, *Dent. Mater.*, 2009, **25**, 1603.
19. S. Asmussen, G. Arenas, W. D. Cook and C. I. Vallo, *Eur. Polym. J.*, 2009, **45**, 515.
20. W. Schroeder, W. D. Cook and C. I. Vallo, *Dent. Mater.*, 2008, **24**, 686.
21. S. Asmussen and C. I. Vallo, *J. Photochem. Photobiol., A*, 2009, **202**, 228.
22. G. Terrones and A. J. Pearlstein, *Macromolecules*, 2001, **34**, 3195.
23. C. I. Vallo, *J. Biomed. Mater. Res.*, 2000, **B53**, 717.
24. J. Jakubiak, X. Allonas, J. P. Fouassier, A. Sionkowska, E. Andrzejewska, L. A. Linden, *et al.*, *Polymer*, 2003, **44**, 5219.
25. M. Trujillo, S. M. Newman and J. W. Stansbury, *Dent. Mater.*, 2004, **20**, 766.
26. C. N. Bowman and C. J. Kloxin, *AIChE J.*, 2008, **54**, 2775.
27. M. D. Goodner and C. N. Bowman, *Macromolecules*, 1999, **32**, 6552.
28. V. Mucci, W. D. Cook and C. I. Vallo, *Polym. Eng. Sci.*, 2009, **49**, 2225.
29. V. Mucci, G. Arenas, R. Duchowicz, W. D. Cook and C. I. Vallo, *Dent. Mater.*, 2009, **25**, 103.
30. S. Loshaek and T. G. Fox, *J. Am. Chem. Soc.*, 1953, **75**, 3544.
31. M. Dewaele, D. Truffier-Boutry, J. Devaux and G. Leloup, *Biomaterials*, 2006, **22**, 359.
32. J. Crivello and J. Lam, *J. Polym. Sci., Part A: Polym. Chem.*, 1978, **16**, 2441.
33. J. Crivello and M. Sangermano, *J. Polym. Sci., Part A: Polym. Chem.*, 2001, **39**, 343.
34. J. Crivello, *J. Polym. Sci., Part A: Polym. Chem.*, 2009, **47**, 866.
35. J. D. Oxman, D. W. Jacobs, M. C. Trom, V. Sipani, B. Fiek and A. B. Scranton, *J. Polym. Sci., Part A: Polym. Chem.*, 2005, **43**, 1747.
36. W. Schroeder, S. V. Asmussen, M. Sangermano and C. I. Vallo, *Polym. Int.*, 2013, **62**, 1368.
37. Y. Bi and D. Neckers, *Macromolecules*, 1994, **27**, 3683.
38. J. Crivello and U. Varlemann, *J. Polym. Sci., Part A: Polym. Chem.*, 1995, **33**, 2473.
39. M. Jang and J. Crivello, *J. Polym. Sci., Part A: Polym. Chem.*, 2003, **41**, 3056.

CHAPTER 16

# Waste Materials Cured and Modified by Irradiating and their Use in Concrete

GONZALO MARTÍNEZ-BARRERA*[a] AND OSMAN GENCEL[b]

[a] Laboratorio de Investigación y Desarrollo de Materiales Avanzados (LIDMA), Facultad de Química, Universidad Autónoma del Estado de México, Km.12 de la carretera Toluca-Atlacomulco, San Cayetano 50200, Mexico; [b] Civil Engineering Department, Faculty of Engineering, Bartin University, 74100 Bartin, Turkey
*Email: gonzomartinez02@yahoo.com.mx

## 16.1 Introduction

Transformation from a conventional consumption-based society to a sustainable society has great importance for several reasons such as lowering the pollution of the natural environment, prevention of exhaustion of natural resources and slowing down filling of final waste-disposal facilities. Recycling has the potential to reduce the amount of wastes disposed of in landfills and to preserve natural resources. Recycling, one of the strategies in minimizing waste offers three benefits: (1) reduces the demand for new resources; (2) cuts down on transport and production energy costs; (3) utilizes waste that would otherwise go into landfill sites. One novel alternative for recycling materials is gamma radiation, which has proved its efficacy for modification and improving of structural and physicochemical properties.

Concrete is one of the principal materials for structures and it is widely used for many applications all over the world. Thus, it is compelling to use life-cycle and sustainable-engineering approaches to concrete technology.[1] Concrete containing waste and recycled materials can support construction sustainability and contribute to the development of the civil engineering areas, minimizing the consumption of natural resources and producing more efficient materials.[2]

This chapter is divided into three sections, in the first one the modifications of waste materials by using gamma radiation are described; such materials are PET bottles, Tetra Pak packages and rubber tires. The second section describes the effects of these waste materials on the mechanical properties of concrete; and in the last section those effects provoked by gamma radiation in concrete containing these waste materials.

## 16.2 Waste Materials Cured and Modified by using Gamma Radiation

When using gamma radiation on polymers, three main processes occur: crosslinking, scission, and grafting of chains. The likelihood of each process occurring depends on the specific polymer properties. The gamma-rays have neither mass nor charge, and they are emitted from different sources, $^{137}$Cs or $^{60}$Co with 0.66 MeV and 1.33 MeV of energy, respectively.[3] The gamma-radiation exposure can provoke either lateral chain fracture and produce radicals combining with each other in order to bond adjacent molecules (crosslinking) or can break the main chain (scission), reducing the mean molecular weight (degradation).[4] Whether both of these processes occur in similar proportions or whether one or the other dominates is dependent on the chemical composition of the polymer. The application of gamma radiation for recycling polymers has increased its acceptation as a current technology due to the ecologic and economical features and mainly to the capacity of ionizing radiation to alter and improve the structure and properties of practically any polymeric material.

The Tetra Pak packages recycling is focused on their components: cellulose (75%), polyethylene (20%), and aluminum (5%); which are recovered in three different ways: (i) generation of energy through incineration, (ii) the recovery of aluminum in pyrolysis ovens, and (iii) the processing of the mixture of PE and aluminum to obtain high-end plastic lumber products.

With the exception of cotton, the main components of natural fibers are cellulose, hemicellulose, and lignin.[5] Linear cellulose molecules are linked laterally by hydrogen bonds to form linear bundles, giving rise to a crystalline structure. The degree of crystallinity is one of the most important structural parameters in cellulose; because physical and mechanical properties depend on it. The crystalline regions are interrupted every 60 nm with noncrystalline amorphous regions; such crystalline regions may contain occasional kinks or folds in the polymer chain, called "defects". When increasing the crystalline/amorphous regions ratio the rigidity of cellulose fibers increases, but flexibility decreases.[6,7]

Gamma radiation causes break-up of cellulose to shorter chains, which are water soluble, and it most likely leads to an "opening of additional microcracks", in which water molecules can easily penetrate. The softening temperature ($T_s$), is 235 °C for untreated cellulose, which decreases linearly up to doses of about 300 kGy, where any changes in the degree of crystallinity are detected, but degree of polymerization value decreases according to the dose increase. Therefore, it is assumed that the $T_s$ shift to lower temperature range would be due to the scission of chemical bonds such as a glucosidic bond. On the other hand, the $T_s$ shift in the range exceeding 300 kGy would be due to degradation of crystalline structure.[8]

At 500 kGy a higher solubility is obtained due to depolymerization and destruction of hemicellulose. The side-chain constituents of hemicelluloses (galactans) are affected sooner and to a greater extent than xylans or mannans that represent the primary backbone of the hemicelluloses. Degradation of cellulose or lingocellulosic material produces huge numbers of water-soluble or insoluble oxygenated compounds. Although the water content of aqueous phases is high, it contains some valuable chemicals. One of them is phenolic compounds.

The two highest-volume polymers HDPE and LDPE are classified as materials that primarily undergo crosslinking when irradiated under an inert atmosphere; however, they are very susceptible to radiation-induced oxidative degradation (chain scission) at relatively low doses in the presence of air. The melting temperature of neat HDPE is 138.2 °C, with a crystallinity of 49.1%. No significant variations of these values are observed after irradiation, except for samples having received the highest irradiation dose (100 kGy), for which the crystallinity decreased to 40% due to crosslinking. PE is known to suffer from degradation and loss of mechanical properties after gamma irradiation that is induced crosslinking leads to a strong decrease of elongation at break and impact strength values, counter balanced by an increase of Young's modulus and yield stress values.

Many efforts for recovering recycled-LDPE samples through the improvement of properties have been made. The recycled PE required a higher dose than virgin PE to achieve crosslinking. Moreover, exploratory research in which the radiation-induced breakdowns of PE scrap was carried out to generate lower molecular weight micropowders.[9]

In the case of PET some authors report changes due to the chain scission process at low dose (up to 10 kGy), while others note them at high dose (above 120 kGy). PET irradiated at 25 and 50 kGy shows a small but meaningful difference in ester ethyl of terephthalic acid generation, less than 1 mg kg$^{-1}$ for nonirradiated PET samples and 2 mg kg$^{-1}$ after irradiating at 50 kGy; measurable quantities of low molecular mass are formed and they resist gamma irradiation. Irradiated PET (from 100 to 300 kGy) was studied; at 100 kGy it shows a low molecular weight and an increment in crystallinity, which influence the aliphatic chain longitude and the aromatic ring (which increases the resistance to the harm by radiation); these phenomena are attributed to the breakage of polymer chains. Moreover, the estimated average size of defects in the crystalline zones and crosslinking are detected, as well as improvement on the stability after irradiating.

For gamma-irradiated PET at doses higher than 135 kGy the bands at 871 cm$^{-1}$ and 1303 cm$^{-1}$ disappear and the absorption increases in the 320–370 nm range, which is attributed to the free radicals produced in PET during the process, which react with air oxygen to form carbonyl and hydroxyl groups. In another study, fair stability in the physicochemical properties at high doses (900 kGy), with changes from crosslinking processes up to 35% from the starting values were seen.

In early studies, recycled PET was compared to virgin PET when irradiating at different doses. It was found that random scission reactions are produced mostly in the main chain, with the consequent decrease in the polymer molecular mass; this was proved by means of fluidity and intrinsic viscosity level tests, performing by DSC. Gamma irradiation effects on thermal properties and the dyeing capacity of virgin and waste PET mixtures show an increase in thermal stability. In addition, color intensity was studied, for a mixture with 80/20% of virgin and recycled PET, the color intensity was improved 53 and 98% after submitting at 30 and 50 kGy dose, respectively.

A broad range of possibilities should be pursued using gamma radiation and recovered scrap rubber in an effort to create useful structural engineering materials. Some studies include ground tire rubber (GTR) and its mix with other materials, where gamma irradiation can vulcanize such thermoplastic/elastomer blends and to evaluate the possibility of recycling GTR and to obtain a product with specific properties.

Some work on GTR mixed with different materials has been conducted, including: (a) Recycling end-of-life GTR powder as functional filler in a thermoplastic matrix that offers an opportunity to design second-generation materials that would be recyclable; (b) GTR and nonrecycled resins, to avoid phase separation and lead to improvement of mechanical properties; (c) GTR and HDPE, allowing compatibilization of both recycled materials and improvement of their mechanical properties. Elongation at break and Charpy impact strength of the blends are significantly increased for irradiation doses ranging from 25 to 50 kGy, thanks to the involvement of GTR dispersed phase in material deformation. The moduli of elasticity of the blends are only slightly decreased by this compatibilization due to the fact that in the same time, the radiation induced crosslinking of the HDPE matrix. Nevertheless, for higher irradiation doses (100 kGy) such crosslinking leads to lower mechanical values.

In a GTR/PE blend gamma-radiation-induced chain scissions within the rubber phase will reduce the incompatibility with the thermoplastic matrix; and provokes crosslinking of polyethylene when using an inert atmosphere, nevertheless, in the presence of air oxidation occurs, leading to lower crosslinking. In general, for rubbers, chain scission dominates and it is enhanced under an air atmosphere.

Blend compatibilization between GTR and HDPE was confirmed by SEM, observations performed during a microtensile test. For a strain of 13% along the vertical axis, a lack of cohesion is observed below the GTR particles, while for 17%, the GTR particle does not participate in blend deformation. After irradiating HDPE/GTR blends at 25 kGy the elongation of the GTR particles increases.

## 16.3 Recycled and Waste Materials used in Concrete

Separation of the paper, aluminum and polyethylene used in packaging cartons is based on plasma technology. The heat transforms the plastic into paraffin and the aluminum is recovered in its pure form.[10] During the separation the first devolatilization takes place between 200 and 400 °C (maximum peak = 360 °C), it is associated with the decomposition of the cardboard layer, including degradation of lignocellulosic material (up to 95% of total degradation). The second devolatilization occurs between 400 and 510 °C (maximum peak = 475 °C) that corresponds to decomposition of polyethylene. Above 700 °C light paraffins and olefins are obtained.

Degradation of cellulose and PE are independent during pyrolysis, the presence of carbonyl group in wax may be related to both interaction between degradation products from cellulose and PE and the presence of some tarry impurities from cellulose degradation. The pyrolysis of PE at moderate temperature gives the mainly waxy product, yields of oil (C4–C15 hydrocarbons) and gases (C1–C4 hydrocarbons) are low. Moreover, the aluminum foil has no effect on the thermal degradation of both PE and cellulose.

Another polymeric waste material is polyethylene terephthalate (PET) from bottles (Figure 16.1), which represents one of the most common plastics in solid urban waste.

Some investigations already confirmed the potential of PET waste in replacing aggregates in concrete (Figure 16.2), which represents a better option than landfill.[11–13] Concrete with PET has a compressive strength up to 28 MPa,[14] increases its thermal performance up to 18%,[15] the unit weight decreases and flow ability increases, which means a better workability. The use of lightweight concrete (desirable in earthquake-prone areas) with a lower density can result in significant benefits such as superior load-bearing capacity of elements, smaller cross sections and reduced foundation sizes.[16]

Fatigue failure is a common problem of asphaltic concrete that can lead to pavement damage. The use of low concentration of PET in stone mastic mixture increases the stiffness modulus of the mixture; conversely, high concentration diminishes stiffness. PET-reinforced mixtures exhibit significantly higher fatigue lives compared to the mixtures without PET.[17]

PET can be returned to fibers that are obtained from melted PET waste to form a roll-type sheet. Then, the sheet is cut into 50 mm long fibers and a deforming machine is used to change the fiber surface geometry. When

**Figure 16.1** PET wastes after milling.

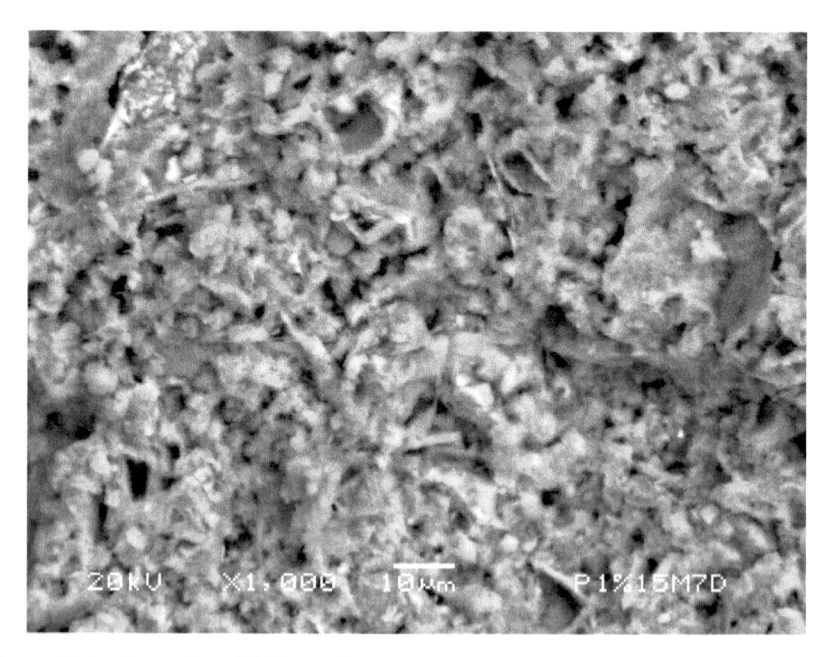

**Figure 16.2** Granules of PET used as aggregate.

using PET-based fibers in concrete a better control on the plastic shrinkage is done. Moreover, depending on the surface treatment of PET fibers, these show a better performance of dispersion and bonding (Figure 16.3).[18]

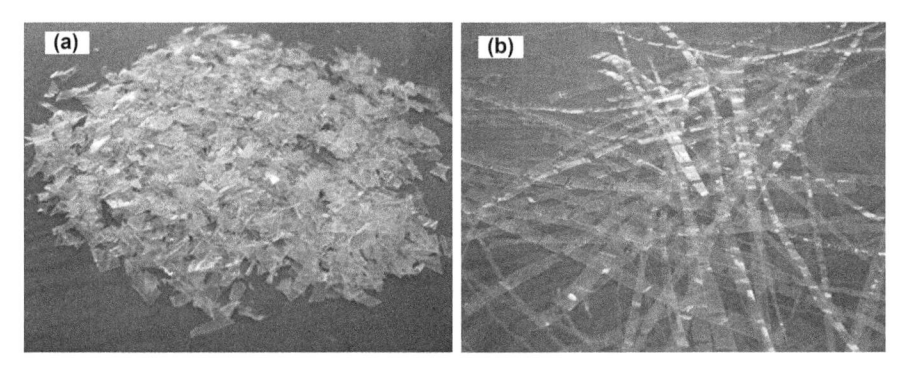

**Figure 16.3** Recycled PET: (a) straight type, and (b) crimped type.

**Figure 16.4** (a) crumb rubber aggregate, (b) ground tire rubber aggregate, and (c) mixed into concrete.

The environmental damage caused by improper management of waste tires increased over the past years creating a relevant problem to be solved. In the field of civil engineering results show that it is possible to reutilize the rubber of the waste tires. The main markets for recycling currently are energy recovery (as kiln fuel in the cement industry) and raw materials recovery. Mechanical and chemical processes, such as tire shredding, pyrolysis and cryogenic reduction are used by the tire-recycling industry.[19] The current applications of recycling waste tires in civil engineering practices mainly are as follows: (a) Modifiers to asphalt paving mixtures, (b) as an additive to Portland cement concrete, (c) Light-weight fillers, and (d) in whole tires as crash barriers, bumpers, and artificial reefs.

Waste tire rubber is produced and classified according its size as follows:[20] (a) Fiber rubber aggregate (8.5–21.5 mm); Shredded or chipped rubber aggregate (>4.75 mm); (b) Crumb rubber (0.425–4.75 mm); (c) Ground tire rubber aggregate (0.425 mm), (Figure 16.4).

Some studies have been conducted on tire-modified concrete and mortars. Results have indicated that rubberized concrete mixtures show lower density, increased toughness and ductility, higher impact resistance, enough compressive and splitting tensile strength, more efficient sound insulation and increased thermal performance.[21]

Self-compacting concrete has a high potential for replacement of natural sand with tire rubber, and can be used in quantities of up to 180 kg m$^{-3}$ of 64 mm crumb rubber aggregate replacement while providing acceptable fresh and hardened properties.[20] Resistance to chloride-ion penetration decreases with increasing waste tire content in concrete containing rubber aggregate. Chloride-ion penetration in concrete mixes after 28 day curing increases approximately 57% for a replacement ratio of the overall volume of aggregate of 25%.[22] Concrete containing rubber waste up to 15% has a high resistance to sulfuric acid attack.[23] The performance of concrete produced with waste tire aggregate is better than that of ordinary concrete under freeze–thaw cycles. High-strength concrete (HSC) shows a higher tendency of explosive spalling when subjected to rapid heating, as in the case of fire. This behavior is mainly due to the lower permeability of cement paste. HSC filled with solid fiber-shaped particles of recycled tire rubber reduce both the risk of HSC explosive spalling and its stiffness without a high loss of strength, enlarging the compatibility of deformation with other building elements.[24]

Bituminous materials have been used in buildings, construction, tunnels, on bridges, for hydraulic structures, and other purposes in civil engineering. The best known and important role of bituminous materials is in road construction and road safety. The accelerated deterioration and eventual failure of bituminous pavements has been partially solved by the addition of tire rubber to an asphalt concrete binder.[25] Waste tires provide a good mechanical behavior under static and dynamic actions. Also, recycled tire rubber-filled concrete can be used as a rigid pavement for roads on elastic subgrade when fatigue loads are applied.[26] A good compatibility and interaction between rubber particles and asphalt binder is achieved, leading to various improved properties of asphalt mixtures. Advantages include high resistance to temperature and to freeze–thaw cycles; better drainage; and diminution of cracking due to the elastic properties of rubber.[27,28]

## 16.4 Irradiated Concrete Containing Waste and Recycled Materials

The use of gamma radiation as a mechanism for reaction initiation and as an accelerator of the polymerization of a monomer in a ceramic matrix can bring considerable advantages to the process. Some studies cover the effects of ionizing radiation on polymer–ceramic composite materials, where the yield of polymerization increased with increasing radiation dose.[29] As this process is completed at room temperature there is much economy of heat energy, besides the reduction in costs to keep the system under pressure. The pressure is used to keep much of the monomer (usually with a high volatility) filling the interstices of the ceramic matrix during the conversion to ceramic–polymer composites.

Virgin materials are modifiers that can be used to improve the properties of road surfaces, nevertheless they are uneconomical and difficult to find.

In the last decade many studies have focused on using waste polymers as a modifier in road surfaces; they can potentially help reduce material wastage and improve the performance of road surfaces at the same time.

For concrete containing recycled cellulose (from packaging cartons) the compressive strength values gradually decrease when more concentration of cellulose is added; the strength depends on the amount of waste cellulose and water cement ratio (w/c). When irradiating concrete the highest compressive strength value is observed at 300 kGy at 28 days of curing time, which means an improvement of 45% respect to control concrete (Figure 16.5); moreover the values increase when the radiation dose increases too. As is known, many types of chemical reactions take place during gamma irradiation of polymeric materials, crosslinking and degradation by chain scission among others, but one or the other of these effects may be predominant in some materials.

Application of high-energy irradiation to cellulose creates free radicals by the scission of the weakest bonds; such radicals can react with some molecules in the cement matrix. The interaction between calcium silicate hydrate (formed during the hydration process) and the cellulose present in the pores during irradiation polymerization that enhances the interphase bonding, and as a result, an improvement of the mechanical strength takes place.

Improvements in modulus of elasticity values of concrete point out a predominant domain of crosslinking of polymer chains in cellulose. However, some shorter chains are produced which are water soluble and in consequence an increment in the solubility is obtained. In general terms, irradiated cellulose cover the sand particles, thus the zone around them is affected by a stress concentration. Therefore, if the distance between

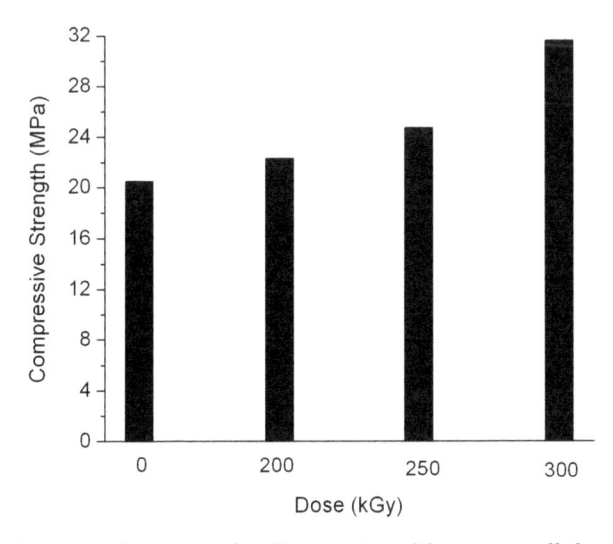

**Figure 16.5** Compressive strength of concrete with waste cellulose at different irradiation doses.

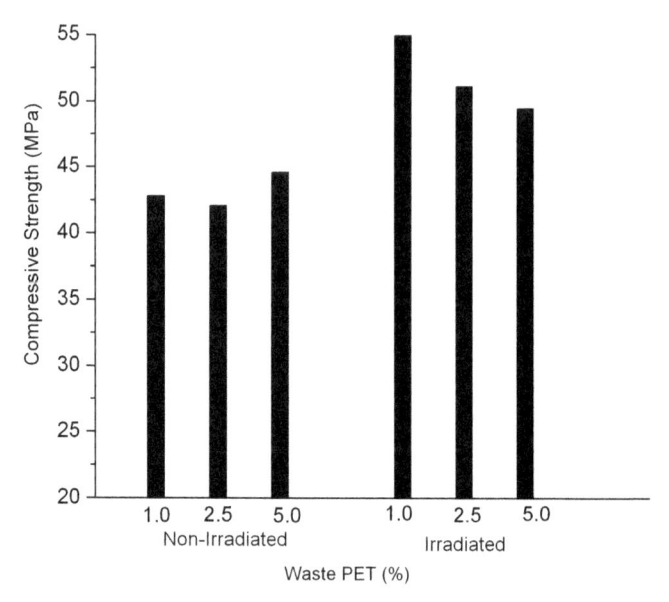

**Figure 16.6** Compressive strength of concrete at different PET volume concentration.

particles is small enough, these zones join together and form a percolation network, which generates good adhesion between the cement matrix and the cellulose and in consequence an increment in the modulus of elasticity is obtained.[8]

The compressive strength of concrete containing waste PET is shown in Figure 16.6. For nonirradiated concrete the values increase when PET concentration increases from 1.0 to 5.0 vol%. The highest value is 44 MPa for concrete with 5.0 vol% of PET. For irradiated concrete the values are up to 35% higher when compared to those of nonirradiated concrete, nevertheless they decrease progressively when increasing the PET concentration.

Compressive strain values of concrete containing waste tire particles decrease progressively according to the particle concentrations increase. Concrete with 10% of particles of 2.8 mm in size is 11% lower than those for control concrete (24 MPa). According to the particle size, the compressive strength values are higher for concretes with particles of 2.8 mm size than those with a size of 0.85 mm. For concrete with waste irradiated tire particles the compressive strength values decrease when increasing the particle concentrations. It is more convenient using bigger size particles instead of smaller ones.

The modulus of elasticity values of concrete are shown in Figure 16.7. They decrease on increasing the concentration of particles; also, concrete with particles of 2.8 mm size have higher modulus of elasticity values when compared to those with a size of 0.85 mm. For concrete with irradiated particles, moduli of elasticity values are higher than those for concrete with nonirradiated particles. Two "stages" are identified in concrete with

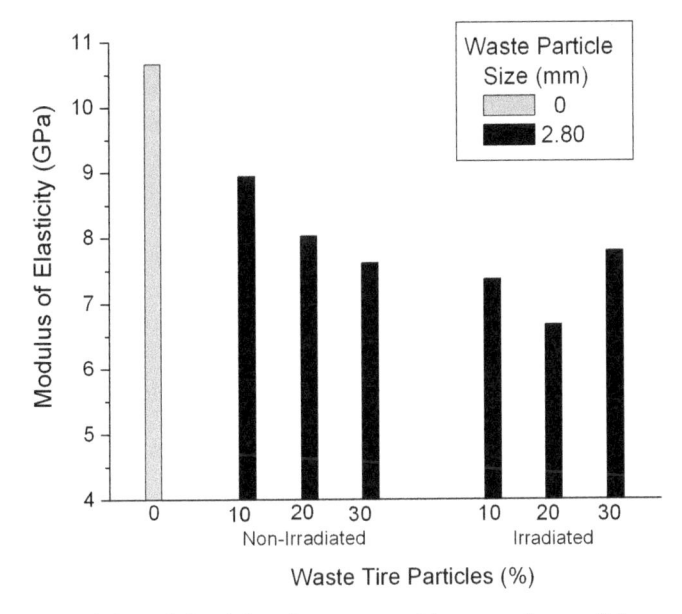

**Figure 16.7** Modulus of elasticity of concrete with waste-tire particles.

irradiated particles, the first one consists in a diminution of the values for concretes with 10% and 20% of particles, and the second increment for concrete with 30% of particles.

## 16.5 Conclusions

As is known, environmental problems caused by waste materials are in constant growth and as consequence different methods have been development, nevertheless some are consuming money and time. One novel alternative consists in using ionizing radiation to modify the physicochemical properties of the wastes and after using them as fillers or modifiers in composite materials as concrete. In this chapter, we have shown that some properties can be improved when using gamma radiation. As concrete's compressive strength is one of the most important structural design parameters used by engineers, waste PET particles can be a suitable material for construction. A small amount of PET could be used for substituting fine aggregate in the mix design to increase strength, and diminish strain. Thus, irradiation can become a useful tool and a suitable method for recycling waste PET. Moreover, in other cases both waste cellulose concentration and gamma radiation are adequate tools for improvement of the mechanical properties of concrete, where sand is substituted by waste cellulose. In general, compressive strength and modulus of elasticity values show an improvement when adding waste cellulose and applying certain gamma dose. Conversely, diminutions on the mechanical properties are seen for nonirradiated concrete.

## Acknowledgements

Financial support of the Autonomous University of the State of Mexico (UAEM), Toluca by Grant UAEM 3408/2013M (Megaproyecto) is acknowledged.

## References

1. O. Gencel, F. Koksal, C. Ozel and W. Brostow, *Constr. Build. Mater.*, 2012, **29**, 633.
2. F. Pelisser, N. Zavarise, T. A. Longo and A. M. Bernardin, *J. Cleaner Prod.*, 2011, **19**, 757.
3. G. Martínez-Barrera and W. Brostow, in *Fiber-reinforced Polymer Concrete: Property Improvement by Gamma Irradiation*, ed. C. Barrera-Díaz and G. Martínez-Barrera, Research Signpost, Kerala, India, 2009, ch. 3, pp. 27–44.
4. G. Martínez-Barrera, C. Menchaca Campos and F. Ureña-Nuñez, in *Gamma Radiation as a Novel Technology for Development of New Generation Concrete*, ed. F. Adrovic, InTech, Rijeka Croatia, 2012, ch. 6, pp. 91–114.
5. A. K. Bledzki and J. Gassan, *Prog. Polym. Sci.*, 1999, **24**, 221.
6. M. Akerholm, B. Hinterstoisser and L. Salmen, *Carbohydr. Res.*, 2004, **399**, 569.
7. E. Gumuskaya, M. Usta and H. Kirci, *Polym. Degrad. Stab.*, 2003, **81**, 559.
8. R. Despot, M. Hasan, A. O. Rapp, C. Brischke, M. Humar, C. R. Welzbacher and D. Razem, in *Changes in Selected Properties of Wood Caused by Gamma Radiation*, ed. F. Adrovic, InTech, Rijeka Croatia, 2012, ch. 14, pp. 281–304.
9. G. Burillo, R. L. Clough, T. Czvikovszky, O. Guven, A. Le-Moel, W. Liu, A. Singh, J. Yang and T. Zaharescu, *Radiat. Phys. Chem.*, 2002, **64**, 41.
10. A. Korkmaz, J. Yanik, M. Brebu and C. Vasile, *Waste Manag.*, 2009, **29**, 2836.
11. F. Pacheco-Torgal, Y. Ding and S. Jalali, *Constr. Build. Mater.*, 2012, **30**, 714.
12. N. L. Modro, N. Modro, N. R. Modro and A. Oliveira, *Rev. Mater.*, 2009, **14**, 725.
13. S. Akcaozoglu, K. Akcaozoglu and C. D. Atis, *Composites: Part B*, 2013, **45**, 721.
14. F. Mahdi, H. Abbas and A. A. Khan, *Constr. Build. Mater.*, 2010, **24**, 25.
15. F. Fraternali, V. Ciancia, R. Chechile, G. Rizzano, L. Feo and L. Incarnato, *Compos. Struct.*, 2011, **93**, 2368.
16. Y. W. Choi, D. J. Moon, Y. J. Kim and M. Lachemi, *Constr. Build. Mater.*, 2009, **23**, 2829.
17. T. B. Moghaddam, M. R. Karim and T. Syammaun, *Constr. Build. Mater.*, 2012, **34**, 236.
18. J. Kim, C. Park, S. Lee, S. Lee and J. Won, *Composites, Part B*, 2008, **39**, 442.

19. G. Centonze, M. Leone and M. A. Aiello, *Constr. Build. Mater.*, 2012, **36**, 46.
20. K. B. Najim and M. R. Hall, *Constr. Build. Mater.*, 2010, **24**, 2043.
21. W. H. Yung, L. C. Yung and L. H. Hua, *Constr. Build. Mater.*, 2013, **41**, 665.
22. M. Gesoglu and E. Guneyisi, *Mater. Struct.*, 2007, **40**, 953.
23. F. Azevedo, F. Pacheco-Torgal, C. Jesus, J. L. Barroso de Aguiar and A. F. Camões, *Constr. Build. Mater.*, 2012, **34**, 186.
24. F. Hernández-Olivares and G. Barluenga, *Cem. Concr. Res.*, 2004, **34**, 109.
25. A. Tortum, C. Celik and A. C. Aydin, *Build. Environ.*, 2005, **40**, 1492.
26. F. Hernandez-Olivares, G. Barluenga, B. Parga-Landa, M. Bollati and B. Witoszek, *Constr. Build. Mater.*, 2007, **21**, 1918.
27. M. Bravo and J. de Brito, *J. Cleaner Prod.*, 2012, **25**, 42.
28. M. H. Paranhos Gazineu, W. A. dos Santos, C. A. Hazin, W. E. de Vasconcelos and C. C. Dantas, *Prog. Nucl. Energy*, 2011, **53**, 1140.
29. Z. N. Kalantar, M. R. Karim and A. Mahrez, *Constr. Build. Mater.*, 2012, **33**, 55.

# Subject Index

Page numbers in *italics* refer to tables and figures.

1-hydroxycyclohexyl phenyl ketone (HPK) (Irgacure 184) 259, *260*, *261*, 266, 267, *293*, 298, 299, 300, *302*, 306, 307, *308*, 309, 310, 311, *312*
1-oxo-4-hydroxymethyl-2,6,7-trioxa-1-phosphabicyclo[2.2.2]octane (PEPA) 174
1-phenyl-1,2-propanedione (PPD) 329, 331
1,4-butanedioldiacrylate (BDDA) 28
1,6-hexanedioldiacrylate (HDDA) 28
2-benzyl-2-dimethylamino-4-morpholino butyrophenone (Irgacure 369) *293*, 298, 306
2-ethyl hexyl acrylate (EHA) 28
2-ethyl hexyl-4-dimethylamino benzoate (EHA) 298, 299, 300, 301, *302*, 306, 307, *308*, 309, 310, 311, *312*
2-hydroxy benzophenone (2-HBP) 298, 299, *302*, 311, *312*
2-hydroxy ethyl acrylate (HEA) 28, 173
2-hydroxy propyl acrylate (HPA) 28
2-hydroxy-2-methyl propiophenone (Irgacure 1173) 294, *302*, 306, *312*
2-ITX 299, 300, 305–6, 311
2-methacryloyloxyethyl phosphorylcholine (MPC) 139
2-methyl benzophenone (2-MBP) *295*, 311

2,2-dimethoxy-2-phenyl acetophenone (DMPA) (Irgacure 651) 32, *293*, 298, 299, *302*, 306, 307, 309, 310, 311, *312*, 328
2,2-dimethyl-1,3-propanediol acryloyloxyethyl phosphate (DPHA) 172, *173*
2,4-diethyl-thioxanthone (DETX) *295*, *302*, 306, 311, *312*
3-methyl benzophenone (3-MBP) *295*, 311
3D pens 75
4-(4-methylphenylthio) benzophenone (BMS) 311
4-(dimethylamino)benzophenone (DMAB) 298, 300, 307
4-4-methyl benzophenone (4-MBP) *295*, 298, 300, *302*, 306, 310, 311, *312*
4-hydroxybenzophenone (4-HBP) 299, 300, 311, *312*
4-ITX 305, 306, 311
4-phenyl benzophenone (PBZ) *296*, 299, 300, *302*, 306, 311, *312*
4,4′-bis(allyloxyphenyl) phenyl phosphine oxide (DAPPO) 157
4,4′-bis(diethylamino)-benzophenone (BDP/DEAB) 259, *260*, 266, 267, 268–9, 271, *272*, 273, 275–6, 277, 279, 281–2, 285, *296*, 298, 299, 300, 306, 307, *312*

*Subject Index* 361

9,10-dihydro-9-oxa 10-phosphaphenanthrene-10-oxide (DOPO) 159, *160*, 175

acetonitrile 303, 305
acrylated benzene phosphonates 174
acrylated epoxies 24
acrylated phenylphosphine oxide (APPO) 163–4, 180
acrylated polyesters *24*, 25–6
acrylated polyethers 24–5
acrylated polyurethanes *24*, 26
acrylated silicones 26
acrylates 27
acrylic acid (AA) 257
acryloyloxyethyl phenyl benzenephosphonate (APBP) 174
active volumes 92
additive manufacturing technologies (AMTs) 75
additives 31
adhesives 2–3
Agilent 307, 310
aircraft coatings 4–5
aliphatic amines 33
alkyd paints 6
allergy diagnosis 142
allyldiphenyl phosphine oxide (ADPPO) 157–8
aluminum 351
amphiphilic polymers 136, 137–9
analytical solutions, multicomponent curing 94–7, *98*
antennas 248–9
anti-Stokes shifts 61
antibodies 142, 145–7, 230
artificial joints 8
autoimmune diseases 142–4
automotive refinish 4, 44
AZ4620 photoresist 235

baby milk 291, 298
ball valves 117
barrier coatings 3
BDEEP (tetra acrylate) 167–8
BDS Hypersil 304

Beer–Lambert law/model 330, 332
benzene 314
benzoin ethers 32
benzophenone (BP) 32, *33*, 79, 136, *295*, 298, 299, 300, 301, *302*, 306, 307, *308*, 309, 310, 311, *312*
benzophenone acrylate (BPAcr) 299, *302*
benzoyl-1-hydroxyl-cyclohexanol (Irgacure 184) 156
beverage cans 5
bimolecular photoinitiators 32–3, 78
biological force measurement 231
biomedical applications 225
biomolecules 135
biosensors 230–1
bis-(triethoxysilylpropyl) phenyl phosphamide (BESPPA) 177–8
bis(4-diphenylsulfonium)-phenylsulfide-bis(hexafluoro-phosphate) (BIS) *297*, 298
BISGMA monomers 326–7
BisGMA/TEGDMA resin 332, *333*, *334*, 336, 337, *338*
bituminous materials 354
blasted holes 116
blood 142, 145–7
blue LEDs *323*, 324–5, 329, *333*, 335, 338, 342
boron methacrylate monomer (BM-M) 180, 181
boron methacrylate oligomer (BM-O) 180, 181
boron phosphate (BPO$_4$) 178
boron-containing flame retardants 178–83
bottom-up approach 123
bovine serum albumin (BSA) 140
Bragg condition 201, 205
Bragg DFB structures 210
Bragg resonators 213
breakfast cereals 311
Brønsted acids 340

bulk micromachining processes (BMPs) 113
byproducts 313–14

$C^{13}$-NMR spectra 52, 72
camphorquinone (CQ) 324, 331, 332, 333, 341, 342, 343–4
camphorquinone/amine systems 329, 332, *333*, *343*
cantilever-based sensors 230–1
carbene 136
carbobetaine 136
carbon fibers 256–7, 281–3
  mechanical properties 283–6
cardboard 298
cationic photopolymerization 34–9, 46, 340–5
  *vs* free-radical photopolymerization 39
cell-specific antibodies 145–7
cells (biological) 114
cellulose 348, 349, 351, 355
chalcones 50
Chimassorb-90 *312*
chitosan 174–5
chloride-ion penetration (concrete) 354
cholesteric liquid crystals (CLCs) 192–3, 210–14
cholesteric phase, (liquid crystals) 189
chromatography 292
"chrome look" 5
chromophores 79, 83
Chrompack 307
circular POLICRYPS gratings 194–7
Class I/II organic–inorganic hybrids 122
compressive strength (concrete) 356
computer-aided drawing (CAD) 111
concrete 348, 351–7
concrete coatings 6
confocal Raman microscopy 41
coplanar waveguides (CPWs) 245, *247*
corrosion protection 4

cotton fabrics 171, 172
covalent immobilization 135
Cramer classification 291, 315
curing intensity *97*
curtain coating 21
curved surfaces 235, 236
cyclic voltammetry (CV) 312
cycloaliphatic epoxides 37–8, 39
cyclobutane ring formation 67
cyclohexane dimethanol divinyl ether (CHDMDE) 29, 30
cyclohexyl epoxides 39
cylindrical specimens 334

D–π–A–π–D photoinitiators 79, 83
D-bulbs 19, *20*
dairy products 301
dark cures 39
dark reaction 341
DART/TOF-MS method 313
DEBP *302*
deep reactive ion etching (DRIE) 244
deformable mirrors 233–5
dentistry 8, 324
Department of Health and Human Services (USA) 30
devolatilization 351
di(acryloyloxyethyl) benzenephosphonate (DABP) 174, 183
di(acryloyloxyethyl) ethyl phosphate (DAEEP) 158, 161–2
diacrylates 27
diallylphenylphosphine oxide (DAPPO) 164–5
diaryliodonium salt 34, 35, 342–4
diazirine 136
dibutyl tin dilaurate (DBTDL) 159
dielectrophoretic (ODEP) switches 229
diethyleneglycol diacrylate (DEGDA) *28*
diethyleneglycol divinyl ether (DEGDE) 29, *30*
differential pulse voltammetry (DPV) 312

## Subject Index

differential scanning calorimetry (DSC) (*see also* photo-DSC) 41–2, 264
diluents 326
dimethacrylate monomers 326–7, 337
dimethyl sulfoxide (DMSO) 259, 260
diphenyl-(2,4,6-trimethylbenzoyl) phosphine oxide (TPO) 306
diphenyl[(phenylthio) phenyl] sulfonium hexafluorophosphate) (THIO) *297*, 298
direct laser writing (DLW) 111, 239
Discovery HS F5 305
Discovery ZR-PS 305
distributed feedback (DFB) 210
divinyl ethers 29–30
DMPE (phosphate-based diacrylate) 161, *162*
DNA detection 230
DNA microarrays 135
Dorocure 1173 (photoinitiator) 159, 165, 169, 211
dry-film lamination 221, 223

EB600 epoxy acrylates 155
Eclipse XDB 299
electric discharge machining (EDM) 245
electrodeless lamps 20–1
electron transfer 258
electrostatic actuators 226–7
electrostatic micromotors *112*
energy levels 104
Environmental Protection Agency (EPA) (USA) 15
enzyme immunoassays (EIAs) 144
Epon SU-8 epoxy 221
epoxides 36, 37
epoxidized linseed oil (ELO) 37
epoxidized soybean oil (ESBO) 153–4
epoxidized vegetable oils 42
epoxy acrylate flame retardants 155–7, 173–6, 183
epoxy monomers 339–40, 341–5

epoxy norbornane linseed oils (ENLOs) 37, 42
equipment for UV curing 18–21
ETA-TTA formulation 80, 81, 82
ethyl-4-(dimethylamino)benzoate (EDB) *297*, 300, *302*, 306, 307, *308*, 309, 310, 311, *312*
European Food Safety Agency (EFSA) 291
European Printing Ink Association (EuPIA) 292, *293–7*

fatigue failure (concrete) 351
ferroelectric liquid crystals (FLCs) 193–4
fetal human hepatocytes (FHHs) 114
fibronectin 140
filter effect 23
final polymer concentration 97
finish defects 43
fire hazards 150
flame retardancy 150–84
fluorescence 61–2
fluorescence imaging method 116
fluorescence spectra 62–6
fluorinated materials 165–7
food packaging 2, 290
food-contact material (FCM) 290–316
Fourier transform infrared spectroscopy (FT-IR) 263
free-form objects 104–5, *111*
free-moving microparts 223
free-radical intermediates 258
free-radical photopolymerization 31–4, 46, 327–8
  *vs* cationic photopolymerization 39
full-color imaging 6
full-width half-maximum (FWHM) intensity 265, 277
FUSION LC-6B Curing System 19

gallium arsenide 322
gallium arsenide phosphide (GaAsP) 322

gallium nitride 322, 324
gamma irradiation 348–51
  concrete 354–7
gammabutyrolactone (GBL) 221
gas chromatography (GC) 292, 307–12, 315
GC-FID method 307
GC-MS method 307, 309–10, *312*, 313
glass-transition temperature 337
graphene oxide (GOx) 129
green LEDs *323*
grid towers 107
ground tire rubber (GTR) 350–1

H-bulbs 19, *20*
$H^1$-NMR spectra 50–2, 72
HACP/HECP (acrylated phosphazene derivative) 173
halogen light curing 324
halogenated flame retardants 151
HDPE polymers 349, 350, 351
hemicellulose 348, 349
Hep G2 cells 114
high-energy EB curing 8
high-performance thin-layer chromatography (HPTLC) 312
high-$Q$ inductors 242–3
high-resolution nanolithography 7
high-strength concrete (HSC) 354
holographic lithography (HL) 111, 239
holographic polymer-dispersed liquid crystals (HPDLCs) (*see also* HPDLC gratings) 200–1, 214
homolytic cleavage 258
hot metal adhesives 3
HPDLC gratings 88, *98*, 101, 204–6
HPLC-DAD method 300–1, 307, 309, 313
HPLC-DAD-FLD method 300
HPLC-MS method 301, 303, 307, 309, 310, 315
HPLC-MS/MS method 301, 303, 305, 307, *308*, 309, 310, 315
HPLC-MS/TOF method 301
hydrogen abstraction 258

hyperbranched PU acrylates 155
Hypersil GOLD 303

ignition stage, radical polymerization 88
image forming/transferring devices 6–7
imide azogroup 50
immobilization, biomolecules 135
immunoglobulin G (IgE) assay 142, 143–4
*in vivo* pressure monitoring 231
inclined photolithography 237–8
indium tin oxide (ITO) 205, 207, 211
Inertsil ODS-3 304
infrared LEDs 322, *323*, 324
inkjet printing (IJP) 2, 111–12, 113, 115, 116
inks 2
inorganic fillers 122
inorganic oxides 122
interference 94
International Agency for Research on Cancer (IARC) 314
inverse kinetic chain length 90
Irgacure 2100 211
isobornyl acrylate (IBA) *28*
isodecyl acrylate (IDA) *28*
isopropylthioxanthone (ITX) 291, *294*, 298, 299, 300, 301, *302*, 304, 305, 306, 307, *308*, 309, 310, 311, 312
itaconic acid (IA) 257

Jablonsky diagrams 22, 61
Jones matrices 207

Ka band filter 245
ketyl 136
KHPDH (phosphorus/silicon-containing monomer) 176–7
Kromasil-100 299
Kusumgar, Nerlfi & Growney Inc 1

ladder structure 257
layer-by-layer lamination 239

Subject Index

LC-MS/MS method 307, 310
LDPE polymers 349
LED light curing 324–5, 345
LIGA-like processes 220
light 103
light sources, UV curing 19–21
light stereolithography (LS) 110–11, 113
light-emitting composite materials 128–9
light-emitting diodes (LEDs) 322–6
light-emitting hybrid materials 127
lignin 348
lignocellulosic material 351
liquid chromatography (LC) 292, 298–307, 315
liquid crystals (LCs) 188–98
liquid-crystal displays 8
liquid-crystalline phases 67–73
liquid-crystalline polymers (LCPs) 49, 50
liquids 103–4
LOI measurement technique (flame retardancy) 151
low-energy EB curing 8
low-gloss coatings 5
low-intensity fringes 97
Lucirin TPO 294
luminescent materials 127

M$_2$CMK photoinitiator 79–80, 81, 82, 84
Mach–Zehnder interferometers (MZIs) 230
"Maltese cross" (liquid crystals) 196
mammalian cells 139
margarine 312
Matte finish coatings 5
medical devices 116
medium-density fiberboard (MDF) 44
medium-energy EB curing 8
MEK rubbing 179
MEMS technology 234, 235
mercury lamps 19–20
mesogens 49–50

mesophases 188
metal-transfer micromolding (MTM) 245–6, 247
methacrylate monomers 326–7, 335
methacrylated phenolic melamine (MAMP) 159
methyl acrylate (MA) 257
methyl-1-(4-methylthio)phenyl-2-morpholinopropan-1-one (Irgacure 907) 293, 298, 299, 300, 301, 309, 311, 312
methyl-2-benzoylbenzoate (MBB) 296, 299, 300, 302, 306, 307, 308, 311, 312
methylene blue 129, 130
Michler's ketone (MK) 298, 300, 307, 309
microactuators 112–13, 225–6
microarray technologies 134–47
microchannels 229–30
microchips (microelectronic chips) 7, 114
microcylindrical lenses 234–5
microfabrication schemes 108–12
microflow cytometers 228–9
microfluidic applications 114, 115, 227–30
microfluidic chips 227–8
microlenses 234–6
microscale optical devices 114–16
microsensors 113–14
microstampers 117
microstructures 107–8
microsystem integration 223
milk 301, 305, 310, 311, 213
mirrors 237
modulated radiation 94
modulus of elasticity (concrete) 355, 356–7
molecular directors 206–7
molecular weights 52–3
monoacrylates 27, 28
monomers 50, 326–7
monomolecular photoinitiators 78
morphological analysis 264
mouse fibroblast STO cells 141–2

366  Subject Index

multicomponent curing  87–101
multifunctional coatings  124–5
multifunctionality  121–2
multilayered stacked inductors (MLSIs)  *246*
multiple exposing beams  240

*n*-butyl acrylate (BA)  *28*
*n*-vinyl pyrrolidone  30–1
nail polish  8
nanocomposite gearwheels  117
nanoparticles (NPs)  122
National Fire Protection Association (USA)  150
National Toxicological Program (NTP)  30
natural fibers  348–9
Nd:YAG lasers  214, 215
near-IR spectroscopy (NIR)  336
negative photoresists (NPRs)  109
nematic liquid crystals (NLCs)  189, 191, 195, 196, 201–4, 207–10
neopentylglycoldiacrylate (NPGDA)  *28*
neutron shielding  178
nitrene  136
nonintentionally added substances (NIAS)  292, 313–15, 316
noninterferometric sensors  232
nonlinear optical (NLO) applications  50
Norrish type-I/II photoinitiators  328
numerical solutions, multicomponent curing  97–100

oligomers  23–7, 326
    cationic photopolymerization  36–9
on-chip antennas  248
one-photon/single-photon polymerization (1PP)  79, 111
onium salts  34, 36
optical accelerometers  232
optical applications  7–8, 191
optical fibers  7, 229
optical holographs  *203*

Optima Delta-6  309
optoelectronic applications  7–8
optofluidics  115–16
orange juice  300
orange LEDs  *323*
Ormocer pillars  116, *117*
OrmoClear polymer  115, *116*
out-of-plane microassembled devices  225
overprint varnishings  2
oxygen inhibition  33–4, 39, 335

P/B-containing additives  151
packaging  351
PAN (polyacrylonitrile) polymer chains  257
PAN-based copolymer  269–75
PAN-based fibers (*see also* carbon fibers)  256–7, 275–83, 286–7
PAN-based precursors  257, 258, 260, 265–9, 271, 273, 286–7
paper (packaging)  298
paraffin  351
passive volumes  92, 93
patch antennas  248, *249*
PDHA (phosphorous monomer)  172
PDLC gratings (*see also* HPDLC gratings)  97, 100
pentaerythritoltetraacrylate  29
pentaerythritoltriacrylate (PETA)  27, *29*
peroxy radicals  335
Phenomenex  311
phenyl azide  136
phenyl boronic acid  182
PHOENIX single-filament tensile testing  264
phosphate-based acrylate  162–5
phosphobetaine  136
phosphorus-based flame retardants  152–73, 183, 184
photo-DSC  336
photocatalysis  129–30
photocrosslinkable polymers  49, 72–3
    synthesis  50

Subject Index

photocured materials
  manufacture 1
  uses 1–2
photodegradation 129–30
photoimmobilization 135–47
photoinduced polymerization 106–7
photoinitiators (PIs) 21–3, 326
  cationic photopolymerization 34–6
  free-radical photopolymerization 31–4
  manufacture of food-contact material 290–312, 313, 314
  two-photon polymerization 78–9
photolithography (PL) 108–9, 113, 115, 116, 221, 223
photoluminescence 127–9
photon-absorption efficiency (PAE) 330–1
photonic bandgaps (PBGs) 210
photonic chip devices 213
photonic crystals 238–41
photopolymerizable formulations 76–8
photopolymerization 321–2
photoresists 7, 104, 108
photoresponsive liquid crystals (photo-LCs) 191–2
photosensitive materials 105–6
photovoltaics 8
piezoresistors 232–3
pink LEDs *323*
piperazine-N,N'-bis(acryloxyethylarylphosphoramidate) (NPBAAP) 168–9, *170*
plasma arc curing (PAC) 324
plasma technology (recycling) 351
plastic 351
polarized optical microscopy (POM) 189, 201
POLICRYPS gratings 88, *98*, 100–1, 190–1, 198, 201–17
poly(2-[3,5-di benzyl oxy benzoyl styryloyloxy] ethyl methacrylate) (P4) 50, *51*, 52, 53, *54*, 55, 59, *60*, 64, *65*, *66*, *68*, *70*, 71, *72*, 73
poly(2-[4-biphenyl oxy benzoyl styryloyloxy] ethyl methacrylate) (P5) 50, *51*, 53, *54*, 55, 59, *60*, 64, *65*, *66*, *68*, *70*, 71, *72*, 73
poly(2-[4-bromo benzoyl styryloyloxy] ethyl methacrylate) (P2) 50, *51*, 53, 54, 55, 56–7, 58, *60*, 62, *63*, *66*, *67*, 68–9, *70*
poly(2-[4-methoxy benzoyl styryloyloxy] ethyl methacrylate) (P3) 50, *51*, 53, 54, 55, 58, *60*, 62, 64, *66*, *67*, 69, *70*, 71, 73
poly(2-[benzoyl styryloyloxy] ethyl methacrylate) (P1) 50, *51*, 53, 54, 55, 56–7, *60*, 62, *63*, *66*, *67*, 68, 69, *70*, 73
poly(acrylic acid) 140
poly(acrylonitrle-co-methylacrylate) 259
poly(ethylene glycol) (PEG) 136, 137
poly(ethylene glycol) acrylate 182
poly(vinyl alcohol) 136, 141
polyacrylonitrile (*see* PAN)
polyclonal antibodies 142
polydimethysiloxane (PDMS) liquid 109, *110*, 114
polydispersity index (PDI) 53
polyethylene 351
polyethylene terephthalate (PET) 349–50, 351–2, 356, 357
polymer-dispersed liquid crystals (PDLCs) 200, 204
polymer–ceramic composites 354
polymers
  $C^{13}$-NMR spectra 52, 72
  fluorescence 61–2
  fluorescence spectra 62–6
  $H^1$-NMR spectra 50–2, 72
  liquid-crystalline phases 67–73
  molecular weight 52–3
  scanning electron microscopy 66–7, 73
  synthesis 50

polymers (continued)
    thermogravimetric
        analysis 53–5, 72
    UV spectra 55–61, 73
POPHA (flame retardant
    monomer) 169, 171
positive photoresists (PPRs) 108–9
potassium bromide (KBr) 40
powder coatings 3–4, 44–5
printing (see inks)
prisms 237–8
prolongation stage, radical
    polymerization 88
proteins 140–2
Purosphers Star RP 300
purple LEDs 323

quartz-tungsten-halogen light
    curing 324
QuEChERS (quick, easy, cheap,
    effective, rugged and safe)
    method 299, 301, 306

radical polymerization 88, 327
radio-frequency identifiers
    (RFIDs) 243
Rapid Alert System for Food and
    Feed (RASFF) (EU/EFTA) 291
rapid prototyping 111
rate of polymerization 41
reactive diluents 27
real-time infrared (RT-IR)
    spectroscopy 40–1
recycling 347
red LEDs 322, 323
refractive microlenses 235
relative humidity 39
release coatings 3
Restek 309
RF components 221, 241–2
RF filters 245–6
RF inductors 242–5
RF MEMS devices 249–50
RF power dividers 250
Rhodamine B 83
roll coating 21

rotary parts 223
Runge–Kutta method 97

scanning electron microscopy
    (SEM) 66–7, 73, 201
Scientific Polymer 259
scratch-resistant coatings 5, 122–4
sealants 3
self-assembled monolayers (SAMs)
    117
self-compacting concrete 354
sensing waveguides 231–2
shark-skin-inspired microscale
    patterns 5
SHEA (silicon-containing
    diacrylate) 165
shear force sensors 232
shrinkage 39, 338–9
Sigma-Aldrich 259
silane-functionalized
    compatibilizer 124
silica ($SiO_2$) nanoparticles
    122, 123–4
silica precursor (tetraethoxysilane,
    TEOS) 123, 124, 127
silica sol–gel ($SiO_2$–$P_2O_5$–$B_2O_3$)
    mixture 179–80
silicate oligomers 45
silicon-based processes 225
silicones 26
siloxane (Si–O–Si) bonds 344
smectic phase, (liquid crystals) 189
Snell's law 325
soft lithography (SL) 109, 113, 115,
    116
soybean oil 312
spatial light modulators
    (SLMs) 214–17
spin coating 221–3
spray guns 21
steady-state distribution 96
stencils 7
stereolithography 7
Stokes, George G. 61, 62
Stokes shifts 61–2, 64, 66, 73
streptavidin detection 231

Subject Index

styrene 30
SU-8 photoresist 220–51
sulfobetaine 136, 139
SunFire C18 305
Supelcosil LC-PAH 299
superacids 36
surface micromachining processes (SMPs) *112*, 113
surface-stabilized ferroelectric liquid crystals (SSFLCs) 193–4
suspended structures 117
switchable holographic gratings 190, 204
Synergi MAX-RP 80A 303
system-on-a-package (SOP) schemes 241, 242

TEGDMA monomers 327
termination stage, radical polymerization 89
Tetra Pak packages 348
tetraacrylates 29
tetrahydrofuran (THF) 191
textiles 172
TGMAP (phosphorous acrylate) 161, *163*
thermo-oxidative reactions 257, 262–3
thermogravimetric analysis (TGA) 53–5, 72
thick systems 321–5
thin films 344
thin-film transistors (TFTs) 117
thiol–ene polymerization 27, 179, *180*
thioxanthone 32
thiyl radicals 27
three-dimensional (3D) microstructures 107–8
three-dimensional (3D) printing *111*
threshold of toxicological concern (TTC) 315
Ti-sapphire lasers 239, 240
tire rubber 353–4
tissue engineering/repair 8–9
titanium dioxide (TiO$_2$) 129

Tokyo Skytree 107, *108*
top-down approach 123
total ossicular replacement prostheses (TORPs) 117
transdermal drug delivery 116
tri(acryloyloxyethyl) phosphate (TAEP) 158–9, 161–2, 171, 176, 184
triacrylates 27
trialkoxy-silyl ammonium salt 124
triaryl-sulfonium salts 34, 221
triethylene glycol divinyl ether (TEGDE) 29, 30
triglycidyl isocyanurate acrylate (TGICA) 171, 172
trimethylolpropane triacrylate (TMPTA) 29
trimethylolpropane trimethacrylate (TMPTMA) 42
triphenylsulfonium hexafluorophosphate 314
tungsten-halogen light curing 324
two-photon absorption (2PA) 79, 82, 239
two-photon polymerization (2PP) 75–84, *110*, 111

UDMA monomers 326
ultrathick SU-8 photoresist 223
ultraviolet LEDs *323*
uniform lying helixes (ULH) 192–3
unimolecular photoinitiators 32
urethane acrylate 161
urethanes 26
UV A radiation 17
UV B radiation 17
UV C radiation 17
UV curing (*see also* cationic photopolymerization, free-radical photopolymerization)
    advantages 16–17
    components 21–31
    disadvantages 17
    equipment 18–21
    evaluation 40–2
    purpose 15–16

UV light/radiation 17–18, 103
UV spectra 55–61, 73
UV-9385C photoinitiator 35
UV-curable coatings 15, 121–31
  use as flame retardants 151
UV-moisture dual-curing 45
UV-thermal dual-curing 45
UVI-6974 photoinitiator 35

V-bulbs 19, *20*
vacuum UV radiation 17
vegetable oils 42
vinyl ether monomers 340
vinyl ethers 29–30, 36, 37
vinyl phosphonic acid (VPA) 152–4, 183–4
violet LEDs *323*
virgin materials 354
viruses (biological) 144–5

viscosity 104, 326
volatile organic compounds (VOCs) 15, 42, 151

waste materials 347–57
water inhibition 39
water-reducible UV-cured coatings 42–3
waveguides 229
wet-etching 245
white LEDs *323*
wide-angle X-ray diffraction (WAXD) 264–5, 277–81
wine 312
wood coatings 6

yellow LEDs *323*
yoghurt 300, 301, 312

zwitterionic groups 136, 139